lectures on classical mechanics

lectures on classical mechanics

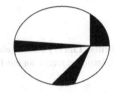

berthold-georg englert
National University of Singapore, Singapore

World Scientific
NEW JERSEY • LONDON • SINGAPORE • BEIJING • SHANGHAI • HONG KONG • TAIPEI • CHENNAI

Published by

World Scientific Publishing Co. Pte. Ltd.
5 Toh Tuck Link, Singapore 596224
USA office: 27 Warren Street, Suite 401-402, Hackensack, NJ 07601
UK office: 57 Shelton Street, Covent Garden, London WC2H 9HE

British Library Cataloguing-in-Publication Data
A catalogue record for this book is available from the British Library.

LECTURES ON CLASSICAL MECHANICS

ISBN 978-981-4678-44-5
ISBN 978-981-4678-45-2 (pbk)

In-house Editor: Christopher Teo

Printed in Singapore

To my teachers, colleagues, and students

Preface

Classical Mechanics is the branch of theoretical physics that deals with the motion of massive bodies under the influence of forces, whereby — one should add — relativistic effects or quantum effects are not important in the physical situation under consideration. The nonrelativistic circumstances require that all speeds are very small compared with the speed of light; but there is, of course, an extension of classical mechanics to the relativistic domain. When disregarding quantum effects we refrain from applying the methods of classical mechanics to phenomena on the scale set by atoms; but there is, of course, quantum mechanics for that purpose. Both relativistic mechanics and quantum mechanics turn into classical mechanics when, respectively, the speeds are small or the scales are large.

To quite some extent, classical mechanics deals with the physical phenomena that are the most familiar and the most intuitive, such as falling stones and swinging pendulums. As a natural tradition, then, the physics student's first exposure to theoretical physics is in lectures on classical mechanics. This has historical reasons because classical mechanics has its roots in the 16th and the 17th century and pre-dates all branches of Physics in maturity, but it also has factual reasons: In classical mechanics the student encounters for the first time universal principles that apply throughout all of physics and learns about concepts of general importance and usefulness. A solid knowledge of classical mechanics is the basis for all subsequent studies of physics.

The course on classical mechanics plays yet another important role in the physics curriculum. It is the student's first serious encounter with physics spoken in its natural tongue, the language of mathematics: A solid command of calculus and linear algebra is indispensable for the physicist, and so is a good knowledge of trigonometry and the basic special functions. While students are expected to be somewhat familiar with these mathematical tools when embarking on classical mechanics, good skills and confidence in using the tools are only developed by lots of practice. It is important that the learning student works out many of the exercises that are included here or can be found in textbooks. You do not learn to play the violin by watching a violinist perform; you must try your own hands on the instrument.

These *Lectures on Classical Mechanics* grew out of a set of lecture notes for a second-year undergraduate course that I taught at the National University of Singapore (NUS) in recent years. The presentation is rather detailed and does not skip intermediate steps that — as experience shows — are not so obvious for the learning student.

Prior to this course, students would have gone through the usual first-year overview of physics, partly a review of pre-university physics and partly a preview of coming attractions, and thus know quite a bit about classical mechanics, albeit in simple contexts and handled with elementary mathematical methods. Accordingly, the reader is expected to know basic facts, concepts, and notions of physics. Some of that material is covered here as well, in particular matters that are central to classical mechanics and deserve a systematic exposition.

A set of lecture notes is not a monograph on the subject and is not meant to be one. Rather, its purpose is to give a solid introduction and prepare the student for further studies on her own. Accordingly, there is no ambition of, and no attempt at, treating each and every aspect of classical mechanics in these notes — they just represent what I could and would deal with in one semester. The material of this book is my personal selection for that one-semester second-year course, presented in full during twenty-three two-hour lectures. Other lecturers will surely omit some of the material of my choice in favor of topics that I did not choose to include. What I selected is, in fact, much material for one semester; one might wish to spread it out over one and a half semesters and cover additional topics in the remaining half of the second semester, perhaps introducing the students to relativistic mechanics or deterministic chaos.

The feedback I received from students in class, from the Ph.D. students and postdoctoral fellows who conducted tutorial sessions, and from colleagues in the NUS Department of Physics and elsewhere was invaluable and led to many improvements of the text. While I am much obliged to all of them, I can only name a few: DAI Jibo, HAN Rui, HU Yu-Xin, LE Huy Nguyen, LEE Kean Loon, LEN Yink Loong, LI Xikun, Hui Khoon NG, SEAH Yi-Lin, SHANG Jiangwei, and YE Luyao represent this large crowd.

I am sincerely grateful for the professional help by the staff of World Scientific Publishing Co., which was crucial for the completion; in particular, I acknowledge Christopher TEO's competent support. CHAI Jing Hao, NGUYEN Duy Quang, and SEAH Zuo Sheng typed the original set of notes and thus provided the electronic version that I could then work on, and DO Thi Xuan Hung helped me prepare the figures; I thank them cordially.

This book would not exist without the outstanding teachers, colleagues, and students who taught me so much. I dedicate these lectures to them.

I wish to thank my dear wife Ola for her continuing understanding and patience by which she is giving me the peace of mind that is the source of all achievements.

Singapore, January 2015 *BG Englert*

Contents

Glossary

Here is a list of the symbols used in the text; the numbers in square brackets indicate the pages of first occurrence or of other significance.

Miscellanea

PSA	Principle of Stationary Action [174]
$\mathbf{1}$; $1_2, 1_3$	unit dyadic [102], 2×2, 3×3 unit matrix [16,3]
$\nu!$; $\lfloor x \rfloor$	factorial of number ν; floor of x [70]
$\mathrm{Re}\,(z)$, $\mathrm{Im}\,(z)$	real, imaginary part of complex number z
$a = \lvert\boldsymbol{a}\rvert$	length of vector $\boldsymbol{a}$ [5]
$\boldsymbol{a}^{\mathrm{T}}$	transposed row version of column vector $\boldsymbol{a}$ [100]
$\boldsymbol{a}\cdot\boldsymbol{b}$, $\boldsymbol{a}\times\boldsymbol{b}$	scalar product [3], vector product [5] of vectors $\boldsymbol{a}$ and $\boldsymbol{b}$
$\boldsymbol{a}\,\boldsymbol{b}$	dyadic product of vectors $\boldsymbol{a}$ and $\boldsymbol{b}$ [101]
$\mathsf{A} \times\!\!\times \mathsf{B}$	two-fold vector product of dyadics A and B [294]
$\dfrac{\partial}{\partial t}$, $\dfrac{\mathrm{d}}{\mathrm{d}t}$	parametric, total time derivative [225]
$\dot{\boldsymbol{r}}$, $\ddot{\boldsymbol{r}}$	first, second time derivative of $\boldsymbol{r}(t)$
$f'(x)$, $f''(x)$	first, second derivative of $f(x)$ with respect to its argument
$f^2(x)$, $f^{-1}(x)$	square, inverse of function f: $f^2(x) = f\big(f(x)\big)$, $f\big(f^{-1}(x)\big) = x$
$f(x)^2$, $f(x)^{-1}$	square, reciprocal of the function value: $f(x)^2 = \big(f(x)\big)^2$, $f(x)^{-1} = 1/f(x)$
$\overline{f(t)}$	long-time average of $f(t)$ [295]
$\boldsymbol{\nabla} V$	gradient of field $V(\boldsymbol{r},t)$ [33]
$\boldsymbol{\nabla}\times\boldsymbol{F}$, $\boldsymbol{\nabla}\cdot\boldsymbol{F}$	curl [90], divergence [105] of vector field $\boldsymbol{F}(\boldsymbol{r},t)$
$\boldsymbol{\nabla}^2$	Laplacian differential operator [105]
$\mathrm{tr}\{\mathsf{A}\}$, $\det\{\mathsf{A}\}$	trace, determinant of a dyadic [293]

$\{F, G\}$	Poisson bracket of phase-space functions $F(X, P, t)$ and $G(X, P, t)$ [226]
$\langle F\rangle$, $\sigma(F)$	mean value [308], variance [308] of phase-space function $F(X, P, t)$

Latin alphabet

a	major half-axis of an ellipse [127]
a_{eff}	effective length [254]
$\boldsymbol{a}$, $\boldsymbol{b}$, $\boldsymbol{c}$, ...	vectors
$\boldsymbol{A}(\boldsymbol{r}, t)$	vector potential [239]
b	impact parameter [141]; minor half-axis of an ellipse [323]
$\boldsymbol{B}(\boldsymbol{r}, t)$	magnetic field [236]
c	speed of light, $c = 2.99792 \times 10^8\,\mathrm{m\,s^{-1}}$ [236]; linear eccentricity [327]
$\cos, \sin, \ldots$	trigonometric functions; $\sin\alpha = \sin(\alpha)$, $\sin\frac{\beta}{2} = \sin(\frac{1}{2}\beta)$
$\cosh, \sinh, \ldots$	hyperbolic functions; $\cosh\vartheta = \cosh(\vartheta)$
$\mathrm{d}X$	differential of quantity X
d	distance from directrix [326]
$\boldsymbol{d}$	electric dipole moment [306]
$(\mathrm{d}\boldsymbol{r})$	volume element [39]
$\mathrm{d}\boldsymbol{S}$	vectorial surface element [37]
$(\mathrm{d}X)(\mathrm{d}P)$	phase-space volume element [232]
D	distance from focus to directrix [327]
e	Euler's number, $\mathrm{e} = 2.71828\ldots$
$\boldsymbol{e}_a$	local unit vector for coordinate $a = x, y, z, s, \varphi, r, \vartheta, \ldots$
E, E_{tot}	energy [74], total energy [110]
$\boldsymbol{E}(\boldsymbol{r}, t)$	electric field [105]
f, $f_\pm$	distance from focus [325]
F, $\boldsymbol{F}$	force [41]
$\boldsymbol{F}^{(\mathrm{ext})}$, $P^{(\mathrm{ext})}$	external force, its power [114]
$\boldsymbol{g}$	gravitational acceleration, $g \cong 9.8\,\mathrm{m\,s^{-2}}$ [43]
G	gravitational constant, $G = 6.67 \times 10^{-11}\,\mathrm{N\,m^2\,kg^{-2}}$ [129]
$G(t - t')$	retarded Green's function [66]
$G(\boldsymbol{r} - \boldsymbol{r}')$	Green's function for the Laplacian [155]
$\hbar$	Planck's constant$/(2\pi)$, $\hbar = 6.62607 \times 10^{-34}\,\mathrm{J\ s}$ [214]
$H(X, P, t)$	Hamilton function [212]
i	imaginary unit, $\mathrm{i}^2 = -1$
I_1, I_2, I_3	principal moments of inertia [247]

I, I_R	inertia dyadic [245], for reference point $\boldsymbol{R}$ [246]
J	joule, SI unit of energy, $1\,\mathrm{J} = 1\,\mathrm{kg\,m^2\,s^{-2}}$
kg	kilogram, basic SI unit of mass
k	spring constant [48]
l, ℓ, L	length of an object
$L(X, \dot{X}, t)$	Lagrange function [174]
$\boldsymbol{l}$, $\boldsymbol{L}$; $\boldsymbol{L}_{\mathrm{tot}}$	angular momentum; total angular momentum [111]
log	natural logarithm
m	meter, basic SI unit of length
m, m_k; $m_\odot$	mass, of the kth particle; mass of the sun [129]
m, M, m_{eff}	reduced [123], total mass [114], effective mass [257]
N	newton, SI unit of force, $1\,\mathrm{N} = 1\,\mathrm{kg\,m\,s^{-2}}$
$\boldsymbol{p}$, $\boldsymbol{P}_{\mathrm{tot}}$	momentum [212], total momentum [108]
p_x, p_y, p_φ, ...	canonical momentum for coordinate x, y, φ, ... [215]
P	collection of canonical momenta [215]
q; q_k	electric charge [105]; of the kth particle
Q	quadrupole moment dyadic [150]
r	radial distance (spherical coordinates) [29]
$r_>$, $r_<$	larger, smaller one of r and r' [152]
$\boldsymbol{r}$, $\boldsymbol{r}_j$	position vector [1], of the jth particle
$\boldsymbol{r}_\parallel$, $\boldsymbol{r}_\perp$	parallel, perpendicular component of $\boldsymbol{r}$ [10]
R	radius of a circle, of a sphere
$\boldsymbol{R}$, $\boldsymbol{V}$	center-of-mass position, velocity [115]
s	second, basic SI unit of time
s	radial distance (polar, cylindrical coordinates) [20]
$\mathrm{sgn}(x)$	sign of x
t	time
T	period [51,77], radial period [135], duration
v, $\boldsymbol{v}$; $\boldsymbol{v}_j$	speed, velocity [18]; of the jth particle
$\boldsymbol{v}_\infty$	terminal velocity [44]
$V(\boldsymbol{r}, t)$, $V_{\mathrm{eff}}(s)$	potential energy [74], effective potential energy [133]
W	watt, SI unit of power, $1\,\mathrm{W} = 1\,\mathrm{kg\,m^2\,s^{-3}}$
W_{12}, δW_{12}	action, its infinitesimal change [174]
x, y, z	cartesian coordinates of $\boldsymbol{r}$ [2]
x_0; x_1, x_2	coordinate of potential-energy minimum; turning points [74]
X, $\dot{X}$	collection of coordinates [176], of velocities [215]

Greek alphabet and Greek-Latin combinations

$\alpha, \beta, \gamma, \ldots$	coefficients
γ	Newtonian friction coefficient [44]
δ_{jk}	delta symbol [4]
δX	infinitesimal change of quantity X
$\delta(t)$, $\delta(\boldsymbol{r})$	one-dimensional, three-dimensional delta function [62,156]
$\delta_\tau(t)$, $\delta'(t)$	model for [63], derivative of $\delta(t)$ [286]
ϵ, $\boldsymbol{\epsilon}$	very small positive quantity, very short vector
ϵ	numerical eccentricity of an ellipse [127]
ϵ_{jkl}	epsilon symbol [7]
$\eta(t)$, $\eta_\tau(t)$	unit step function [61], model for it [63]
ϑ	polar angle (spherical coordinates) [29]; Euler angle [259]
θ, Θ	scattering angle [140], in the lab frame [298]
κ	air-drag friction coefficient [46]
κ	constant of motion (area per unit time) [127]
λ	latitude [29]
λ, $\lambda(t)$	Lagrange multiplier [167], $\sim$ function [183]
λ_1, λ_2	complex frequencies (damped harmonic oscillator) [51]
ξ, η, φ	parabolic coordinates [283]
π	Archimedes's constant, $\pi = 3.14159\ldots$
$\rho(\boldsymbol{r})$; $\rho(X, P, t)$	mass density [145]; phase-space density [232]
σ, $\dfrac{\mathrm{d}\sigma}{\mathrm{d}\Omega}$	total [299], differential cross section [142]
τ	characteristic time [47]
$\boldsymbol{\tau}$, $\boldsymbol{\tau}_j$; $\boldsymbol{\tau}^{(\mathrm{ext})}$	torque, of the jth particle [112]; external torque [114]
φ	rotation angle [13]; azimuth [20]; Euler angle [259]
$\boldsymbol{\delta\varphi}$	infinitesimal rotation vector [12]
ϕ	angle parameter, phase shift [57]
Φ	angular period [136]
$\Phi(\boldsymbol{r})$, $\Phi(\boldsymbol{r}, t)$	gravitational potential [146], scalar potential [239]
ψ	Euler angle [259]
ω, $\boldsymbol{\omega}$	angular velocity [19], angular velocity vector [19]
ω, ω_l	circular frequency [49], characteristic frequency [192]
Ω	circular frequency of periodic force [57]
$\boldsymbol{\Omega}$	angular velocity vector for the rotating earth [267]
$\mathrm{d}\Omega$	solid-angle element [39]

Chapter 1

Kinematics

We begin with Kinematics, that is: How do we describe the physical objects under study? First, there is an idealization that can be removed later, in the simplification that we treat material bodies as if they were point-like objects. This should be understood as focusing on the relations between material bodies that are sufficiently far apart that their individual sizes and shapes do not matter: When dealing with the earth in orbit about the sun, the size of the earth is so tiny compared with its distance from the sun that we get a very good understanding of the physics by describing the earth as a point-like object with position and velocity — one position and one velocity, not many positions and velocities for all the parts that constitute the earth. Other phenomena, such as tides, for example, do require a much more detailed description of the earth, but even then we can think of her as being composed of a large number of essentially point-like parts. In short, it is well justified to begin with a discussion of the kinematics of point-like massive bodies, of *mass points* or *point masses.* The emphasis on mass, that is: on the material nature of the objects, is a reminder that we are not going to deal with immaterial things — the motion of shadows, for example, may be intriguing, but it is not what we are interested in.

1.1 Vectors and all that

1.1.1 *Cartesian coordinates*

At any time t, as measured by the wall clock, say, a point mass has a position that we describe by a *position vector* $\boldsymbol{r}(t)$, usually a function of time unless the mass is at rest and does not move. The position vector is relative to a reference point that we take as the origin of a cartesian coordinate system (named after Descartes*):

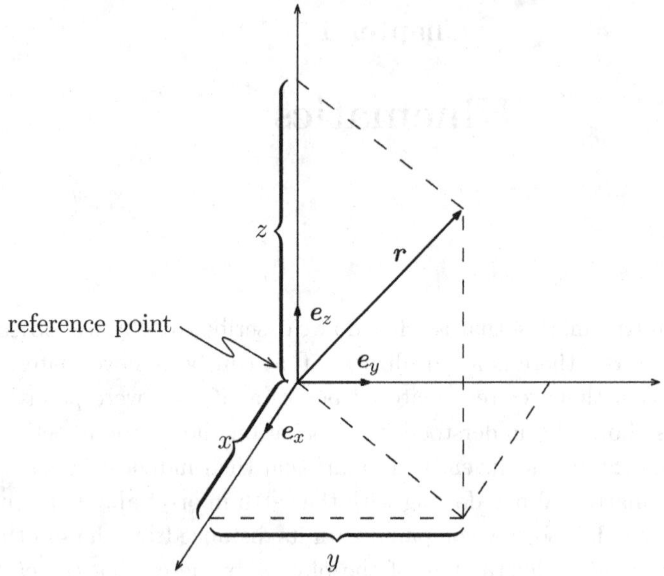

(1.1.1)

This gives us a parameterization of the position vector in terms of three coordinates,

$$r \mathrel{\widehat{=}} \begin{pmatrix} x \\ y \\ z \end{pmatrix}. \tag{1.1.2}$$

We do not write "=" because the vector $\boldsymbol{r}$ is a geometrical object that has a meaning irrespective of the chosen coordinate system. There are many coordinate systems, each of them giving another parameterization of the same position vector. The vector is not equal to its column of coordinates; the column is a coordinate-system dependent representation of the vector.

The actual specification of the coordinate system is done by stating, in addition to the reference point, the *unit vectors* for the coordinate axes, denoted by $\boldsymbol{e}_x$, $\boldsymbol{e}_y$, $\boldsymbol{e}_z$ in (1.1.1). These are vectors with no metrical dimensions and unit length. For a cartesian coordinate system we further require that the three unit vectors are pairwise orthogonal and, by convention, they make up a right-hand system in the order $\boldsymbol{e}_x$, $\boldsymbol{e}_y$, $\boldsymbol{e}_z$, meaning that if you

*René DESCARTES (1596–1650, Latinized name: Renatus Cartesius)

arrange the thumb, the forefinger, and the middle finder of your right hand accordingly, you can align them with $\boldsymbol{e}_x$, $\boldsymbol{e}_y$, $\boldsymbol{e}_z$, respectively, or a cyclic permutation thereof.

In terms of the unit vectors, we then have the equality

$$\boldsymbol{r} = x\boldsymbol{e}_x + y\boldsymbol{e}_y + z\boldsymbol{e}_z\,, \tag{1.1.3}$$

which is an example of vector addition. There are three vectors on the right, each obtained by multiplying the unit vector by the corresponding coordinate. The simple mathematics of multiplying vectors by numbers and adding vectors is familiar and we will not spend time on it.

We can unify the "$\widehat{=}$" of (1.1.2) with the "$=$" of (1.1.3) by writing

$$\boldsymbol{r} = \begin{pmatrix}\boldsymbol{e}_x & \boldsymbol{e}_y & \boldsymbol{e}_z\end{pmatrix}\begin{pmatrix}x\\ y\\ z\end{pmatrix} \widehat{=} \begin{pmatrix}x\\ y\\ z\end{pmatrix}, \tag{1.1.4}$$

where the row of unit vectors $\begin{pmatrix}\boldsymbol{e}_x & \boldsymbol{e}_y & \boldsymbol{e}_z\end{pmatrix}$ is multiplied by the column of coordinates $\begin{pmatrix}x\\ y\\ z\end{pmatrix}$ in the usual way of matrix multiplication to yield the sum of (1.1.3). If we take this one step further by representing the unit vectors themselves by their columns of coordinates, such as $\boldsymbol{e}_x \,\widehat{=}\, \begin{pmatrix}1\\ 0\\ 0\end{pmatrix}$, the row of unit vectors is represented by the 3×3 unit matrix $\mathbf{1}_3$,

$$\begin{pmatrix}\boldsymbol{e}_x & \boldsymbol{e}_y & \boldsymbol{e}_z\end{pmatrix} \widehat{=} \begin{pmatrix}1\,0\,0\\ 0\,1\,0\\ 0\,0\,1\end{pmatrix} = \mathbf{1}_3\,. \tag{1.1.5}$$

1.1.2 *Scalar product*

The *scalar product* of two vectors is familiar as well, for which

$$\boldsymbol{e}_a \cdot \boldsymbol{e}_b = \begin{cases} 1 \text{ if } a = b \\ 0 \text{ if } a \neq b \end{cases} \tag{1.1.6}$$

is the basic example. It expresses what is stated above: The three basic vectors are pairwise orthogonal. The statements

$$x = \boldsymbol{e}_x \cdot \boldsymbol{r}\,, \quad y = \boldsymbol{e}_y \cdot \boldsymbol{r}\,, \quad z = \boldsymbol{e}_z \cdot \boldsymbol{r} \tag{1.1.7}$$

follow immediately. Looking at the x part,

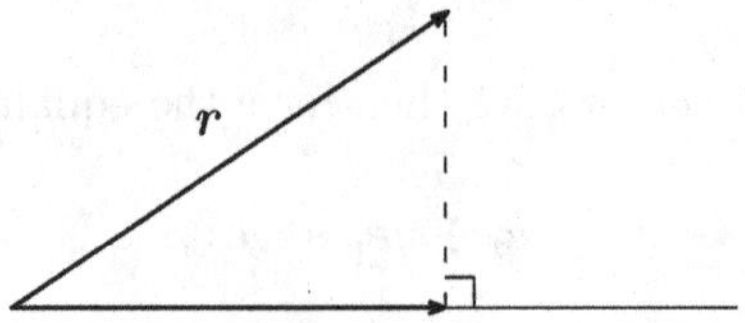

$$x\boldsymbol{e}_x \tag{1.1.8}$$

reminds us of the geometrical meaning: $\boldsymbol{e}_x \boldsymbol{e}_x \cdot \boldsymbol{r}$ is the projection of $\boldsymbol{r}$ onto the direction specified by the unit vector $\boldsymbol{e}_x$; and likewise for any arbitrary unit vector that need not coincide with one of the unit vectors that define the chosen cartesian coordinates.

For the situation on the right-hand side of (1.1.6), we have a useful special symbol, known as *Kronecker's* delta symbol*,

$$\left\{\begin{matrix} 1 \text{ if } a = b \\ 0 \text{ if } a \neq b \end{matrix}\right\} = \delta_{ab} \,. \tag{1.1.9}$$

If we use it, the *orthonormality* statement in (1.1.6) has the compact appearance

$$\boldsymbol{e}_a \cdot \boldsymbol{e}_b = \delta_{ab} \,. \tag{1.1.10}$$

The basic vectors are *orthonormal*, that is: pairwise orthogonal and each of unit length.

There is a distributive law for the scalar product,

$$(\boldsymbol{a} + \boldsymbol{b}) \cdot \boldsymbol{c} = \boldsymbol{a} \cdot \boldsymbol{c} + \boldsymbol{b} \cdot \boldsymbol{c} \,, \tag{1.1.11}$$

otherwise we would not call it a product in the first place, and this is frequently exploited when computing the scalar products of vectors parameterized by cartesian coordinates; the two vectors

$$\boldsymbol{a} = a_x \boldsymbol{e}_x + a_y \boldsymbol{e}_y + a_z \boldsymbol{e}_z \mathrel{\widehat{=}} \begin{pmatrix} a_x \\ a_y \\ a_z \end{pmatrix}$$

$$\text{and} \quad \boldsymbol{b} = b_x \boldsymbol{e}_x + b_y \boldsymbol{e}_y + b_z \boldsymbol{e}_z \mathrel{\widehat{=}} \begin{pmatrix} b_x \\ b_y \\ b_z \end{pmatrix} \tag{1.1.12}$$

*Leopold KRONECKER (1823–1891)

have the scalar product

$$\begin{aligned} \boldsymbol{a}\cdot\boldsymbol{b} &= a_x\boldsymbol{e}_x\cdot\boldsymbol{b} + a_y\boldsymbol{e}_y\cdot\boldsymbol{b} + a_z\boldsymbol{e}_z\cdot\boldsymbol{b} \\ &= a_xb_x + a_yb_y + a_zb_z\,, \end{aligned} \tag{1.1.13}$$

that is: we multiply the corresponding coordinates and add the products. The individual terms in this sum depend on the coordinate system used, whereas the value of $\boldsymbol{a}\cdot\boldsymbol{b}$ does not depend on the choice of coordinate system; it has the geometrical meaning that derives from the projection property depicted in (1.1.8).

In this context, we finally note that the scalar product of a vector with itself,

$$\boldsymbol{a}\cdot\boldsymbol{a} = |\boldsymbol{a}|^2 = a^2\,, \tag{1.1.14}$$

is the square of the length $|\boldsymbol{a}| = a$ of the vector. We use this very often when we need to calculate the length of a vector. For example, the distance between two point masses with position vectors $\boldsymbol{r}$ and $\boldsymbol{r}'$ is the length of the difference vector $\boldsymbol{r}-\boldsymbol{r}'$,

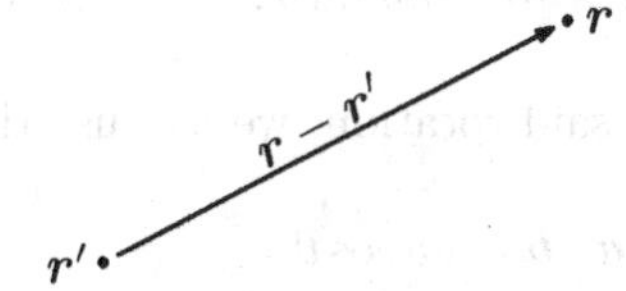

(1.1.15)

$$\begin{aligned} |\boldsymbol{r}-\boldsymbol{r}'| &= \sqrt{(\boldsymbol{r}-\boldsymbol{r}')\cdot(\boldsymbol{r}-\boldsymbol{r}')} \\ &= \sqrt{(x-x')^2+(y-y')^2+(z-z')^2}\,, \end{aligned}$$

fully consistent with what we expect as a consequence of the Pythagorean* theorem about the sides of rectangular triangles.

1.1.3 *Vector product*

Further, we remind ourselves of the *vector product.* For two vectors $\boldsymbol{a}$ and $\boldsymbol{b}$, their vector product $\boldsymbol{a}\times\boldsymbol{b}$ is orthogonal to the plane spanned by $\boldsymbol{a}$ and $\boldsymbol{b}$, such that $\boldsymbol{a}$, $\boldsymbol{b}$, $\boldsymbol{a}\times\boldsymbol{b}$ are a right-handed trio of vectors in this order, and the length of $\boldsymbol{a}\times\boldsymbol{b}$ is the area of the parallelogram formed by $\boldsymbol{a}$ and $\boldsymbol{b}$:

---

*PYTHAGORAS of Samos (510–495 BC)

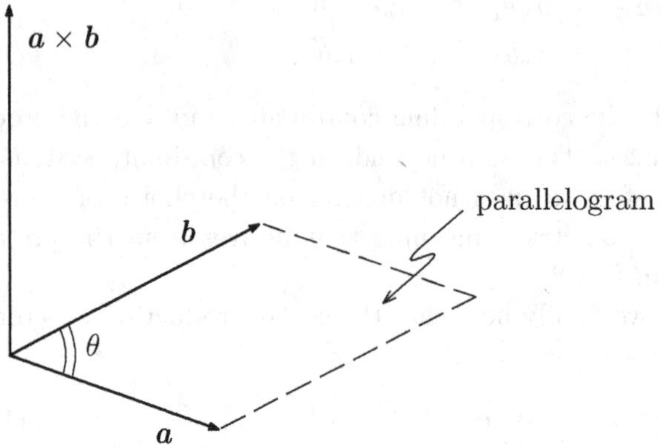

(1.1.16)

If we denote by θ the angle by which we have to rotate $\boldsymbol{a}$ (in the shortest way) to align it with $\boldsymbol{b}$, then that parallelogram has area $ab\sin\theta$, so that

$$|\boldsymbol{a}\times\boldsymbol{b}| = ab\sin\theta\,. \tag{1.1.17}$$

Rather than defining θ by the said rotation, we can use the scalar product

$$\boldsymbol{a}\cdot\boldsymbol{b} = ab\cos\theta \tag{1.1.18}$$

to identify the angle between $\boldsymbol{a}$ and $\boldsymbol{b}$, as follows immediately from the projection meaning.

Combining the two statements, we have a vector version of the Pythagorean theorem,

$$(\boldsymbol{a}\cdot\boldsymbol{b})^2 + (\boldsymbol{a}\times\boldsymbol{b})^2 = a^2b^2\,, \tag{1.1.19}$$

where, as in (1.1.14), squaring a vector means taking its scalar product with itself,

$$(\boldsymbol{a}\times\boldsymbol{b})^2 = (\boldsymbol{a}\times\boldsymbol{b})\cdot(\boldsymbol{a}\times\boldsymbol{b})\,. \tag{1.1.20}$$

Since it is often simpler to calculate $\boldsymbol{a}\cdot\boldsymbol{b}$ than $\boldsymbol{a}\times\boldsymbol{b}$, the very useful identity (1.1.19) offers a convenient way of finding the length of $\boldsymbol{a}\times\boldsymbol{b}$.

For the vector product, we also have a distributive law,

$$(\boldsymbol{a}+\boldsymbol{b})\times\boldsymbol{c} = \boldsymbol{a}\times\boldsymbol{c}+\boldsymbol{b}\times\boldsymbol{c}\,. \tag{1.1.21}$$

Together with the basic vector products of the cartesian unit vectors,

$$e_a \times e_b = \begin{cases} 0 & \text{if } a = b\,, \\ e_c & \text{if } abc \text{ is an even permutation of } xyz\,, \\ -e_c & \text{if } abc \text{ is an odd permutation of } xyz\,, \end{cases} \tag{1.1.22}$$

it enables us to find the coordinates of $\boldsymbol{a} \times \boldsymbol{b}$ in terms of the coordinates of $\boldsymbol{a}$ and $\boldsymbol{b}$,

$$\begin{aligned} \boldsymbol{a} \times \boldsymbol{b} &= (a_x e_x + a_y e_y + a_z e_z) \times (b_x e_x + b_y e_y + b_z e_z) \\ &= (a_y b_z e_y \times e_z + a_z b_y e_z \times e_y) + (a_z b_x e_z \times e_x + a_x b_z e_x \times e_z) \\ &\quad + (a_x b_y e_x \times e_y + a_y b_x e_y \times e_x) \\ &= (a_y b_z - a_z b_y) e_x + (a_z b_x - a_x b_z) e_y + (a_x b_y - a_y b_x) e_z\,, \end{aligned} \tag{1.1.23}$$

which exhibits a cyclic structure that is worth memorizing.

In passing we note that, upon combining (1.1.22) with the orthonormality relation (1.1.6), we have

$$\big(e_a \times e_b\big) \cdot e_c = \left\{ \begin{matrix} 1 & \text{if } abc \text{ is an even permutation of } xyz \\ -1 & \text{if } abc \text{ is an odd permutation of } xyz \\ 0 & \text{else} \end{matrix} \right\} = \epsilon_{abc}\,, \tag{1.1.24}$$

which introduces the so-called *epsilon symbol.* The "else" case includes the "$a = b$" case of (1.1.22) and also the cases of $b = c$ or $a = c$, as one notices immediately. Whereas Kronecker's delta symbol δ_{ab} compares two quantities ("Are they equal or not?"), the epsilon symbol compares three quantities that have a conventional cyclic order ("Are all three present, and is the order cyclic or anticyclic?"). The epsilon symbol can be used, for example, to give a compact expression to (1.1.22),

$$e_a \times e_b = \sum_c \epsilon_{abc} e_c\,, \tag{1.1.25}$$

where the summation covers $c = x, y, z$.

As follows from their geometrical meaning, the order of the factors does not matter in the scalar product,

$$\boldsymbol{a} \cdot \boldsymbol{b} = \boldsymbol{b} \cdot \boldsymbol{a}\,, \tag{1.1.26}$$

whereas the vector product reverses direction when the factors are interchanged,

$$\boldsymbol{a} \times \boldsymbol{b} = -\boldsymbol{b} \times \boldsymbol{a}\,. \tag{1.1.27}$$

Both statements are confirmed by the coordinate version for the two kinds of products of vectors. In particular, it follows that $\boldsymbol{a} \times \boldsymbol{a} = 0$; there is no plane spanned by $\boldsymbol{a}$ alone.

1.1.4 *Dyadic product*

In addition to the scalar product $\boldsymbol{a} \cdot \boldsymbol{b}$ and the vector product $\boldsymbol{a} \times \boldsymbol{b}$ of the vectors $\boldsymbol{a}$ and $\boldsymbol{b}$, there is also their *dyadic product.* We postpone its discussion to Section 3.2.4, where the dyadic product comes up in a typical context.

1.1.5 *Three-vector product*

Two vectors $\boldsymbol{a}$ and $\boldsymbol{b}$ define a parallelogram with area $|\boldsymbol{a} \times \boldsymbol{b}|$, three vectors span a parallelepiped whose volume is $(\boldsymbol{a} \times \boldsymbol{b}) \cdot \boldsymbol{c}$,

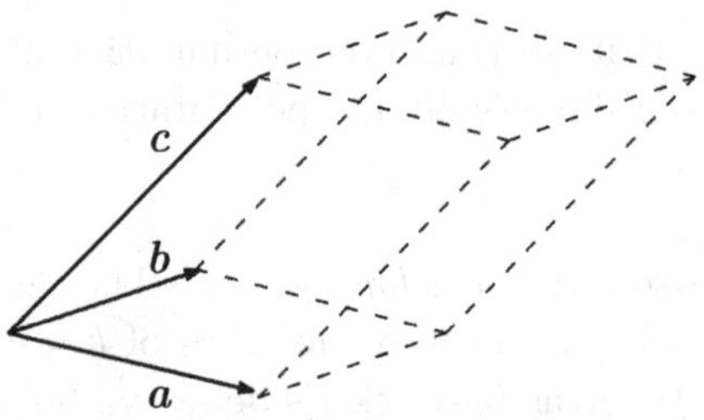

(1.1.28)

which is a *signed volume*: positive if $\boldsymbol{a}$, $\boldsymbol{b}$, $\boldsymbol{c}$ are a right-handed system, negative if they are left-handed. The geometrical meaning of the *three-vector product* $(\boldsymbol{a} \times \boldsymbol{b}) \cdot \boldsymbol{c}$ tells us that the value does not change if the vectors are permuted cyclically,

$$\begin{aligned} (\boldsymbol{a} \times \boldsymbol{b}) \cdot \boldsymbol{c} &= (\boldsymbol{b} \times \boldsymbol{c}) \cdot \boldsymbol{a} = (\boldsymbol{c} \times \boldsymbol{a}) \cdot \boldsymbol{b} \\ &= \boldsymbol{c} \cdot (\boldsymbol{a} \times \boldsymbol{b}) = \boldsymbol{a} \cdot (\boldsymbol{b} \times \boldsymbol{c}) = \boldsymbol{b} \cdot (\boldsymbol{c} \times \boldsymbol{a}) \,, \end{aligned} \tag{1.1.29}$$

where the bottom line exploits the symmetry of the scalar product under factor interchange. In summary, it does not matter where we have the vector product and where the scalar product in the three-vector product but it matters whether the order is cyclic or anticyclic. These observations are confirmed by a look at the expression in terms of cartesian coordinates,

$$(\boldsymbol{a} \times \boldsymbol{b}) \cdot \boldsymbol{c} = a_x b_y c_z + a_y b_z c_x + a_z b_x c_y - a_z b_y c_x - a_x b_z c_y - a_y b_x c_z \,, \tag{1.1.30}$$

which happens to have the form of the determinant of a 3×3 matrix whose columns are the coordinate columns of the three vectors,

$$(\boldsymbol{a} \times \boldsymbol{b}) \cdot \boldsymbol{c} = \begin{vmatrix} a_x & b_x & c_x \\ a_y & b_y & c_y \\ a_z & b_z & c_z \end{vmatrix} . \tag{1.1.31}$$

As an immediate consequence of the geometrical meaning of $(\boldsymbol{a} \times \boldsymbol{b}) \cdot \boldsymbol{c}$, the value of this determinant is the same irrespective of the cartesian coordinate system to which the matrix entries refer.

1.1.6 *Double vector product*

What about a double vector product, $\boldsymbol{a} \times (\boldsymbol{b} \times \boldsymbol{c}) = ?$ We can figure it out by exploiting the geometrical meaning. First, we note that $\boldsymbol{b} \times \boldsymbol{c}$ is orthogonal to the plane spanned by $\boldsymbol{b}$ and $\boldsymbol{c}$. Then $\boldsymbol{a} \times (\boldsymbol{b} \times \boldsymbol{c})$ is orthogonal to $\boldsymbol{b} \times \boldsymbol{c}$, telling us that the resulting vector is in the plane spanned by $\boldsymbol{b}$ and $\boldsymbol{c}$,

$$\boldsymbol{a} \times (\boldsymbol{b} \times \boldsymbol{c}) = \beta \boldsymbol{b} + \gamma \boldsymbol{c} \tag{1.1.32}$$

with currently unknown coefficients β and γ. Now we note that the left-hand side is linear in $\boldsymbol{a}$, $\boldsymbol{b}$, and $\boldsymbol{c}$, so that, in particular, if we multiply any of the three vectors by a number ($\boldsymbol{a} \to 3\boldsymbol{a}$, say) the right-hand side must also be multiplied by the same number. Therefore, β must contain $\boldsymbol{a}$ and $\boldsymbol{c}$ linearly, which is only possible in the form of $\boldsymbol{a} \cdot \boldsymbol{c}$, and we conclude that $\beta = \lambda \boldsymbol{a} \cdot \boldsymbol{c}$ with a numerical constant λ that does not depend on which three vectors are involved. By the same token, we get $\gamma = \mu \boldsymbol{a} \cdot \boldsymbol{b}$ with another numerical constant μ. At this stage, we have

$$\boldsymbol{a} \times (\boldsymbol{b} \times \boldsymbol{c}) = \lambda \boldsymbol{a} \cdot \boldsymbol{c}\, \boldsymbol{b} + \mu \boldsymbol{a} \cdot \boldsymbol{b}\, \boldsymbol{c} \tag{1.1.33}$$

and observe further that interchanging $\boldsymbol{b}$ and $\boldsymbol{c}$ on the left reverses the direction of the resulting vector, which in turn requires $\mu = -\lambda$ on the right, so that

$$\boldsymbol{a} \times (\boldsymbol{b} \times \boldsymbol{c}) = \lambda \boldsymbol{a} \cdot \boldsymbol{c}\, \boldsymbol{b} - \lambda \boldsymbol{a} \cdot \boldsymbol{b}\, \boldsymbol{c} . \tag{1.1.34}$$

What remains is to determine the value of λ. A single special example serves for this purpose. Let us take

$$\boldsymbol{a} = \boldsymbol{e}_y , \quad \boldsymbol{b} = \boldsymbol{e}_x , \quad \boldsymbol{c} = \boldsymbol{e}_y , \tag{1.1.35}$$

for which $\boldsymbol{a} \times (\boldsymbol{b} \times \boldsymbol{c}) = \boldsymbol{e}_y \times (\boldsymbol{e}_x \times \boldsymbol{e}_y) = \boldsymbol{e}_y \times \boldsymbol{e}_z = \boldsymbol{e}_x$ on the left, and $\boldsymbol{a} \cdot \boldsymbol{c}\, \boldsymbol{b} = \boldsymbol{e}_y \cdot \boldsymbol{e}_y\, \boldsymbol{e}_x = \boldsymbol{e}_x$ and $\boldsymbol{a} \cdot \boldsymbol{b}\, \boldsymbol{c} = \boldsymbol{e}_y \cdot \boldsymbol{e}_x\, \boldsymbol{e}_y = 0$ on the right, together we thus have $\boldsymbol{e}_x = \lambda \boldsymbol{e}_x$, so that $\lambda = 1$. In summary, then,

$$\boldsymbol{a} \times (\boldsymbol{b} \times \boldsymbol{c}) = \boldsymbol{a} \cdot \boldsymbol{c}\, \boldsymbol{b} - \boldsymbol{a} \cdot \boldsymbol{b}\, \boldsymbol{c}\,, \tag{1.1.36}$$

and

$$(\boldsymbol{a} \times \boldsymbol{b}) \times \boldsymbol{c} = \boldsymbol{a} \cdot \boldsymbol{c}\, \boldsymbol{b} - \boldsymbol{b} \cdot \boldsymbol{c}\, \boldsymbol{a} \tag{1.1.37}$$

is another, equivalent version of the identity.

This identity for double vector products is worth memorizing; it is frequently used in vector calculations. Here is an application. Given the position vector $\boldsymbol{r}$ and some unit vector $\boldsymbol{e}$, how do we decompose $\boldsymbol{r}$ into its components parallel and orthogonal to $\boldsymbol{e}$? Simply by means of

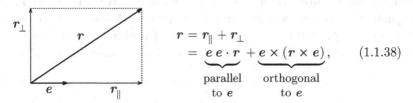

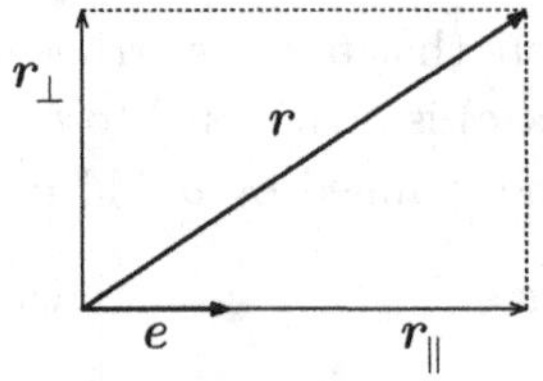

$$\begin{aligned} \boldsymbol{r} &= \boldsymbol{r}_\parallel + \boldsymbol{r}_\perp \\ &= \underbrace{\boldsymbol{e}\, \boldsymbol{e} \cdot \boldsymbol{r}}_{\text{parallel to } \boldsymbol{e}} + \underbrace{\boldsymbol{e} \times (\boldsymbol{r} \times \boldsymbol{e})}_{\text{orthogonal to } \boldsymbol{e}}\,, \end{aligned} \tag{1.1.38}$$

which is immediately verified in view of $\boldsymbol{e} \times (\boldsymbol{r} \times \boldsymbol{e}) = \boldsymbol{e} \cdot \boldsymbol{e}\, \boldsymbol{r} - \boldsymbol{e}\, \boldsymbol{e} \cdot \boldsymbol{r}$.

Another application is the scalar product of two vector products,

$$\begin{aligned} (\boldsymbol{a} \times \boldsymbol{b}) \cdot (\boldsymbol{c} \times \boldsymbol{d}) &= \boldsymbol{a} \cdot \big[\boldsymbol{b} \times (\boldsymbol{c} \times \boldsymbol{d})\big] \\ &= \boldsymbol{a} \cdot (\boldsymbol{c}\, \boldsymbol{b} \cdot \boldsymbol{d} - \boldsymbol{d}\, \boldsymbol{b} \cdot \boldsymbol{c})\,, \end{aligned} \tag{1.1.39}$$

so that

$$(\boldsymbol{a} \times \boldsymbol{b}) \cdot (\boldsymbol{c} \times \boldsymbol{d}) = \boldsymbol{a} \cdot \boldsymbol{c}\, \boldsymbol{b} \cdot \boldsymbol{d} - \boldsymbol{a} \cdot \boldsymbol{d}\, \boldsymbol{b} \cdot \boldsymbol{c}\,, \tag{1.1.40}$$

where the two first and the two second vectors are paired in the first, positive term while the two inner and the two outer vectors are paired in the second, negative term. The Pythagorean theorem (1.1.19) is a special case of (1.1.40).

1.1.7 *Infinitesimal rotations*

We took for granted that the vector product of two vectors is another vector. Let us now justify this, for which purpose we need to consider the response of vectors to rotations because this characterizes vectors. Finite rotations are complicated (see Section 1.1.8), but they are composed of a sequence of very many infinitesimal rotations and, therefore, it is sufficient to examine the response to infinitesimal rotations, that is: to rotations by angles $\delta\varphi$ so small that only the terms linear in $\delta\varphi$ matter, and all contributions of order $(\delta\varphi)^2$, or $(\delta\varphi)^3$, ..., can be consistently ignored.

The situation is as depicted here:

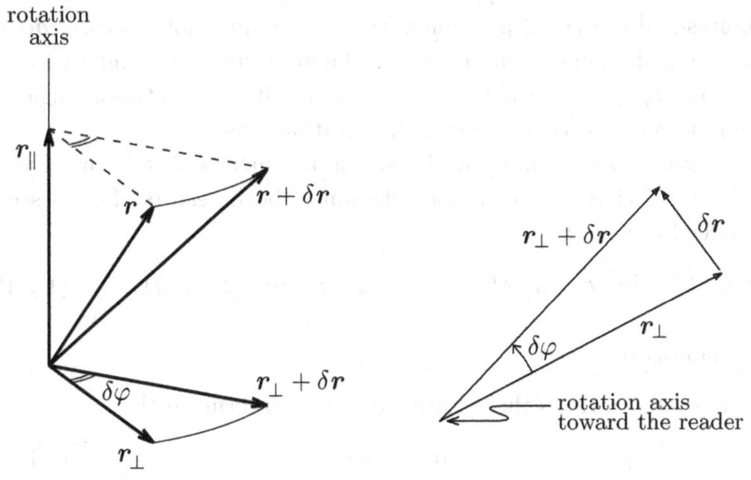

(1.1.41)

The increment $\delta\boldsymbol{r}$ of $\boldsymbol{r}$, the difference between the vector after and before the rotation, is orthogonal to $\boldsymbol{r}$ (this is where the infinitesimal nature of $\delta\boldsymbol{r}$ enters), because the length of $\boldsymbol{r}+\delta\boldsymbol{r}$ is the same as the length of $\boldsymbol{r}$, $(\boldsymbol{r}+\delta\boldsymbol{r})\cdot(\boldsymbol{r}+\delta\boldsymbol{r}) = r^2 + 2\boldsymbol{r}\cdot\delta\boldsymbol{r} = r^2$, and it is also orthogonal to the axis of rotation that is specified by unit vector $\boldsymbol{e}$. The length of $\delta\boldsymbol{r}$ is $r_\perp\delta\varphi$, as follows from the geometrical meaning of a rotation: The distance covered by the tip of the vector is proportional to the length of the orthogonal component $\boldsymbol{r}_\perp = \boldsymbol{e}\times(\boldsymbol{r}\times\boldsymbol{e})$ of the rotated vector, the parallel component $\boldsymbol{r}_\parallel = \boldsymbol{e}\,\boldsymbol{e}\cdot\boldsymbol{r}$ is not affected by the rotation about axis $\boldsymbol{e}$. Now,

$$r_\perp = |\boldsymbol{e}\times(\boldsymbol{r}\times\boldsymbol{e})| = |\boldsymbol{e}\times\boldsymbol{r}|\,, \tag{1.1.42}$$

and in summary, then, we have

$$\delta\boldsymbol{r} = \delta\varphi\,\boldsymbol{e}\times\boldsymbol{r} \tag{1.1.43}$$

for the effect of the rotation where, by convention, the unit vector $\boldsymbol{e}$ of the rotation axis points toward the reader in the situation depicted in (1.1.41) where, owing to the constraints of a drawing, $\delta\boldsymbol{r}$ is not infinitesimal but finite. It is customary to write $\boldsymbol{\delta\varphi}$ for the product $\delta\varphi\,\boldsymbol{e}$ of the infinitesimal rotation angle and the unit vector specifying the rotation axis. We then arrive at the compact expression

$$\delta \boldsymbol{r} = \boldsymbol{\delta\varphi} \times \boldsymbol{r} \tag{1.1.44}$$

for the response of vector $\boldsymbol{r}$ to the infinitesimal rotation associated with $\boldsymbol{\delta\varphi}$.

Note that both $\delta\boldsymbol{r}$ and $\boldsymbol{\delta\varphi}$ are infinitesimal vectors but, whereas $\delta\boldsymbol{r}$ is the infinitesimal change of position vector $\boldsymbol{r}$, we must not interpret $\boldsymbol{\delta\varphi}$ as the infinitesimal change of a vector $\boldsymbol{\varphi}$. There is no such "angle vector," we only have $\boldsymbol{\delta\varphi} = \delta\varphi\, \boldsymbol{e}$, the product of the infinitesimal rotation angle $\delta\varphi$ and the unit vector $\boldsymbol{e}$ that specifies the rotation axis.

The geometrical meaning of the scalar product $\boldsymbol{a}\cdot\boldsymbol{b}$ tells us that it should be unaffected by a common rotation of both vectors. Let us see if this is indeed the case:

$$\delta(\boldsymbol{a}\cdot\boldsymbol{b}) \underset{\substack{\uparrow \\ \text{product rule}}}{=} \delta\boldsymbol{a}\cdot\boldsymbol{b} + \boldsymbol{a}\cdot\delta\boldsymbol{b} = (\boldsymbol{\delta\varphi}\times\boldsymbol{a})\cdot\boldsymbol{b} + \boldsymbol{a}\cdot(\boldsymbol{\delta\varphi}\times\boldsymbol{b})\,, \tag{1.1.45}$$

where the second term is the negative of the first term, so that

$$\delta\,(\boldsymbol{a}\cdot\boldsymbol{b}) = 0\,, \tag{1.1.46}$$

as it should be. This observation is, of course, not a profound insight; it is merely a check of consistency.

We turn to vector products. For $\boldsymbol{a}\times\boldsymbol{b}$ to be a vector, we need

$$\delta(\boldsymbol{a}\times\boldsymbol{b}) = \boldsymbol{\delta\varphi}\times(\boldsymbol{a}\times\boldsymbol{b})\,, \tag{1.1.47}$$

which requires

$$(\boldsymbol{\delta\varphi}\times\boldsymbol{a})\times\boldsymbol{b} + \boldsymbol{a}\times(\boldsymbol{\delta\varphi}\times\boldsymbol{b}) = \boldsymbol{\delta\varphi}\times(\boldsymbol{a}\times\boldsymbol{b}) \tag{1.1.48}$$

after making use of the product rule in

$$\delta(\boldsymbol{a}\times\boldsymbol{b}) = \delta\boldsymbol{a}\times\boldsymbol{b} + \boldsymbol{a}\times\delta\boldsymbol{b}\,. \tag{1.1.49}$$

What is required, is the case indeed, because the *Jacobi* identity*

$$(\boldsymbol{a}\times\boldsymbol{b})\times\boldsymbol{c} + (\boldsymbol{b}\times\boldsymbol{c})\times\boldsymbol{a} + (\boldsymbol{c}\times\boldsymbol{a})\times\boldsymbol{b} = 0 \tag{1.1.50}$$

or, equivalently,

$$\boldsymbol{a}\times(\boldsymbol{b}\times\boldsymbol{c}) + \boldsymbol{b}\times(\boldsymbol{c}\times\boldsymbol{a}) + \boldsymbol{c}\times(\boldsymbol{a}\times\boldsymbol{b}) = 0 \tag{1.1.51}$$

is obeyed by any three vectors, as verified in Exercise 5.

*Carl Gustav Jacob JACOBI (1804–1851)

1.1.8 *Finite rotations*

To see how finite rotations grow from infinitesimal rotations, let us first consider rotations about the z axis. The vector rotated by angle φ has cartesian components

$$\boldsymbol{r}(\varphi) \mathrel{\widehat{=}} \begin{pmatrix} x(\varphi) \\ y(\varphi) \\ z_0 \end{pmatrix} \tag{1.1.52}$$

with

$$\boldsymbol{r}_0 = \boldsymbol{r}(\varphi = 0) \mathrel{\widehat{=}} \begin{pmatrix} x_0 \\ y_0 \\ z_0 \end{pmatrix} \tag{1.1.53}$$

for the unrotated vector, and we have already taken into account that the z component is not affected by a rotation about the z axis. Rotating by another $\delta\varphi$ takes us to

$$\boldsymbol{r}(\varphi + \delta\varphi) \mathrel{\widehat{=}} \begin{pmatrix} x(\varphi + \delta\varphi) \\ y(\varphi + \delta\varphi) \\ z_0 \end{pmatrix} \tag{1.1.54}$$

which we compare with

$$\boldsymbol{r}(\varphi + \delta\varphi) = \boldsymbol{r}(\varphi) + \delta\varphi\, \boldsymbol{e}_z \times \boldsymbol{r}(\varphi) \mathrel{\widehat{=}} \begin{pmatrix} x(\varphi) \\ y(\varphi) \\ z_0 \end{pmatrix} + \delta\varphi \begin{pmatrix} -y(\varphi) \\ x(\varphi) \\ 0 \end{pmatrix} \tag{1.1.55}$$

to find

$$\begin{aligned} x(\varphi + \delta\varphi) &= x(\varphi) - \delta\varphi\, y(\varphi)\,, \\ y(\varphi + \delta\varphi) &= y(\varphi) + \delta\varphi\, x(\varphi)\,. \end{aligned} \tag{1.1.56}$$

Upon noting that $x(\varphi+\delta\varphi) = x(\varphi) + \delta\varphi \dfrac{\mathrm{d}}{\mathrm{d}\varphi} x(\varphi)$ and likewise for $y(\varphi+\delta\varphi)$, these statements become a pair of coupled differential equations.

$$\frac{\mathrm{d}}{\mathrm{d}\varphi} x(\varphi) = -y(\varphi)\,, \qquad \frac{\mathrm{d}}{\mathrm{d}\varphi} y(\varphi) = x(\varphi)\,, \tag{1.1.57}$$

to be solved with the initial conditions $x(\varphi = 0) = x_0$, $y(\varphi = 0) = y_0$. This mathematical problem occurs surprisingly often in all branches of physics and many different contexts. Therefore it is useful to know methods for solving these equations. We will take a look at three methods; they are really three versions of one method.

First, we decouple the equations by another differentiation:

$$\left(\frac{\mathrm{d}}{\mathrm{d}\varphi}\right)^2 x(\varphi) = -\frac{\mathrm{d}}{\mathrm{d}\varphi}y(\varphi) = -x(\varphi) \tag{1.1.58}$$

and

$$\left(\frac{\mathrm{d}}{\mathrm{d}\varphi}\right)^2 y(\varphi) = \frac{\mathrm{d}}{\mathrm{d}\varphi}x(\varphi) = -y(\varphi)\,, \tag{1.1.59}$$

which is the same second-order differential equation twice. We need to solve one of them, say, the equation of $x(\varphi)$, because $y(\varphi) = -\frac{\mathrm{d}}{\mathrm{d}\varphi}x(\varphi)$ gives us the solution of the second equation without additional work. Before proceeding, we should observe that the pair of coupled first-order differential equations has been converted into a pair of uncoupled second-order differential equations. In the original form, the coupling between the equations is (a bit) troublesome, in the new version the coupling is removed at the price of now having second-order differential equations.

Regarding the second-order differential equation for $x(\varphi)$, we note that it is a linear equation (in the unknown function $x(\varphi)$), with the consequence that any linear combination $\lambda x_1(\varphi) + \mu x_2(\varphi)$ of two solutions $x_1(\varphi)$, $x_2(\varphi)$ with φ-independent coefficients λ, μ is also a solution. The basic solutions are

$$x_1(\varphi) = \cos\varphi \quad \text{and} \quad x_2(\varphi) = \sin\varphi\,, \tag{1.1.60}$$

since it is a familiar fact that the second derivative of either one of these trigonometric functions is the negative of the function itself. Further, $\cos\varphi$ and $\sin\varphi$ are linearly independent in the sense that $\lambda\cos\varphi + \mu\sin\varphi = 0$ for all φ is only possible if $\lambda = \mu = 0$. Are two basic solutions of (1.1.58) sufficient to get all solutions by linear superposition? Yes, says the general theory of differential equations, and this answer is easily understood: If we know the values of $x(\varphi = 0)$ and $\frac{\mathrm{d}x}{\mathrm{d}\varphi}(\varphi = 0)$, the differential equation tells us $\frac{\mathrm{d}^2x}{\mathrm{d}\varphi^2}(\varphi = 0)$ and so takes us to $x(\varphi = \epsilon)$ and $\frac{\mathrm{d}x}{\mathrm{d}\varphi}(\varphi = \epsilon)$, for a sufficiently small ϵ; then we repeat and advance to $\varphi = 2\epsilon$, then 3ϵ, and so forth in a unique fashion without any further input in addition to those initial values of $x(\varphi)$ and $\frac{\mathrm{d}x}{\mathrm{d}\varphi}(\varphi)$ at $\varphi = 0$.

Accordingly, $x(\varphi)$ must be of the form

$$x(\varphi) = \lambda \cos\varphi + \mu \sin\varphi\,, \tag{1.1.61}$$

and then

$$y(\varphi) = -\frac{\mathrm{d}}{\mathrm{d}\varphi}x(\varphi) = -\mu\cos\varphi + \lambda\sin\varphi\,, \tag{1.1.62}$$

and, just to protect against slips of the pen, we verify that

$$x(\varphi) = \frac{\mathrm{d}}{\mathrm{d}\varphi}y(\varphi) = \lambda\cos\varphi + \mu\sin\varphi\,, \tag{1.1.63}$$

indeed. Now we incorporate $x(\varphi = 0) = x_0$ and $y(\varphi = 0) = y_0$, which require $\lambda = x_0$ and $\mu = -y_0$, and arrive at the final answer

$$\boldsymbol{r}(\varphi) \mathrel{\widehat{=}} \begin{pmatrix} x_0\cos\varphi - y_0\sin\varphi \\ y_0\cos\varphi + x_0\sin\varphi \\ z_0 \end{pmatrix}. \tag{1.1.64}$$

In the second method we replace the two unknown real functions $x(\varphi)$ and $y(\varphi)$ by one unknown complex function

$$u(\varphi) = x(\varphi) + \mathrm{i}y(\varphi)\,, \tag{1.1.65}$$

which obeys the first-order differential equation

$$\frac{\mathrm{d}}{\mathrm{d}\varphi}u(\varphi) = \frac{\mathrm{d}}{\mathrm{d}\varphi}x(\varphi) + \mathrm{i}\frac{\mathrm{d}}{\mathrm{d}\varphi}y(\varphi) = -y(\varphi) + \mathrm{i}x(\varphi) = \mathrm{i}u(\varphi)\,. \tag{1.1.66}$$

It has the immediate solution

$$u(\varphi) = u_0\,\mathrm{e}^{\mathrm{i}\varphi} = (x_0 + \mathrm{i}y_0)(\cos\varphi + \mathrm{i}\sin\varphi)\,. \tag{1.1.67}$$

We extract $x(\varphi)$ and $y(\varphi)$ by means of

$$\begin{aligned} x(\varphi) &= \mathrm{Re}\,\bigl(u(\varphi)\bigr) = x_0\cos\varphi - y_0\sin\varphi\,, \\ y(\varphi) &= \mathrm{Im}\,\bigl(u(\varphi)\bigr) = y_0\cos\varphi + x_0\sin\varphi\,, \end{aligned} \tag{1.1.68}$$

the same answers as before — hardly a surprise.

The third method treats the coupled first-order differential equation as a single differential equation for the two-component column $\begin{pmatrix} x(\varphi) \\ y(\varphi) \end{pmatrix}$,

$$\frac{\mathrm{d}}{\mathrm{d}\varphi}\begin{pmatrix} x(\varphi) \\ y(\varphi) \end{pmatrix} = \begin{pmatrix} -y(\varphi) \\ x(\varphi) \end{pmatrix} = \begin{pmatrix} 0 & -1 \\ 1 & 0 \end{pmatrix}\begin{pmatrix} x(\varphi) \\ y(\varphi) \end{pmatrix} = M\begin{pmatrix} x(\varphi) \\ y(\varphi) \end{pmatrix} \tag{1.1.69}$$

with a φ-independent 2×2 matrix M. Quite analogous to the first-order differential equation for $u(\varphi)$, this equation has an immediate solution

$$\begin{pmatrix} x(\varphi) \\ y(\varphi) \end{pmatrix} = \mathrm{e}^{M\varphi} \begin{pmatrix} x_0 \\ y_0 \end{pmatrix}, \tag{1.1.70}$$

where the exponential function $\mathrm{e}^{M\varphi}$ of the 2×2 matrix $M\varphi$ is the 2×2 matrix

$$\mathrm{e}^{M\varphi} = \sum_{k=0}^{\infty} \frac{1}{k!} M^k \varphi^k \tag{1.1.71}$$

with

$$\begin{aligned} M^0 &= \begin{pmatrix} 1 & 0 \\ 0 & 1 \end{pmatrix} = \mathbf{1}_2\,, \text{ the } 2 \times 2 \text{ unit matrix}\,, \\ M^1 &= M = \begin{pmatrix} 0 & -1 \\ 1 & 0 \end{pmatrix}, \\ M^2 &= \begin{pmatrix} -1 & 0 \\ 0 & -1 \end{pmatrix} = -\mathbf{1}_2\,, \\ M^3 &= M^2 M = -M\,, \\ M^4 &= M^2 M^2 = \mathbf{1}_2\,, \\ M^5 &= M^2 M^3 = M\,, \end{aligned} \tag{1.1.72}$$

and so forth, giving

$$\begin{aligned} \mathrm{e}^{M\varphi} &= \Bigl(1 - \frac{\varphi^2}{2!} + \frac{\varphi^4}{4!} + \cdots\Bigr)\mathbf{1}_2 + \Bigl(\frac{\varphi}{1!} - \frac{\varphi^3}{3!} + \frac{\varphi^5}{5!} + \cdots\Bigr)M \\ &= \mathbf{1}_2 \sum_{k=0}^{\infty} \frac{(-\varphi^2)^k}{(2k)!} + M \sum_{k=0}^{\infty} \frac{(-\varphi^2)^k\,\varphi}{(2k+1)!} \\ &= \mathbf{1}_2 \cos\varphi + M \sin\varphi \\ &= \begin{pmatrix} \cos\varphi & -\sin\varphi \\ \sin\varphi & \cos\varphi \end{pmatrix} \end{aligned} \tag{1.1.73}$$

after we recognize the power series of $\cos\varphi$ and $\sin\varphi$. Thus, we have

$$\begin{pmatrix} x(\varphi) \\ y(\varphi) \end{pmatrix} = \begin{pmatrix} \cos\varphi & -\sin\varphi \\ \sin\varphi & \cos\varphi \end{pmatrix} \begin{pmatrix} x_0 \\ y_0 \end{pmatrix} = \begin{pmatrix} x_0 \cos\varphi - y_0 \sin\varphi \\ y_0 \cos\varphi + x_0 \sin\varphi \end{pmatrix}, \tag{1.1.74}$$

the now familiar expressions for $x(\varphi)$ and $y(\varphi)$, first seen in (1.1.64).

In summary, the rotated vector is

$$
\begin{aligned}
\boldsymbol{r}(\varphi) &\mathrel{\widehat{=}} \begin{pmatrix} x_0\cos\varphi - y_0\sin\varphi \\ y_0\cos\varphi + x_0\sin\varphi \\ z_0 \end{pmatrix} \\
&= \begin{pmatrix} 0 \\ 0 \\ z_0 \end{pmatrix} + \begin{pmatrix} x_0 \\ y_0 \\ 0 \end{pmatrix}\cos\varphi + \begin{pmatrix} -y_0 \\ x_0 \\ 0 \end{pmatrix}\sin\varphi
\end{aligned} \tag{1.1.75}
$$

for rotation about the z axis, $\boldsymbol{e} \mathrel{\widehat{=}} \begin{pmatrix} 0 \\ 0 \\ 1 \end{pmatrix}$. We recognize that the three terms have particular geometrical significance:

$$
\begin{aligned}
&\begin{pmatrix} 0 \\ 0 \\ z_0 \end{pmatrix} \mathrel{\widehat{=}} \boldsymbol{e}\,\boldsymbol{e}\cdot\boldsymbol{r}_0 && \text{is the component of } \boldsymbol{r}_0 \text{ parallel to the axis of rotation;} \\
&\begin{pmatrix} x_0 \\ y_0 \\ 0 \end{pmatrix} \mathrel{\widehat{=}} \boldsymbol{e}\times(\boldsymbol{r}_0\times\boldsymbol{e}) && \text{is the component of } \boldsymbol{r}_0 \text{ orthogonal to the axis of rotation;} \\
&\begin{pmatrix} -y_0 \\ x_0 \\ 0 \end{pmatrix} \mathrel{\widehat{=}} \boldsymbol{e}\times\boldsymbol{r}_0 && \text{is the tangential vector for } \varphi = 0.
\end{aligned} \tag{1.1.76}
$$

We infer that the corresponding vector statement is

$$
\boldsymbol{r}(\varphi) = \boldsymbol{e}\,\boldsymbol{e}\cdot\boldsymbol{r}_0 + \boldsymbol{e}\times(\boldsymbol{r}_0\times\boldsymbol{e})\cos\varphi + \boldsymbol{e}\times\boldsymbol{r}_0\sin\varphi \tag{1.1.77}
$$

for rotation about axis $\boldsymbol{e}$ by angle φ. Let us verify that this is correct, for which purpose we consider

$$
\begin{aligned}
\boldsymbol{r}(\varphi+\delta\varphi) &= \boldsymbol{e}\,\boldsymbol{e}\cdot\boldsymbol{r}_0 + \boldsymbol{e}\times(\boldsymbol{r}_0\times\boldsymbol{e})\cos(\varphi+\delta\varphi) + \boldsymbol{e}\times\boldsymbol{r}_0\sin(\varphi+\delta\varphi) \\
&= \boldsymbol{r}(\varphi) + \bigl(-\boldsymbol{e}\times(\boldsymbol{r}_0\times\boldsymbol{e})\sin\varphi + \boldsymbol{e}\times\boldsymbol{r}_0\cos\varphi\bigr)\delta\varphi \\
&= \boldsymbol{r}(\varphi) + \delta\varphi\,\boldsymbol{e}\times\boldsymbol{r}(\varphi) \\
&= \boldsymbol{r}(\varphi) + \boldsymbol{\delta\varphi}\times\boldsymbol{r}(\varphi)\,;
\end{aligned} \tag{1.1.78}
$$

indeed, we obtain what we should get.

1.1.9 *An application: Motion on a circle*

As an application, we have the motion of a point mass on a circle of radius R in the xy plane, with the circle centered at $\begin{pmatrix} x \\ y \\ z \end{pmatrix} = \begin{pmatrix} 0 \\ 0 \\ 0 \end{pmatrix}$. The position vector $\boldsymbol{r}(t)$ has the cartesian coordinates indicated in this sketch:

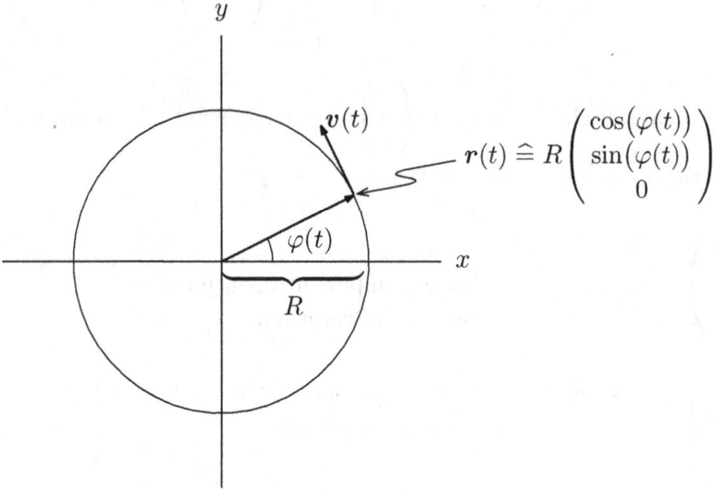

(1.1.79)

where $\varphi(t)$ is the angle between $\boldsymbol{r}(t)$ and the x axis at time t. The *velocity vector*

$$\boldsymbol{v}(t) = \frac{\mathrm{d}}{\mathrm{d}t}\boldsymbol{r}(t) \mathrel{\widehat{=}} R\frac{\mathrm{d}\varphi}{\mathrm{d}t}(t)\begin{pmatrix} -\sin\bigl(\varphi(t)\bigr) \\ \cos\bigl(\varphi(t)\bigr) \\ 0 \end{pmatrix} \tag{1.1.80}$$

is always tangential to the circle, with the *speed*

$$v(t) = |\boldsymbol{v}(t)| = R\left|\frac{\mathrm{d}\varphi}{\mathrm{d}t}(t)\right| = R|\dot{\varphi}| \tag{1.1.81}$$

changing in time only if $\frac{\mathrm{d}}{\mathrm{d}t}\varphi(t) \equiv \dot{\varphi}$ changes in time. Here, we use the over-dot notation for time derivatives, and we suppress the time argument for simplicity of notation, which is permissible in contexts, in which it is clear which quantities depend on time (here: $\varphi(t)$) and which do not (here: R).

Motion on the circle with constant speed (*not* constant velocity) is thus the case when

$$\varphi(t) = \omega t\,, \tag{1.1.82}$$

where ω is the *angular velocity*, which is a misnomer — "angular speed" would be better but is not used. There is, however, also the *angular velocity vector*, here

$$\boldsymbol{\omega} = \omega \boldsymbol{e}_z\,, \tag{1.1.83}$$

corresponding to $\boldsymbol{\delta\varphi} = \omega\delta t\, \boldsymbol{e}_z$, that is: $\boldsymbol{\delta\varphi} = \boldsymbol{\omega}\delta t$.

More generally, we recognize that $\boldsymbol{v}(t) \propto \boldsymbol{e}_z \times \boldsymbol{r}(t)$ in (1.1.80). This invites us to write

$$\boldsymbol{v}(t) = \boldsymbol{\omega}(t) \times \boldsymbol{r}(t) \tag{1.1.84}$$

with the instantaneous angular velocity vector

$$\boldsymbol{\omega}(t) = \dot{\varphi}(t)\boldsymbol{e}_z\,. \tag{1.1.85}$$

In the form

$$\boldsymbol{\omega}(t) = \omega(t)\boldsymbol{e}_z \tag{1.1.86}$$

it identifies the instantaneous angular velocity $\omega = \dot{\varphi}$, which can be positive or negative.

Further, the *acceleration vector* is

$$\begin{aligned}\left(\frac{\mathrm{d}}{\mathrm{d}t}\right)^2 \boldsymbol{r}(t) &= \frac{\mathrm{d}}{\mathrm{d}t}\boldsymbol{v}(t) = \dot{\boldsymbol{v}} = \dot{\boldsymbol{\omega}} \times \boldsymbol{r} + \boldsymbol{\omega} \times \boldsymbol{v} \\ &\mathrel{\widehat{=}} R\ddot{\varphi}\begin{pmatrix} -\sin\varphi \\ \cos\varphi \\ 0 \end{pmatrix} - R\dot{\varphi}^2 \begin{pmatrix} \cos\varphi \\ \sin\varphi \\ 0 \end{pmatrix}.\end{aligned} \tag{1.1.87}$$

It has a tangential component parallel to $\boldsymbol{v}$,

$$\dot{\boldsymbol{v}}_{\text{tangential}} = \dot{\boldsymbol{\omega}} \times \boldsymbol{r} \mathrel{\widehat{=}} R\ddot{\varphi}\begin{pmatrix} -\sin\varphi \\ \cos\varphi \\ 0 \end{pmatrix}, \tag{1.1.88}$$

and a normal component that is antiparallel to $\boldsymbol{r}$,

$$\begin{aligned}\dot{\boldsymbol{v}}_{\text{normal}} &= \boldsymbol{\omega} \times \boldsymbol{v} = \boldsymbol{\omega} \times (\boldsymbol{\omega} \times \boldsymbol{r}) = -\omega^2 \boldsymbol{r} = -\dot{\varphi}^2 \boldsymbol{r} \\ &\widehat{=} -R\dot{\varphi}^2 \begin{pmatrix} \cos\varphi \\ \sin\varphi \\ 0 \end{pmatrix} .\end{aligned} \tag{1.1.89}$$

This normal component is always nonzero; it keeps the point mass on the circle. By contrast, the tangential component can vanish (constant speed) or be positive or negative, depending on whether the speed increases or decreases, whereby positive/negative refer to the direction of $\boldsymbol{v}$.

This remark is based on considering the time derivative of the speed $v = |\boldsymbol{v}|$. See,

$$\frac{\mathrm{d}}{\mathrm{d}t} v^2 = 2v\dot{v} = \frac{\mathrm{d}}{\mathrm{d}t} \boldsymbol{v}^2 = 2\boldsymbol{v} \cdot \dot{\boldsymbol{v}} = 2\boldsymbol{v} \cdot \dot{\boldsymbol{v}}_{\text{tangential}} , \tag{1.1.90}$$

or

$$\dot{v} = \frac{\boldsymbol{v}}{v} \cdot \dot{\boldsymbol{v}}_{\text{tangential}} = \pm R\ddot{\varphi} = \pm R\dot{\omega} , \tag{1.1.91}$$

where we use that

$$\frac{\boldsymbol{v}}{v} \widehat{=} \pm \begin{pmatrix} -\sin\varphi \\ \cos\varphi \\ 0 \end{pmatrix} , \tag{1.1.92}$$

that is: $\boldsymbol{v}/v$ is parallel or antiparallel to the tangential unit vector, depending on the sign of $\dot{\varphi} = \omega$ in (1.1.80). We observe, in particular, that the acceleration $\frac{\mathrm{d}}{\mathrm{d}t}\boldsymbol{v}$ is never zero (unless the point mass is at rest on the circle), but the speed only changes if the angular velocity does.

1.1.10 *Polar coordinates, cylindrical coordinates*

Motion on a circle, or more generally motion in the xy plane with a center singled out, is often more conveniently described by *polar coordinates* in the plane. For the three-dimensional space, where we keep the cartesian component for the z direction, we have *cylindrical coordinates* s, φ, z related to the cartesian coordinates x, y, z by

$$\begin{aligned} x &= s\cos\varphi , \\ y &= s\sin\varphi , \\ z &= z . \end{aligned} \tag{1.1.93}$$

Since $x^2+y^2 = s^2$, the coordinate s is the distance from the z axis, provided that we adopt the usual convention that restricts s to positive values: $s \geq 0$. Regarding φ, we note that it is the azimuthal angle around the z axis, with $\varphi = 0$ for points in the half-plane with $x > 0$, $y = 0$; $\varphi = \frac{1}{2}\pi$ for the half-plane with $x = 0$, $y > 0$; $\varphi = \pi$ for the half-plane with $x < 0$, $y = 0$; and $\varphi = \frac{3}{2}\pi$ for the half-plane with $x = 0$, $y < 0$. The azimuth φ is a periodic coordinate, points with the same s, z coordinates and φ values differing by a multiple of 2π are the same. If we want a unique φ value for each x, y pair of cartesian coordinates, we must restrict φ to a 2π-range of values, such as $0 \leq \varphi < 2\pi$ or $-\pi < \varphi \leq \pi$ or similar, but it is often more convenient to just make no difference between φ and $\varphi + 2\pi$. What matters are $\cos\varphi$ and $\sin\varphi$, after all.

When $x = y = 0$, the value of φ is undetermined, which is to say that the cylindrical coordinates are singular on the z axis. Such singularities are found for many coordinate systems, they are usually harmless but one must be aware of them: Given x, y, and z, we can determine $s \geq 0$, φ (periodic or restricted to a 2π interval), and z (now the cylindrical coordinate), except when $x = y = 0$, in which case $s = 0$ and φ is both undetermined and irrelevant.

The picture to keep in mind is this:

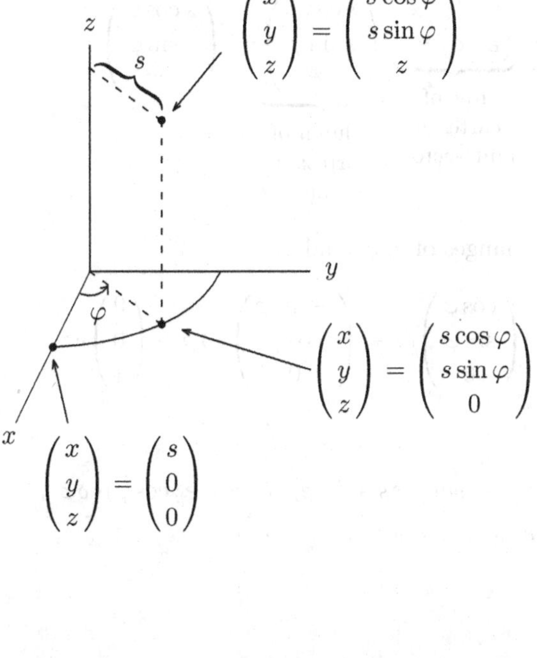

$$\begin{pmatrix} x \\ y \\ z \end{pmatrix} = \begin{pmatrix} s\cos\varphi \\ s\sin\varphi \\ z \end{pmatrix}$$

$$\begin{pmatrix} x \\ y \\ z \end{pmatrix} = \begin{pmatrix} s\cos\varphi \\ s\sin\varphi \\ 0 \end{pmatrix}$$

$$\begin{pmatrix} x \\ y \\ z \end{pmatrix} = \begin{pmatrix} s \\ 0 \\ 0 \end{pmatrix} \tag{1.1.94}$$

We mentioned above the special cases of $\varphi = 0, \frac{\pi}{2}, \pi, \frac{3\pi}{2}$; more generally, the points with the same φ values are on a half-plane that has the z axis ($s = 0$) as the border and is orthogonal to the xy plane. Points with the same z values are on a plane parallel to the xy plane. And points with the same s value are on a circular cylinder around the z axis with radius s; if $s = 0$, the cylinder degenerates into the z axis, and we just have a one-dimensional line rather than a two-dimensional surface.

For cartesian coordinates, we have the standard unit vectors that define the coordinate system together with the $x = y = z = 0$ point of reference,

$$\boldsymbol{r} = \boldsymbol{e}_x\, x + \boldsymbol{e}_y\, y + \boldsymbol{e}_z\, z = \begin{pmatrix}\boldsymbol{e}_x & \boldsymbol{e}_y & \boldsymbol{e}_z\end{pmatrix}\begin{pmatrix} x \\ y \\ z \end{pmatrix} \mathrel{\widehat{=}} \begin{pmatrix} x \\ y \\ z \end{pmatrix}, \tag{1.1.95}$$

which are the statements in (1.1.3) and (1.1.4). Are there also such unit vectors for the cylindrical coordinates? Yes, there are, and we find them by considering the analog of

$$\delta\boldsymbol{r} = \boldsymbol{e}_x\, \delta x + \boldsymbol{e}_y\, \delta y + \boldsymbol{e}_z\, \delta z\,, \tag{1.1.96}$$

that is: the response of $\boldsymbol{r}$ to small changes of the coordinates, here the cartesian coordinates. For cylindrical coordinates we proceed from

$$\begin{aligned}\boldsymbol{r} &= \boldsymbol{e}_x\, s\cos\varphi + \boldsymbol{e}_y\, s\sin\varphi + \boldsymbol{e}_z\, z \\ &= \underbrace{\begin{pmatrix}\boldsymbol{e}_x & \boldsymbol{e}_y & \boldsymbol{e}_z\end{pmatrix}}_{\substack{\text{row of}\\ \text{cartesian}\\ \text{unit vectors}}} \underbrace{\begin{pmatrix} s\cos\varphi \\ s\sin\varphi \\ z \end{pmatrix}}_{\substack{\text{column of}\\ \text{cartesian}\\ \text{coordinates}}} \mathrel{\widehat{=}} \begin{pmatrix} s\cos\varphi \\ s\sin\varphi \\ z \end{pmatrix},\end{aligned} \tag{1.1.97}$$

and consider small changes of s, φ, and z:

$$\delta\begin{pmatrix} s\cos\varphi \\ s\sin\varphi \\ z \end{pmatrix} = \begin{pmatrix} \cos\varphi \\ \sin\varphi \\ 0 \end{pmatrix}\delta s + \begin{pmatrix} -\sin\varphi \\ \cos\varphi \\ 0 \end{pmatrix} s\,\delta\varphi + \begin{pmatrix} 0 \\ 0 \\ 1 \end{pmatrix}\delta z\,. \tag{1.1.98}$$

Accordingly, we have

$$\begin{aligned}\delta\boldsymbol{r} &= (\boldsymbol{e}_x \cos\varphi + \boldsymbol{e}_y \sin\varphi)\delta s + (-\boldsymbol{e}_x \sin\varphi + \boldsymbol{e}_y \cos\varphi) s\,\delta\varphi + \boldsymbol{e}_z\, \delta z \\ &= \boldsymbol{e}_s\, \delta s + \boldsymbol{e}_\varphi\, s\,\delta\varphi + \boldsymbol{e}_z\, \delta z\end{aligned} \tag{1.1.99}$$

and so identify the unit vectors associated with the cylindrical coordinates:

$$\begin{aligned} \boldsymbol{e}_s &= \boldsymbol{e}_x \cos\varphi + \boldsymbol{e}_y \sin\varphi \mathrel{\widehat{=}} \begin{pmatrix} \cos\varphi \\ \sin\varphi \\ 0 \end{pmatrix} \quad \text{for } s \text{ increments,} \\ \boldsymbol{e}_\varphi &= -\boldsymbol{e}_x \sin\varphi + \boldsymbol{e}_y \cos\varphi \mathrel{\widehat{=}} \begin{pmatrix} -\sin\varphi \\ \cos\varphi \\ 0 \end{pmatrix} \quad \text{for } \varphi \text{ increments,} \\ \boldsymbol{e}_z &\mathrel{\widehat{=}} \begin{pmatrix} 0 \\ 0 \\ 1 \end{pmatrix} \quad \text{for } z \text{ increments.} \end{aligned} \tag{1.1.100}$$

Since the z coordinate is still the cartesian coordinate, let us focus on s and φ and just look at a plane with constant z, in which we have the polar coordinates s and φ:

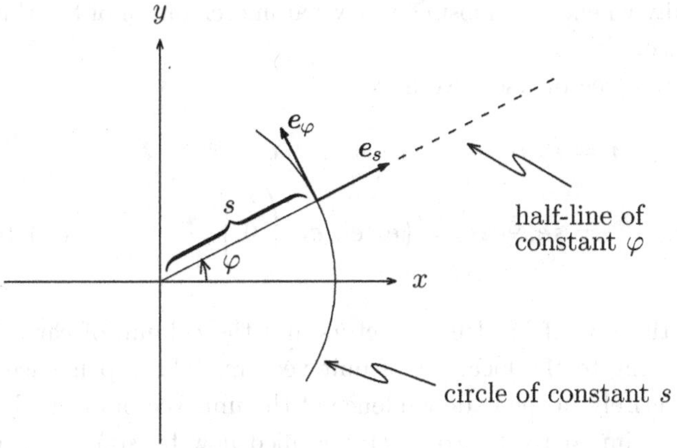

(1.1.101)

The respective unit vectors are normal to the curves of points with constant coordinate value, here: circles for s, half-lines for φ. If we remember the z coordinate, the unit vectors are actually normal to the two-dimensional surfaces of constant coordinate values (cylinders for s, half-planes for φ, planes for z).

Note the term $\boldsymbol{e}_\varphi s\,\delta\varphi$, where we put the factor s aside to ensure that the remaining vector is of unit length. Again, we are confronted with the $s = 0$ singularity of the cylindrical coordinate system, inasmuch as a change of φ has no effect on $\boldsymbol{r}$ when $s = 0$.

It is important to note that the unit vectors $\boldsymbol{e}_s \mathrel{\widehat{=}} \begin{pmatrix} \cos\varphi \\ \sin\varphi \\ 0 \end{pmatrix}$ and $\boldsymbol{e}_\varphi \mathrel{\widehat{=}} \begin{pmatrix} -\sin\varphi \\ \cos\varphi \\ 0 \end{pmatrix}$ depend on the coordinate φ, and in other coordinate systems the respective unit vectors can depend on all coordinates. This is to say that we have a *local* set of unit vectors, one set for each point in space. In the case of cylindrical coordinates, matters are still relatively simple because there is only a dependence of the unit vectors on one coordinate, the azimuth φ. Furthermore, the local unit vectors are an orthogonal right-handed system in the order $\boldsymbol{e}_s$, $\boldsymbol{e}_\varphi$, $\boldsymbol{e}_z$, as we verify by confirming that

$$\boldsymbol{e}_s \times \boldsymbol{e}_\varphi = \boldsymbol{e}_z \quad \text{and} \quad \boldsymbol{e}_s \cdot \boldsymbol{e}_\varphi = 0\,. \tag{1.1.102}$$

Most coordinate systems used in practical calculations are *systems of orthogonal coordinates* in this sense, namely that the directions of infinitesimal changes are pairwise orthogonal. But there are also other coordinate systems for which this is not the case. One needs to examine the situation carefully whenever choosing a new parameterization of the three-dimensional space.

For the position vector itself, we have

$$\begin{aligned} \boldsymbol{r} &= \boldsymbol{e}_x\, s\cos\varphi + \boldsymbol{e}_y\, s\sin\varphi + \boldsymbol{e}_z\, z \\ &= s\boldsymbol{e}_s + z\boldsymbol{e}_z = \begin{pmatrix} \boldsymbol{e}_s & \boldsymbol{e}_\varphi & \boldsymbol{e}_z \end{pmatrix} \begin{pmatrix} s \\ 0 \\ z \end{pmatrix}, \end{aligned} \tag{1.1.103}$$

where we have the row of local unit vectors and the column of cartesian coordinates referring to the local set of unit vectors. The dependence on the azimuth is entirely in the φ dependence of the unit vectors $\boldsymbol{e}_s$ and $\boldsymbol{e}_\varphi$. When taking the time derivative of $\boldsymbol{r}(t)$, specified now by $s(t)$, $\varphi(t)$, and $z(t)$, this φ dependence of the unit vectors must, of course, be taken into account. We get

$$\begin{aligned} \boldsymbol{v} = \dot{\boldsymbol{r}} &= \begin{pmatrix} \boldsymbol{e}_s & \boldsymbol{e}_\varphi & \boldsymbol{e}_z \end{pmatrix} \begin{pmatrix} \dot{s} \\ 0 \\ \dot{z} \end{pmatrix} + \begin{pmatrix} \dfrac{\partial \boldsymbol{e}_s}{\partial\varphi} & \dfrac{\partial \boldsymbol{e}_\varphi}{\partial\varphi} & 0 \end{pmatrix} \begin{pmatrix} \dot{\varphi}s \\ 0 \\ \dot{\varphi}z \end{pmatrix} \\ &= \dot{s}\boldsymbol{e}_s + \dot{z}\boldsymbol{e}_z + \dot{\varphi}s\frac{\partial}{\partial\varphi}\boldsymbol{e}_s \end{aligned} \tag{1.1.104}$$

for the velocity vector, which we could, of course, also find by differentiating $\boldsymbol{r} = s\boldsymbol{e}_s + z\boldsymbol{e}_z$ without using the column-times-row expression of (1.1.103).

To proceed further, we need to recognize that

$$\frac{\partial}{\partial\varphi}\boldsymbol{e}_s = \frac{\partial}{\partial\varphi}(\boldsymbol{e}_x \cos\varphi + \boldsymbol{e}_y \sin\varphi) = -\boldsymbol{e}_x \sin\varphi + \boldsymbol{e}_y \cos\varphi = \boldsymbol{e}_\varphi \quad (1.1.105)$$

and

$$\frac{\partial}{\partial\varphi}\boldsymbol{e}_\varphi = \frac{\partial}{\partial\varphi}(-\boldsymbol{e}_x \sin\varphi + \boldsymbol{e}_y \cos\varphi) = -\boldsymbol{e}_x \cos\varphi - \boldsymbol{e}_y \sin\varphi = -\boldsymbol{e}_s\,, \quad (1.1.106)$$

which are summarized in

$$\frac{\partial}{\partial\varphi}\begin{pmatrix}\boldsymbol{e}_s & \boldsymbol{e}_\varphi & \boldsymbol{e}_z\end{pmatrix} = \begin{pmatrix}\boldsymbol{e}_s & \boldsymbol{e}_\varphi & \boldsymbol{e}_z\end{pmatrix}\begin{pmatrix}0 & -1 & 0\\ 1 & 0 & 0\\ 0 & 0 & 0\end{pmatrix}, \quad (1.1.107)$$

where the 3×3 matrix tells us how the derivatives of the local unit vectors are expressed as linear combinations of themselves. This matrix notation may seem excessive in

$$\begin{aligned}\frac{\mathrm{d}}{\mathrm{d}t}\boldsymbol{r} &= \begin{pmatrix}\boldsymbol{e}_s & \boldsymbol{e}_\varphi & \boldsymbol{e}_z\end{pmatrix}\left[\frac{\mathrm{d}}{\mathrm{d}t}\begin{pmatrix}s\\0\\z\end{pmatrix} + \begin{pmatrix}0 & -1 & 0\\ 1 & 0 & 0\\ 0 & 0 & 0\end{pmatrix}\frac{\mathrm{d}\varphi}{\mathrm{d}t}\begin{pmatrix}s\\0\\z\end{pmatrix}\right]\\ &= \begin{pmatrix}\boldsymbol{e}_s & \boldsymbol{e}_\varphi & \boldsymbol{e}_z\end{pmatrix}\left[\begin{pmatrix}1&0&0\\0&1&0\\0&0&1\end{pmatrix}\frac{\mathrm{d}}{\mathrm{d}t} + \frac{\mathrm{d}\varphi}{\mathrm{d}t}\begin{pmatrix}0 & -1 & 0\\ 1 & 0 & 0\\ 0 & 0 & 0\end{pmatrix}\right]\begin{pmatrix}s\\0\\z\end{pmatrix}, \quad (1.1.108)\end{aligned}$$

and it is indeed somewhat overblown in the present simple context of cylindrical coordinates, but it is a useful tool for more complicated cases.

In passing, we note the obvious danger of keeping too many important details implicit. We *could* write

$$\boldsymbol{r} \mathrel{\widehat{=}} \begin{pmatrix}s\\0\\z\end{pmatrix} \quad (1.1.109)$$

as the coordinate representation of $\boldsymbol{r}$ with respect to the local cartesian coordinates defined by the local set of unit vectors, as suggested by (1.1.103), but we *should not*, because sooner or later we are bound to forget that there is a hidden φ dependence in the now invisible set of unit vectors. We should not, and we will not. A column of coordinates will always refer to the fixed, standard cartesian system, relative to which the other coordinates are defined, here:

$$\boldsymbol{r} \mathrel{\widehat{=}} \begin{pmatrix} s\cos\varphi \\ s\sin\varphi \\ z \end{pmatrix}, \tag{1.1.110}$$

which repeats (1.1.93).

Returning to the velocity, we put the ingredients together and arrive at

$$\boldsymbol{v} = \dot{s}\boldsymbol{e}_s + s\dot{\varphi}\boldsymbol{e}_\varphi + \dot{z}\boldsymbol{e}_z\,. \tag{1.1.111}$$

For motion along a circle of radius R in the xy plane, that is: $s = R$, $z = 0$, this is simply

$$\boldsymbol{v} = R\dot{\varphi}\boldsymbol{e}_\varphi \mathrel{\widehat{=}} R\dot{\varphi}\begin{pmatrix} -\sin\varphi \\ \cos\varphi \\ 0 \end{pmatrix}, \tag{1.1.112}$$

which is exactly the expression found earlier in (1.1.80) and we recognize that the unit vector $\boldsymbol{v}/v$ of (1.1.92) is just $\boldsymbol{e}_\varphi$, the unit vector tangential to the circle at the current position of the moving point mass.

The acceleration is obtained by another differentiation,

$$\ddot{\boldsymbol{r}} = \dot{\boldsymbol{v}} = \ddot{s}\boldsymbol{e}_s + \dot{s}\dot{\boldsymbol{e}}_s + (\dot{s}\dot{\varphi} + s\ddot{\varphi})\boldsymbol{e}_\varphi + s\dot{\varphi}\dot{\boldsymbol{e}}_\varphi + \ddot{z}\boldsymbol{e}_z\,, \tag{1.1.113}$$

where, as we know, $\dot{\boldsymbol{e}}_s = \dot{\varphi}\boldsymbol{e}_\varphi$ and $\dot{\boldsymbol{e}}_\varphi = -\dot{\varphi}\boldsymbol{e}_s$, so that

$$\dot{\boldsymbol{v}} = (\ddot{s} - s\dot{\varphi}^2)\boldsymbol{e}_s + (s\ddot{\varphi} + 2\dot{s}\dot{\varphi})\boldsymbol{e}_\varphi + \ddot{z}\,\boldsymbol{e}_z\,. \tag{1.1.114}$$

Again, let us consider $s = R$, $z = 0$, for which

$$\dot{\boldsymbol{v}} = -R\dot{\varphi}^2\boldsymbol{e}_s + R\ddot{\varphi}\boldsymbol{e}_\varphi\,. \tag{1.1.115}$$

The two terms on the right are the normal and the tangential components, respectively, that we found in (1.1.87)–(1.1.89).

When evaluating expressions such as $\boldsymbol{r}\cdot\boldsymbol{v}$ or $\boldsymbol{r}\times\boldsymbol{v}$ where both vectors are naturally given by coefficients that refer to the *same* local set of unit vectors, the calculation is very similar to that in cartesian coordinates. See,

$$\boldsymbol{r}\cdot\boldsymbol{v} = (s\boldsymbol{e}_s + z\boldsymbol{e}_z)\cdot(\dot{s}\boldsymbol{e}_s + s\dot{\varphi}\boldsymbol{e}_\varphi + \dot{z}\boldsymbol{e}_z) = s\dot{s} + z\dot{z}\,, \tag{1.1.116}$$

which we recognize as

$$\frac{1}{2}\frac{\mathrm{d}}{\mathrm{d}t}(s^2 + z^2) = \frac{1}{2}\frac{\mathrm{d}}{\mathrm{d}t}r^2 = \frac{1}{2}\frac{\mathrm{d}}{\mathrm{d}t}\boldsymbol{r}^2\,, \tag{1.1.117}$$

since $\boldsymbol{r}^2 = r^2 = s^2 + z^2$. Likewise,

$$\begin{aligned}\boldsymbol{r} \times \boldsymbol{v} &= (s\boldsymbol{e}_s + z\boldsymbol{e}_z) \times (\dot{s}\boldsymbol{e}_s + s\dot{\varphi}\boldsymbol{e}_\varphi + \dot{z}\boldsymbol{e}_z) \\ &= -s\dot{\varphi}z\,\underbrace{\boldsymbol{e}_\varphi \times \boldsymbol{e}_z}_{=\boldsymbol{e}_s} + (\dot{s}z - s\dot{z})\,\underbrace{\boldsymbol{e}_z \times \boldsymbol{e}_s}_{=\boldsymbol{e}_\varphi} + s^2\dot{\varphi}\,\underbrace{\boldsymbol{e}_s \times \boldsymbol{e}_\varphi}_{=\boldsymbol{e}_z} \\ &= -s\dot{\varphi}z\boldsymbol{e}_s + (\dot{s}z - s\dot{z})\boldsymbol{e}_\varphi + s^2\dot{\varphi}\,\boldsymbol{e}_z\,. \end{aligned} \tag{1.1.118}$$

In particular, for the point mass moving on the circle of radius R in the xy plane ($s = R$, $z = 0$), these are

$$\boldsymbol{r} \cdot \boldsymbol{v} = 0 \quad \text{and} \quad \boldsymbol{r} \times \boldsymbol{v} = R^2\dot{\varphi}\boldsymbol{e}_z\,, \tag{1.1.119}$$

as they should be.

More care is required for two vectors that have different natural sets of local unit vectors. As an example, we consider two position vectors,

$$\boldsymbol{r} \mathrel{\widehat{=}} \begin{pmatrix} s\cos\varphi \\ s\sin\varphi \\ z \end{pmatrix} \quad \text{and} \quad \boldsymbol{r}' \mathrel{\widehat{=}} \begin{pmatrix} s'\cos\varphi' \\ s'\sin\varphi' \\ z' \end{pmatrix} \tag{1.1.120}$$

or

$$\boldsymbol{r} = s\boldsymbol{e}_s + z\boldsymbol{e}_z \quad \text{and} \quad \boldsymbol{r}' = s'\boldsymbol{e}'_s + z'\boldsymbol{e}_z \tag{1.1.121}$$

with $\boldsymbol{e}_s \mathrel{\widehat{=}} \begin{pmatrix} \cos\varphi \\ \sin\varphi \\ 0 \end{pmatrix}$ as before, and $\boldsymbol{e}'_s \mathrel{\widehat{=}} \begin{pmatrix} \cos\varphi' \\ \sin\varphi' \\ 0 \end{pmatrix}$. This gives

$$\boldsymbol{r} \cdot \boldsymbol{r}' = ss'\,\boldsymbol{e}_s \cdot \boldsymbol{e}'_s + zz' \tag{1.1.122}$$

with

$$\boldsymbol{e}_s \cdot \boldsymbol{e}'_s = \cos\varphi\cos\varphi' + \sin\varphi\sin\varphi' = \cos(\varphi - \varphi') \tag{1.1.123}$$

correctly stating that the angle between $\boldsymbol{e}_s$ and $\boldsymbol{e}'_s$ is the difference in azimuth:

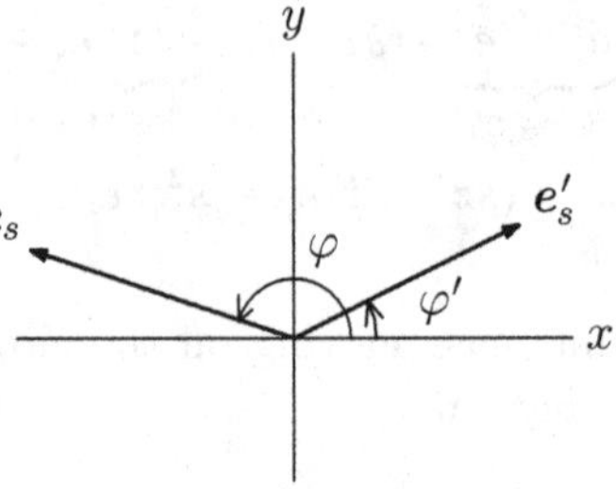

(1.1.124)

We thus have

$$\boldsymbol{r}\cdot\boldsymbol{r}' = ss'\cos(\varphi-\varphi') + zz'\,. \tag{1.1.125}$$

For $\boldsymbol{r}=\boldsymbol{r}'$, this repeats what we know, $r^2 = s^2+z^2$, and more generally, we can use it to find the distance from position $\boldsymbol{r}$ to position $\boldsymbol{r}'$ in cylindrical coordinates,

$$\begin{aligned}|\boldsymbol{r}-\boldsymbol{r}'| &= \sqrt{(\boldsymbol{r}-\boldsymbol{r}')^2} = \sqrt{\boldsymbol{r}^2 - 2\boldsymbol{r}\cdot\boldsymbol{r}' + \boldsymbol{r}'^2}\\ &= \sqrt{s^2 - 2ss'\cos(\varphi-\varphi') + s'^2 + (z-z')^2}\,.\end{aligned} \tag{1.1.126}$$

As a basic check of consistency, one verifies immediately that this distance vanishes for $\boldsymbol{r}=\boldsymbol{r}'$, that is: $s=s'$, $\varphi=\varphi'$, $z=z'$.

1.1.11 *Spherical coordinates*

Another important coordinate system is that of spherical coordinates. One can understand it as the result of introducing polar coordinates in the half-plane of constant φ that we have for cylindrical coordinates:

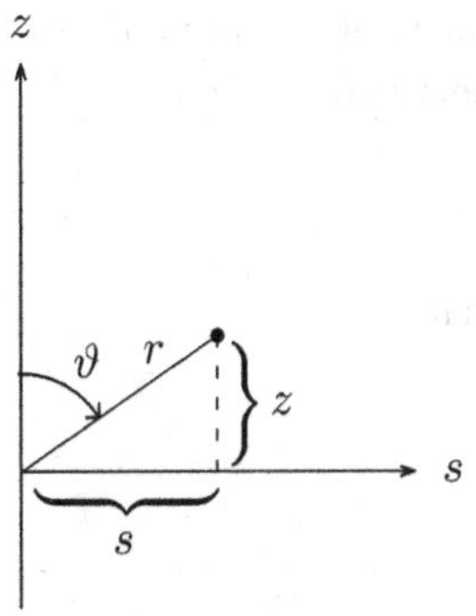

$$s = r\sin\vartheta\,,\quad z = r\cos\vartheta\,,\quad r \geq 0\,, \tag{1.1.127}$$

where the polar angle ϑ is restricted to the range $0 \leq \vartheta \leq \pi$ since s cannot be negative. There is another connection, used for example in geography when specifying the latitude λ of a site:

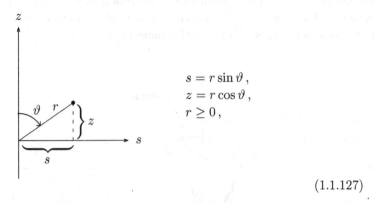

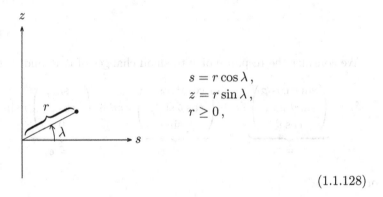

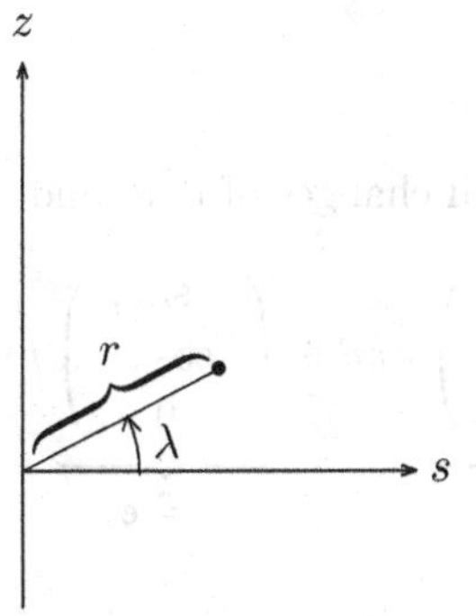

$$s = r\cos\lambda\,,\quad z = r\sin\lambda\,,\quad r \geq 0\,, \tag{1.1.128}$$

with $-\frac{\pi}{2} \leq \lambda \leq \frac{\pi}{2}$. Clearly, ϑ and λ are related to each other by $\vartheta + \lambda = \frac{\pi}{2}$. In physics, we use the latitude version very rarely and the polar-angle version very often.

In the polar-angle version, then, we have

$$\begin{aligned} x &= r\sin\vartheta\cos\varphi\,,\\ y &= r\sin\vartheta\sin\varphi\,,\\ z &= r\cos\vartheta \end{aligned} \tag{1.1.129}$$

with $r \geq 0$, $0 \leq \vartheta \leq \pi$, $0 \leq \varphi < 2\pi$ (or any other 2π interval, or we regard the azimuth φ as a periodic variable with no restriction). As before, the two-dimensional surfaces of constant φ are half-planes with the z axis as the boundary line. The two-dimensional surfaces of constant r are spheres centered at $x = y = z = 0$; and the surfaces of constant ϑ are cones with the z axis as the axis of rotational symmetry:

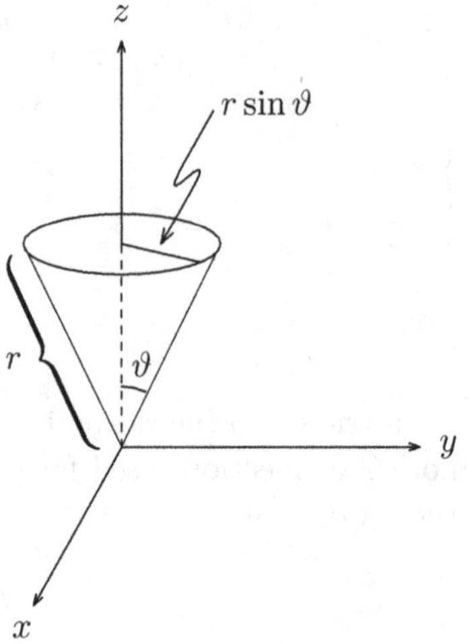

(1.1.130)

We consider the response of $\boldsymbol{r}$ to small changes of r, ϑ, and φ, that is

$$\delta\boldsymbol{r} \mathrel{\widehat{=}} \underbrace{\begin{pmatrix} \sin\vartheta\cos\varphi \\ \sin\vartheta\sin\varphi \\ \cos\vartheta \end{pmatrix}}_{\widehat{=}\,\boldsymbol{e}_r} \delta r + \underbrace{\begin{pmatrix} \cos\vartheta\cos\varphi \\ \cos\vartheta\sin\varphi \\ -\sin\vartheta \end{pmatrix}}_{\widehat{=}\,\boldsymbol{e}_\vartheta} r\,\delta\vartheta + \underbrace{\begin{pmatrix} -\sin\varphi \\ \cos\varphi \\ 0 \end{pmatrix}}_{\widehat{=}\,\boldsymbol{e}_\varphi} r\sin\vartheta\,\delta\varphi \tag{1.1.131}$$

or

$$\delta\boldsymbol{r} = \boldsymbol{e}_r\,\delta r + \boldsymbol{e}_\vartheta\, r\,\delta\vartheta + \boldsymbol{e}_\varphi\, r\sin\vartheta\,\delta\varphi\,, \tag{1.1.132}$$

where $\boldsymbol{e}_\varphi$ is the same unit vector as the one we found for cylindrical coordinates in (1.1.99) with $s\,\delta\varphi$ correctly replaced by $r\sin\vartheta\,\delta\varphi$. The local unit vectors $\boldsymbol{e}_r$ and $\boldsymbol{e}_\vartheta$ depend on both ϑ and φ, so all the unit vectors of a local set are position dependent.

The singularity of the cylindrical coordinates on the z axis is inherited by the spherical coordinates. For $x = y = 0$, $z \geq 0$ we have $r = z$ and $\vartheta = 0$ while φ is undetermined, and for $x = y = 0$, $z \leq 0$ we have $r = |z| = -z$ and $\vartheta = \pi$ with φ undetermined as well, and for $x = y = z = 0$ we have $r = 0$ and both ϑ and φ are undetermined.

The local set of unit vectors are a right-handed orthogonal set in the order $\boldsymbol{e}_r$, $\boldsymbol{e}_\vartheta$, $\boldsymbol{e}_\varphi$,

$$\boldsymbol{e}_r \times \boldsymbol{e}_\vartheta = \boldsymbol{e}_\varphi\,, \quad \boldsymbol{e}_\vartheta \times \boldsymbol{e}_\varphi = \boldsymbol{e}_r\,, \quad \boldsymbol{e}_\varphi \times \boldsymbol{e}_r = \boldsymbol{e}_\vartheta\,. \tag{1.1.133}$$

If you picture yourself on the surface of the earth, idealized as a sphere of radius 6738 km, $\boldsymbol{e}_r$ is pointing "up" (away from the center of the earth), $\boldsymbol{e}_\vartheta$ is pointing "south" (the north pole is at $\vartheta = 0$, the equator has $\vartheta = \frac{1}{2}\pi$, the south pole is at $\vartheta = \pi$), and $\boldsymbol{e}_\varphi$ is pointing "east" (the direction of the rising sun and, therefore, the direction of the velocity vector for the surface points that follow the 24 h rotation of the earth).

Following the example of cylindrical coordinates above, we establish

$$\frac{\partial}{\partial\vartheta}\big(\boldsymbol{e}_r\ \boldsymbol{e}_\vartheta\ \boldsymbol{e}_\varphi\big) = \big(\boldsymbol{e}_\vartheta\ \ {-\boldsymbol{e}_r}\ \ 0\big) = \big(\boldsymbol{e}_r\ \boldsymbol{e}_\vartheta\ \boldsymbol{e}_\varphi\big)\begin{pmatrix} 0 & -1 & 0 \\ 1 & 0 & 0 \\ 0 & 0 & 0 \end{pmatrix} \tag{1.1.134}$$

and

$$\begin{aligned}\frac{\partial}{\partial\varphi}\big(\boldsymbol{e}_r\ \boldsymbol{e}_\vartheta\ \boldsymbol{e}_\varphi\big) &= \big(\boldsymbol{e}_\varphi \sin\vartheta\ \ \boldsymbol{e}_\varphi\cos\vartheta\ \ {-(\boldsymbol{e}_r\sin\vartheta + \boldsymbol{e}_\vartheta\cos\vartheta)}\big) \\ &= \big(\boldsymbol{e}_r\ \boldsymbol{e}_\vartheta\ \boldsymbol{e}_\varphi\big)\begin{pmatrix} 0 & 0 & -\sin\vartheta \\ 0 & 0 & -\cos\vartheta \\ \sin\vartheta & \cos\vartheta & 0 \end{pmatrix}.\end{aligned} \tag{1.1.135}$$

Then, proceeding from the position vector

$$\boldsymbol{r} = r\,\boldsymbol{e}_r\,, \tag{1.1.136}$$

we first get the velocity vector

$$\boldsymbol{v} = \dot{r}\boldsymbol{e}_r + r\dot{\vartheta}\,\boldsymbol{e}_\vartheta + r\sin\vartheta\,\dot{\varphi}\,\boldsymbol{e}_\varphi \tag{1.1.137}$$

and then the acceleration vector

$$\begin{aligned}\ddot{\boldsymbol{r}} = \dot{\boldsymbol{v}} &= \ddot{r}\,\boldsymbol{e}_r + \dot{r}\dot{\vartheta}\,\frac{\partial}{\partial\vartheta}\boldsymbol{e}_r + \dot{r}\dot{\varphi}\frac{\partial}{\partial\varphi}\boldsymbol{e}_r \\ &\quad + \big(r\ddot{\vartheta} + \dot{r}\dot{\vartheta}\big)\,\boldsymbol{e}_\vartheta + r\dot{\vartheta}^2\frac{\partial}{\partial\vartheta}\boldsymbol{e}_\vartheta + r\dot{\vartheta}\dot{\varphi}\,\frac{\partial}{\partial\varphi}\boldsymbol{e}_\vartheta \\ &\quad + \big(r\ddot{\varphi}\sin\vartheta + \dot{r}\dot{\varphi}\sin\vartheta + r\dot{\vartheta}\dot{\varphi}\cos\vartheta\big)\,\boldsymbol{e}_\varphi + r\dot{\varphi}^2\sin\vartheta\,\frac{\partial}{\partial\varphi}\boldsymbol{e}_\varphi \\ &= \big(\ddot{r} - r\dot{\vartheta}^2 - r(\dot{\varphi}\sin\vartheta)^2\big)\,\boldsymbol{e}_r + \big(r\ddot{\vartheta} + 2\dot{r}\dot{\vartheta} - r\dot{\varphi}^2\sin\vartheta\cos\vartheta\big)\,\boldsymbol{e}_\vartheta \\ &\quad + \big(r\ddot{\varphi}\sin\vartheta + 2\dot{r}\dot{\varphi}\sin\vartheta + 2r\dot{\vartheta}\dot{\varphi}\cos\vartheta\big)\,\boldsymbol{e}_\varphi\,,\end{aligned} \tag{1.1.138}$$

which has quite a few terms.

Let us look at two special cases: motion on the sphere with radius R, and motion on the cone with polar angle θ. For the motion on the sphere, we have $r = R =$ constant, $\dot{r} = 0$, $\ddot{r} = 0$, and get

$$\begin{aligned} \boldsymbol{v} &= R\dot{\vartheta}\, \boldsymbol{e}_\vartheta + R\dot{\varphi} \sin\vartheta\, \boldsymbol{e}_\varphi\,, \\ \dot{\boldsymbol{v}} &= -R\bigl(\dot{\vartheta}^2 + (\dot{\varphi}\sin\vartheta)^2\bigr)\, \boldsymbol{e}_r + R\bigl(\ddot{\vartheta} - \dot{\varphi}^2 \sin\vartheta\cos\vartheta\bigr)\, \boldsymbol{e}_\vartheta \\ &\quad + R\bigl(\ddot{\varphi}\sin\vartheta + 2\dot{\vartheta}\dot{\varphi}\cos\vartheta\bigr)\, \boldsymbol{e}_\varphi \end{aligned} \tag{1.1.139}$$

for the velocity and the acceleration, and we note that $\boldsymbol{e}_r \cdot \dot{\boldsymbol{v}}$, the radial component of the acceleration is always negative and equal to v^2/R in magnitude.

For the motion on the cone, we have $\vartheta = \theta =$ constant, $\dot{\vartheta} = 0$, $\ddot{\vartheta} = 0$ and get

$$\begin{aligned} \boldsymbol{v} &= \dot{r}\, \boldsymbol{e}_r + r\dot{\varphi}\, \sin\theta\, \boldsymbol{e}_\varphi\,, \\ \dot{\boldsymbol{v}} &= \bigl(\ddot{r} - r(\dot{\varphi}\sin\theta)^2\bigr)\, \boldsymbol{e}_r - r\dot{\varphi}^2 \sin\theta\cos\theta\, \boldsymbol{e}_\vartheta \\ &\quad + \bigl(r\ddot{\varphi} + 2\dot{r}\dot{\varphi}\bigr) \sin\theta\, \boldsymbol{e}_\varphi \end{aligned} \tag{1.1.140}$$

for the velocity and the acceleration, with $\vartheta \to \theta$ in the expressions for $\boldsymbol{e}_r$ and $\boldsymbol{e}_\vartheta$ in (1.1.131). We note that $\boldsymbol{r} \times \boldsymbol{v} = -r^2\dot{\varphi} \sin\theta\, \boldsymbol{e}_\vartheta$ is always normal to the cone, as it must be because $\boldsymbol{r}$ points from the tip of the cone to the position of the point mass whose velocity is necessarily tangential to the cone, $\boldsymbol{v} \cdot \boldsymbol{e}_\vartheta = 0$.

The spherical-coordinate analog of (1.1.116) is

$$\boldsymbol{r} \cdot \boldsymbol{v} = r\boldsymbol{e}_r \cdot (\dot{r}\, \boldsymbol{e}_r + r\dot{\vartheta}\, \boldsymbol{e}_\vartheta + r\dot{\varphi}\, \sin\vartheta\, \boldsymbol{e}_\varphi) = r\dot{r}\,, \tag{1.1.141}$$

which, of course, is simply the time derivative of $\frac{1}{2}r^2 = \frac{1}{2}\boldsymbol{r}^2$. Similarly,

$$\begin{aligned} \boldsymbol{r} \times \boldsymbol{v} &= r\boldsymbol{e}_r \times (\dot{r}\, \boldsymbol{e}_r + r\dot{\vartheta}\, \boldsymbol{e}_\vartheta + r\dot{\varphi}\, \sin\vartheta\, \boldsymbol{e}_\varphi) \\ &= -r^2\dot{\varphi} \sin\vartheta\, \boldsymbol{e}_\vartheta + r^2\dot{\vartheta} \boldsymbol{e}_\varphi \end{aligned} \tag{1.1.142}$$

is the analog of (1.1.118).

For the scalar product of two position vectors, we have

$$\begin{aligned} \boldsymbol{r} \cdot \boldsymbol{r}' = rr'\, \boldsymbol{e}_r \cdot \boldsymbol{e}_r' &= rr' \begin{pmatrix} \sin\vartheta\cos\varphi \\ \sin\vartheta\sin\varphi \\ \cos\vartheta \end{pmatrix} \cdot \begin{pmatrix} \sin\vartheta'\cos\varphi' \\ \sin\vartheta'\sin\varphi' \\ \cos\vartheta' \end{pmatrix} \\ &= rr' \bigl[\cos\vartheta\, \cos\vartheta' + \sin\vartheta\, \sin\vartheta' \cos(\varphi - \varphi')\bigr] \end{aligned} \tag{1.1.143}$$

in spherical coordinates. It follows that the distance between the points with position vectors $\boldsymbol{r}$ and $\boldsymbol{r}'$ is

$$\begin{aligned}|\boldsymbol{r}-\boldsymbol{r}'| &= \sqrt{r^2+r'^2-2\boldsymbol{r}\cdot\boldsymbol{r}'}\\ &= \sqrt{r^2+r'^2-2rr'\bigl[\cos\vartheta\,\cos\vartheta'+\sin\vartheta\,\sin\vartheta'\cos(\varphi-\varphi')\bigr]}\,.\end{aligned} \tag{1.1.144}$$

1.2 Fields and their gradients

Let us now turn to another question, also with a definite touch of geometry. Consider a position-dependent physical quantity, such as the temperature distribution within a room, the mass distribution in an extended object, the charge distribution inside a battery, or similar. Such a quantity is described by a function of position, $f(\boldsymbol{r})$ say, what the physicist calls a *field*. As always, we are interested in the small changes of the field value that come about because parameters change a bit. In the present content, the question is: How does $f(\boldsymbol{r}+\delta\boldsymbol{r})$ differ from $f(\boldsymbol{r})$? In other words, we compare the field values of neighboring positions.

For a function of a single variable, this question has a familiar answer in terms of the derivative,

$$f(x+\delta x) = f(x) + \delta x\, f'(x) \tag{1.2.1}$$

with $f'(x) = \dfrac{\mathrm{d}}{\mathrm{d}x}f(x)$. The three-dimensional analog pays attention to the vectorial nature of $\delta\boldsymbol{r}$ and needs a vectorial derivative of $f(\boldsymbol{r})$, the so-called *gradient* of $f(\boldsymbol{r})$, denoted by $\boldsymbol{\nabla}f(\boldsymbol{r})$,

$$f(\boldsymbol{r}+\delta\boldsymbol{r}) = f(\boldsymbol{r}) + \delta\boldsymbol{r}\cdot\boldsymbol{\nabla}f(\boldsymbol{r})\,. \tag{1.2.2}$$

The gradient has a fundamental geometrical significance: It points in the direction of fastest growth of $f(\boldsymbol{r})$, and its length $|\boldsymbol{\nabla}f(\boldsymbol{r})|$ is the derivative in this direction. To justify, or to illustrate, these remarks let us consider $\delta\boldsymbol{r} = \epsilon\boldsymbol{e}$ with the unit vector $\boldsymbol{e}$ and infinitesimal $\epsilon > 0$, and regard $f(\boldsymbol{r}+\delta\boldsymbol{r}) = f(\boldsymbol{r}+\epsilon\boldsymbol{e})$ as a function of ϵ. Then, to first order in ϵ we have

$$f(\boldsymbol{r}+\epsilon\boldsymbol{e}) = f(\boldsymbol{r}) + \epsilon\boldsymbol{e}\cdot\boldsymbol{\nabla}f(\boldsymbol{r})\,, \tag{1.2.3}$$

which confirms that we get the largest increase if $\boldsymbol{e}$ is parallel to $\boldsymbol{\nabla}f(\boldsymbol{r})$, and we get the largest decrease if $\boldsymbol{e}$ is antiparallel to $\boldsymbol{\nabla}f(\boldsymbol{r})$; and there is no change if $\delta\boldsymbol{r} = \epsilon\boldsymbol{e}$ is orthogonal to $\boldsymbol{\nabla}f(\boldsymbol{r})$. This last observation, put differently, states that $\boldsymbol{\nabla} f(\boldsymbol{r})$ is normal to the surface of constant f that contains $\boldsymbol{r}$.

We find explicit expressions for $\boldsymbol{\nabla} f(\boldsymbol{r})$ by considering parameterizations of $\boldsymbol{r}$ in terms of coordinates, first looking at cartesian coordinates. For these we have

$$\delta\boldsymbol{r} = \delta x\, \boldsymbol{e}_x + \delta y\, \boldsymbol{e}_y + \delta z\, \boldsymbol{e}_z \tag{1.2.4}$$

so that

$$f(\boldsymbol{r}+\delta\boldsymbol{r}) = f(\boldsymbol{r}) + \delta x\, \boldsymbol{e}_x \cdot \boldsymbol{\nabla} f(\boldsymbol{r}) + \delta y\, \boldsymbol{e}_y \cdot \boldsymbol{\nabla} f(\boldsymbol{r}) + \delta z\, \boldsymbol{e}_z \cdot \boldsymbol{\nabla} f(\boldsymbol{r}), \tag{1.2.5}$$

which we compare with

$$f(\boldsymbol{r}+\delta\boldsymbol{r}) = f(\boldsymbol{r}) + \delta x \frac{\partial}{\partial x} f(\boldsymbol{r}) + \delta y \frac{\partial}{\partial y} f(\boldsymbol{r}) + \delta z \frac{\partial}{\partial z} f(\boldsymbol{r}), \tag{1.2.6}$$

where $\boldsymbol{r} \mathrel{\widehat{=}} \begin{pmatrix} x \\ y \\ z \end{pmatrix}$ is understood, of course. We read off that

$$\boldsymbol{e}_x \cdot \boldsymbol{\nabla} f(\boldsymbol{r}) = \frac{\partial}{\partial x} f(\boldsymbol{r}) \Big|_{\boldsymbol{r} = x\boldsymbol{e}_x + y\boldsymbol{e}_y + z\boldsymbol{e}_z} \tag{1.2.7}$$

and likewise for the y and z components, and so conclude that the cartesian form of the gradient is

$$\boldsymbol{\nabla} = \boldsymbol{e}_x \frac{\partial}{\partial x} + \boldsymbol{e}_y \frac{\partial}{\partial y} + \boldsymbol{e}_z \frac{\partial}{\partial z} \mathrel{\widehat{=}} \begin{pmatrix} \dfrac{\partial}{\partial x} \\ \dfrac{\partial}{\partial y} \\ \dfrac{\partial}{\partial z} \end{pmatrix}. \tag{1.2.8}$$

If not obvious earlier, it is clear now that the gradient is a differential-operator vector.

Quite analogously, we have for cylindrical coordinates

$$\begin{aligned} f(\boldsymbol{r}+\delta\boldsymbol{r}) &= f(\boldsymbol{r}) + (\delta s\, \boldsymbol{e}_s + \delta\varphi\, s\, \boldsymbol{e}_\varphi + \delta z \boldsymbol{e}_z) \cdot \boldsymbol{\nabla} f(\boldsymbol{r}) \\ &= f(\boldsymbol{r}) + \left(\delta s \frac{\partial}{\partial s} + \delta\varphi \frac{\partial}{\partial \varphi} + \delta z \frac{\partial}{\partial z}\right) f(\boldsymbol{r}), \end{aligned} \tag{1.2.9}$$

which tells us first that

$$\boldsymbol{e}_s \cdot \boldsymbol{\nabla} = \frac{\partial}{\partial s}, \quad s\boldsymbol{e}_\varphi \cdot \boldsymbol{\nabla} = \frac{\partial}{\partial \varphi}, \quad \boldsymbol{e}_z \cdot \boldsymbol{\nabla} = \frac{\partial}{\partial z}, \tag{1.2.10}$$

and then

$$\nabla = e_s \frac{\partial}{\partial s} + e_\varphi \frac{1}{s} \frac{\partial}{\partial \varphi} + e_z \frac{\partial}{\partial z}, \tag{1.2.11}$$

where it helps that the local unit vectors are pairwise orthogonal. In coordinate systems for which this is not the case (we use them very rarely) the nonorthogonality must be accounted for.

In this cylindrical-coordinate version of the gradient, we carefully write the unit vectors and other factors to the left of the differential operators, because these differential operators are meant to differentiate *everything* that stands to the right of them, and there is a difference between the gradient component

$$e_\varphi \frac{1}{s} \frac{\partial}{\partial \varphi} f(\boldsymbol{r}) \tag{1.2.12}$$

and

$$\frac{1}{s} \frac{\partial}{\partial \varphi} \bigl(e_\varphi f(\boldsymbol{r})\bigr) = e_\varphi \frac{1}{s} \frac{\partial}{\partial \varphi} f(\boldsymbol{r}) - \frac{1}{s} e_s f(\boldsymbol{r}), \tag{1.2.13}$$

where $\dfrac{\partial}{\partial \varphi} e_\varphi = -e_s$ has entered; we must not forget the coordinate dependence of the unit vectors.

Finally, for spherical coordinates, we have

$$\begin{aligned} \delta \boldsymbol{r} \cdot \nabla &= \bigl(\delta r\, e_r + \delta\vartheta\, r\, e_\vartheta + \delta\varphi\, r \sin\vartheta\, e_\varphi\bigr) \cdot \nabla \\ &= \delta r \frac{\partial}{\partial r} + \delta\vartheta \frac{\partial}{\partial \vartheta} + \delta\varphi \frac{\partial}{\partial \varphi} \end{aligned} \tag{1.2.14}$$

and, therefore,

$$\nabla = e_r \frac{\partial}{\partial r} + e_\vartheta \frac{1}{r} \frac{\partial}{\partial \vartheta} + e_\varphi \frac{1}{r \sin\vartheta} \frac{\partial}{\partial \varphi}. \tag{1.2.15}$$

The singularity of $\frac{1}{r}$ and $\frac{1}{r \sin\vartheta}$ when $r = 0$ or $\sin\vartheta = 0$ reminds us of the singularity of the spherical coordinates on the z axis where $\vartheta = 0$ or $\vartheta = \pi$. Similarly, the factor $\frac{1}{s}$ in the cylindrical coordinate version of ∇ in (1.2.11) originates in the respective singularity of that coordinate system. These singularities are usually harmless but one should not ignore them.

Here is an example. Consider

$$f(\boldsymbol{r}) = \boldsymbol{r}^2 = \underbrace{x^2 + y^2 + z^2}_{\text{cartesian}} = \underbrace{s^2 + z^2}_{\text{cylindrical}} = \underbrace{r^2}_{\text{spherical}} \tag{1.2.16}$$

for which

$$\begin{aligned}\boldsymbol{\nabla} f(\boldsymbol{r}) &= \underbrace{2x\,\boldsymbol{e}_x + 2y\,\boldsymbol{e}_y + 2z\,\boldsymbol{e}_z}_{\text{cartesian}} \mathrel{\widehat{=}} \begin{pmatrix} 2x \\ 2y \\ 2z \end{pmatrix} \\ &= \underbrace{2s\,\boldsymbol{e}_s + 2z\,\boldsymbol{e}_z}_{\text{cylindrical}} = \underbrace{2r\,\boldsymbol{e}_r}_{\text{spherical}}\end{aligned} \tag{1.2.17}$$

are equivalent ways of stating

$$\boldsymbol{\nabla} r^2 = 2\boldsymbol{r}\,. \tag{1.2.18}$$

We have least work here when employing spherical coordinates, which are also best suited in the more general situation of a function that depends only on the length r of the position vector $\boldsymbol{r}$,

$$\boldsymbol{\nabla} f(r) = \boldsymbol{e}_r \frac{\partial}{\partial r} f(r) = \frac{\boldsymbol{r}}{r}\frac{\partial}{\partial r} f(r)\,. \tag{1.2.19}$$

The intermediate expression refers to spherical coordinates, but the final expression is just a statement about $\boldsymbol{r}$, r, and the derivative of $f(r)$ with no reference to a particular coordinate system.

Another application concerns the change of $f(\boldsymbol{r})$ in time as we follow the trajectory $\boldsymbol{r}(t)$ of a point mass. We compare $f\big(\boldsymbol{r}(t+\delta t)\big)$ with $f\big(\boldsymbol{r}(t)\big)$, the values at time $t + \delta t$ and the slightly earlier time t. After first noting that

$$\boldsymbol{r}(t+\delta t) = \boldsymbol{r} + \delta t\,\boldsymbol{v}(t) \tag{1.2.20}$$

we have

$$f\big(\boldsymbol{r}(t+\delta t)\big) = f\big(\boldsymbol{r}(t)\big) + \delta t\,\boldsymbol{v}(t)\cdot\boldsymbol{\nabla} f\big(\boldsymbol{r}(t)\big) \tag{1.2.21}$$

and then conclude that

$$\frac{\mathrm{d}}{\mathrm{d}t} f\big(\boldsymbol{r}(t)\big) = \frac{f\big(\boldsymbol{r}(t+\delta t)\big) - f\big(\boldsymbol{r}(t)\big)}{\delta t}\Bigg|_{\delta t\,\to\,0} = \boldsymbol{v}(t)\cdot\boldsymbol{\nabla} f\big(\boldsymbol{r}(t)\big)\,. \tag{1.2.22}$$

This is, of course, nothing but the chain rule of differentiation in this vectorial context: The derivative of $f\big(\boldsymbol{r}(t)\big)$ is the derivative of the function with respect to its argument (that is $\boldsymbol{\nabla} f$) multiplied by the derivative of the argument (here $\boldsymbol{v}(t) = \frac{\mathrm{d}}{\mathrm{d}t}\boldsymbol{r}(t)$), and the correct kind of multiplication is the dot product.

As a simple illustration, we evaluate the time derivative of $f(\boldsymbol{r}) = \frac{1}{2}r^2$ that we have met earlier in (1.1.117). Upon combining (1.2.22) with (1.2.18), we get

$$\frac{\mathrm{d}}{\mathrm{d}t}\frac{1}{2}r^2 = \boldsymbol{v}\cdot\boldsymbol{\nabla}\frac{1}{2}r^2 = \boldsymbol{v}\cdot\boldsymbol{r} = \boldsymbol{r}\cdot\boldsymbol{v}\,. \tag{1.2.23}$$

Of course, we obtain exactly what we had found earlier.

1.3 Surface and volume elements

1.3.1 *Surface elements*

A point $\boldsymbol{r} + \mathrm{d}\boldsymbol{r}$ that is infinitesimally close to $\boldsymbol{r}$ is reached by the displacement

$$\mathrm{d}\boldsymbol{r} = \boldsymbol{a}\,\mathrm{d}\epsilon\,, \tag{1.3.1}$$

where the vector $\boldsymbol{a}$ specifies the direction of the displacement and $\mathrm{d}\epsilon$ is the infinitesimal increment of a parameter ϵ. We can think of ϵ as the coordinate associated with the direction $\boldsymbol{a}$. A second infinitesimally close point is displaced by

$$\mathrm{d}\boldsymbol{r}' = \boldsymbol{b}\,\mathrm{d}\epsilon', \tag{1.3.2}$$

with $\boldsymbol{b}$ and ϵ' analogous to $\boldsymbol{a}$ and ϵ.

The two displacements together give us a little parallelogram with the corners at $\boldsymbol{r}$, $\boldsymbol{r} + \mathrm{d}\boldsymbol{r}$, $\boldsymbol{r} + \mathrm{d}\boldsymbol{r}'$, and $\boldsymbol{r} + \mathrm{d}\boldsymbol{r} + \mathrm{d}\boldsymbol{r}'$, as depicted in (1.1.16) if we replace $\boldsymbol{a}$ by $\boldsymbol{a}\,\mathrm{d}\epsilon$ and $\boldsymbol{b}$ by $\boldsymbol{b}\,\mathrm{d}\epsilon'$. This parallelogram is an infinitesimal area element with its size equal to the length of the normal vector

$$\mathrm{d}\boldsymbol{S} = \mathrm{d}\boldsymbol{r} \times \mathrm{d}\boldsymbol{r}' = \boldsymbol{a} \times \boldsymbol{b}\,\mathrm{d}\epsilon\,\mathrm{d}\epsilon'. \tag{1.3.3}$$

The direction of $\mathrm{d}\boldsymbol{S}$ endows the area element with an orientation in accordance with the right-hand rule for $\boldsymbol{a}$, $\boldsymbol{b}$, and $\mathrm{d}\boldsymbol{S}$. For a window in a wall, say, the orientation of the window area tells us on which side is the room and on which the street. Then we know if wind blowing with a velocity $\boldsymbol{v}$ is carrying air into the room or out of the room, depending on the sign of $\boldsymbol{v}\cdot\mathrm{d}\boldsymbol{S}$. If the orientation is such that $\mathrm{d}\boldsymbol{S}$ points from the room to the street, air leaves the room for $\boldsymbol{v}\cdot\mathrm{d}\boldsymbol{S} > 0$, enters the room for $\boldsymbol{v}\cdot\mathrm{d}\boldsymbol{S} < 0$, and moves parallel to the window surface for $\boldsymbol{v}\cdot \mathrm{d}\boldsymbol{S}=0$:

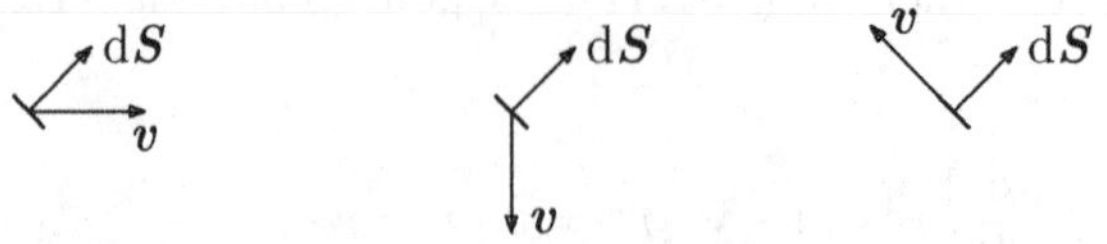

$$\boldsymbol{v}\cdot \mathrm{d}\boldsymbol{S}>0 \qquad \boldsymbol{v}\cdot \mathrm{d}\boldsymbol{S}<0 \qquad \boldsymbol{v}\cdot \mathrm{d}\boldsymbol{S}=0 \tag{1.3.4}$$

The various coordinate systems — cartesian, cylindrical, spherical, and others — have particular surfaces associated with them, the surfaces on which one of the coordinates is constant. These surfaces are parameterized by the other two coordinates and their area elements, the *surface elements* for short, are needed when we integrate over such a surface, or over parts of it.

For cartesian coordinates, we have

$$\begin{aligned}
&\text{the planes of constant } x \text{ with } \mathrm{d}\boldsymbol{S}=\boldsymbol{e}_y\,\mathrm{d}y\times\boldsymbol{e}_z\,\mathrm{d}z=\boldsymbol{e}_x\,\mathrm{d}y\,\mathrm{d}z\,,\\
&\text{the planes of constant } y \text{ with } \mathrm{d}\boldsymbol{S}=\boldsymbol{e}_z\,\mathrm{d}z\times\boldsymbol{e}_x\,\mathrm{d}x=\boldsymbol{e}_y\,\mathrm{d}z\,\mathrm{d}x\,,\\
&\text{the planes of constant } z \text{ with } \mathrm{d}\boldsymbol{S}=\boldsymbol{e}_x\,\mathrm{d}x\times\boldsymbol{e}_y\,\mathrm{d}y=\boldsymbol{e}_z\,\mathrm{d}x\,\mathrm{d}y\,,
\end{aligned}\tag{1.3.5}$$

and the respective area elements $\mathrm{d}y\,\mathrm{d}z$, $\mathrm{d}z\,\mathrm{d}x$, and $\mathrm{d}x\,\mathrm{d}y$ are familiar. For cylindrical coordinates, there are

$$\begin{aligned}
\text{the cylinders of constant } s \text{ with } \mathrm{d}\boldsymbol{S}&=\boldsymbol{e}_\varphi s\,\mathrm{d}\varphi\times\boldsymbol{e}_z\,\mathrm{d}z\\
&=\boldsymbol{e}_s s\,\mathrm{d}\varphi\,\mathrm{d}z\,,\\
\text{the half-planes of constant } \varphi \text{ with } \mathrm{d}\boldsymbol{S}&=\boldsymbol{e}_z\,\mathrm{d}z\times\boldsymbol{e}_s\,\mathrm{d}s\\
&=\boldsymbol{e}_\varphi\,\mathrm{d}z\,\mathrm{d}s\,,\\
\text{the planes of constant } z \text{ with } \mathrm{d}\boldsymbol{S}&=\boldsymbol{e}_s\,\mathrm{d}s\times\boldsymbol{e}_\varphi s\,\mathrm{d}\varphi\\
&=\boldsymbol{e}_z s\,\mathrm{d}s\,\mathrm{d}\varphi\,.
\end{aligned}\tag{1.3.6}$$

Since the planes of constant z, each of them parallel to the xy plane, are the same for cartesian and cylindrical coordinates, we note that $\mathrm{d}x\,\mathrm{d}y=s\,\mathrm{d}s\,\mathrm{d}\varphi$ for corresponding area elements in the xy plane.

For spherical coordinates, there are

$$\begin{aligned}
\text{the spheres of constant } r \text{ with } \mathrm{d}\boldsymbol{S}&=\boldsymbol{e}_\vartheta r\,\mathrm{d}\vartheta\times\boldsymbol{e}_\varphi r\sin\vartheta\,\mathrm{d}\varphi\\
&=\boldsymbol{e}_r r^2\,\mathrm{d}\Omega\,,\\
\text{the cones of constant } \vartheta \text{ with } \mathrm{d}\boldsymbol{S}&=\boldsymbol{e}_\varphi r\sin\vartheta\,\mathrm{d}\varphi\times\boldsymbol{e}_r\,\mathrm{d}r\\
&=\boldsymbol{e}_\vartheta r\sin\vartheta\,\mathrm{d}r\,\mathrm{d}\varphi\,,\\
\text{the half-planes of constant } \varphi \text{ with } \mathrm{d}\boldsymbol{S}&=\boldsymbol{e}_r\,\mathrm{d}r\times\boldsymbol{e}_\vartheta r\,\mathrm{d}\vartheta\\
&=\boldsymbol{e}_\varphi r\,\mathrm{d}r\,\mathrm{d}\vartheta\,,
\end{aligned}\tag{1.3.7}$$

where

$$d\Omega = \sin\vartheta \, d\vartheta \, d\varphi \tag{1.3.8}$$

is the standard notation for the *solid-angle element*, the area element on the unit sphere. It is the two-dimensional analog of the arc element $d\varphi$, the length element on the unit circle. Just as the total arc

$$\int_0^{2\pi} d\varphi = 2\pi \tag{1.3.9}$$

is the circumference of the unit circle, the total solid angle

$$\int_0^{\pi} d\vartheta \, \sin\vartheta \int_0^{2\pi} d\varphi = 4\pi \tag{1.3.10}$$

is the area of the unit sphere.

We note that $dz\, ds = r\, dr\, d\vartheta$ relates the cylindrical-coordinates version and the spherical-coordinates version of the area elements on the half-planes of constant azimuth to each other. This is much like the $dx\, dy = s\, ds\, d\varphi$ relation above, except that the coordinates have different ranges for their values: $-\infty < x, y < \infty$ and $s \geq 0$, $0 \leq \varphi < 2\pi$ for the planes of constant z, whereas $-\infty < z < \infty$, $s \geq 0$ and $r \geq 0$, $0 \leq \vartheta \leq \pi$ for the half-planes of constant φ.

1.3.2 *Volume elements*

If we supplement the two infinitesimal displacements by a third one,

$$d\boldsymbol{r}'' = \boldsymbol{c}\, d\epsilon'' , \tag{1.3.11}$$

then $d\boldsymbol{r}$, $d\boldsymbol{r}'$, $d\boldsymbol{r}''$ make up a little parallelepiped at position $\boldsymbol{r}$, as depicted in (1.1.28) with $\boldsymbol{a}$, $\boldsymbol{b}$, and $\boldsymbol{c}$ scaled by $d\epsilon$, $d\epsilon'$, and $d\epsilon''$, respectively. The volume of this parallelepiped is the *volume element*

$$(d\boldsymbol{r}) = (d\boldsymbol{r} \times d\boldsymbol{r}') \cdot d\boldsymbol{r}'' = (\boldsymbol{a} \times \boldsymbol{b}) \cdot \boldsymbol{c}\, d\epsilon\, d\epsilon'\, d\epsilon'', \tag{1.3.12}$$

where we take for granted that $(\boldsymbol{a} \times \boldsymbol{b}) \cdot \boldsymbol{c} > 0$, that is: the trio $\boldsymbol{a}$, $\boldsymbol{b}$, $\boldsymbol{c}$ is right-handed.

The local unit vectors of the systems of cartesian, cylindrical, or spherical coordinates are right-handed and orthonormal,

$$(\boldsymbol{e}_x \times \boldsymbol{e}_y) \cdot \boldsymbol{e}_z = 1\,, \quad (\boldsymbol{e}_s \times \boldsymbol{e}_\varphi) \cdot \boldsymbol{e}_z = 1\,, \quad (\boldsymbol{e}_r \times \boldsymbol{e}_\vartheta) \cdot \boldsymbol{e}_\varphi = 1\,, \tag{1.3.13}$$

so that

$$(\mathrm{d}\boldsymbol{r}) = \mathrm{d}x\,\mathrm{d}y\,\mathrm{d}z = s\,\mathrm{d}s\,\mathrm{d}\varphi\,\mathrm{d}z = r^2\,\mathrm{d}r\,\mathrm{d}\Omega \tag{1.3.14}$$

are the three respective versions of the volume element. We obtain them as the scalar products of the vector products for the various surface elements $\mathrm{d}\boldsymbol{S}$ in (1.3.5), (1.3.6), or (1.3.7) with the respective third displacement vector,

$$\begin{aligned}(\mathrm{d}\boldsymbol{r}) &= (\boldsymbol{e}_x\,\mathrm{d}x \times \boldsymbol{e}_y\,\mathrm{d}y) \cdot \boldsymbol{e}_z\,\mathrm{d}z \\ &= (\boldsymbol{e}_s\,\mathrm{d}s \times \boldsymbol{e}_\varphi s\,\mathrm{d}\varphi) \cdot \boldsymbol{e}_z\,\mathrm{d}z \\ &= (\boldsymbol{e}_r\,\mathrm{d}r \times \boldsymbol{e}_\vartheta r\,\mathrm{d}\vartheta) \cdot \boldsymbol{e}_\varphi r \sin\vartheta\,\mathrm{d}\varphi\,. \end{aligned} \tag{1.3.15}$$

Chapter 2

Dynamics

2.1 Newton's equation of motion

Having dealt with the basic examples of kinematics (How do we describe masses in motion?), we now move on to dynamics (How does the state of motion change in the course of time?). The fundamental law is *Newton's* equation of motion*

$$\frac{\mathrm{d}}{\mathrm{d}t}(m\boldsymbol{v}) = \boldsymbol{F}\,, \tag{2.1.1}$$

the time derivative of the kinetic momentum — the product $m\boldsymbol{v}$ of mass m and velocity $\boldsymbol{v}$ — is equal to the force $\boldsymbol{F}$. There is one such equation for each mechanical object, for each point mass, and the force on one of them will usually depend on the positions, and perhaps the velocities, of the others. But for a start, we consider the situation of a single point mass.

The force $\boldsymbol{F}$ results from the physical interactions to which the point mass is exposed, such as the gravitational pull of the earth, the Lorentz† force of the electric and magnetic fields if the point mass carries electric charge, or perhaps the force exerted by a string attached to the point mass with someone pulling at the other end. Exceptional situations aside, the force is a function of the position $\boldsymbol{r}$ of the point mass, of its velocity $\boldsymbol{v} = \frac{\mathrm{d}}{\mathrm{d}t}\boldsymbol{r} = \dot{\boldsymbol{r}}$, and possibly some time dependent parameters (externally manipulated strength of an electric field, say), that is

$$\boldsymbol{F} = \boldsymbol{F}\bigl(\underbrace{\boldsymbol{r}(t)}_{\text{position}}, \underbrace{\boldsymbol{v}(t)}_{\text{velocity}}, \underbrace{t}_{\text{parametric time dependence}}\bigr) \tag{2.1.2}$$

(position, velocity at time t; parametric time dependence)

---

*Sir Isaac Newton (1643–1727) †Hendrik Antoon Lorentz (1853–1929) but there is a very large class of important force laws that only have a position dependence,

$$\boldsymbol{F} = \boldsymbol{F}\big(\boldsymbol{r}(t)\big)\,. \tag{2.1.3}$$

In either case, Newton's equation of motion

$$\frac{\mathrm{d}}{\mathrm{d}t}\Big(m\frac{\mathrm{d}}{\mathrm{d}t}\boldsymbol{r}\Big) = \boldsymbol{F}\Big(\boldsymbol{r}, \frac{\mathrm{d}}{\mathrm{d}t}\boldsymbol{r}, t\Big) \tag{2.1.4}$$

is a second-order differential equation for the position vector $\boldsymbol{r}$ as a function of time. As a rule, it has a unique solution when the position and velocity are specified at an initial time t_0: Given $\boldsymbol{r}_0 = \boldsymbol{r}(t_0)$ and $\boldsymbol{v}_0 = \dot{\boldsymbol{r}}(t_0)$, we find $\boldsymbol{r}(t)$ by solving the equation of motion. There is no general expression for $\boldsymbol{r}(t)$ that we could use for all force laws, it is necessary to consider each physical situation separately.

2.2 Elementary examples

We will first consider a couple of elementary examples. These will serve for the illustration of important aspects in simple, and perhaps familiar, contexts. We will also have the opportunity to introduce a few technical tools when studying these elementary examples.

2.2.1 *Force-free motion*

We begin with the simplest example: force-free motion, $\boldsymbol{F} = 0$,

$$\frac{\mathrm{d}}{\mathrm{d}t}(m\boldsymbol{v}) = 0\,, \tag{2.2.1}$$

and since the mass of the point mass is usually not time-dependent, this simplifies further to

$$\frac{\mathrm{d}}{\mathrm{d}t}\boldsymbol{v} = 0\,. \tag{2.2.2}$$

The solution is immediate,

$$\boldsymbol{v}(t) = \boldsymbol{v}_0\,, \tag{2.2.3}$$

which incorporates the value of the constant velocity, equal to the initial velocity, of course. We then find $\boldsymbol{r}(t)$ as the solution of

$$\frac{\mathrm{d}}{\mathrm{d}t}\boldsymbol{r}(t) = \boldsymbol{v}(t) = \boldsymbol{v}_0\,, \tag{2.2.4}$$

that is

$$\boldsymbol{r}(t) = \boldsymbol{r}_0 + \boldsymbol{v}_0(t - t_0)\,. \tag{2.2.5}$$

This states simply what we all know: If no force is acting, the point mass moves along a straight line with constant velocity.

2.2.2 *Constant force*

The next example is that of a constant force,

$$m\frac{\mathrm{d}}{\mathrm{d}t}\boldsymbol{v} = \boldsymbol{F} = \text{constant}\,. \tag{2.2.6}$$

This equation is solved by

$$\boldsymbol{v}(t) = \boldsymbol{v}_0 + \frac{\boldsymbol{F}}{m}(t - t_0)\,, \tag{2.2.7}$$

and then

$$\frac{\mathrm{d}}{\mathrm{d}t}\boldsymbol{r}(t) = \boldsymbol{v}(t) \tag{2.2.8}$$

gives

$$\boldsymbol{r}(t) = \boldsymbol{r}_0 + \boldsymbol{v}_0(t - t_0) + \frac{\boldsymbol{F}}{2m}(t - t_0)^2\,. \tag{2.2.9}$$

This is motion along a parabola, very familiar from all those high-school examples about stones thrown in the presence of the gravitational pull, which gives a constant force $m\boldsymbol{g}$ near the surface of the earth with the universal gravitational acceleration $\boldsymbol{g}$ — universal because it is the same for all point masses: In the absence of the frictional forces that result from air drag, all bodies fall equally fast, as was first stated by Galilei* who extrapolated from the real experimental situations with air drag to the ideal circumstances without these frictional forces. It was only after Galilei's death in 1642 that his contemporary Torricelli† produced, in 1643, a low-pressure volume with so little residual gas (mostly mercury vapor) that close-to ideal circumstances could be realized in the laboratory.

*Galileo GALILEI (1564–1642) †Evangelista TORRICELLI (1608–1647)

2.2.3 *Frictional forces*

We can model the frictional force of the air drag by a force that is proportional to the velocity and opposed to the direction of $\boldsymbol{v}$,

$$\boldsymbol{F}_{\text{friction}} = -m\gamma\boldsymbol{v} \quad \text{with} \quad \gamma > 0\,, \tag{2.2.10}$$

where the inclusion of the mass m in the prefactor is for convenience. Such a frictional force is called *Newtonian friction* for historical reason, it is the simplest modeling of friction that one can have. Very often, however, frictional forces have a more complicated velocity dependence, and air drag is no exception inasmuch as the actual force is approximately proportional to the square of the speed. For now, nevertheless, we shall be content with the model of Newtonian friction. The equation of motion is then

$$m\frac{\mathrm{d}}{\mathrm{d}t}\boldsymbol{v} = -m\gamma\boldsymbol{v} + m\boldsymbol{g} \tag{2.2.11}$$

after we write $\boldsymbol{F} = m\boldsymbol{g}$ for the force of the gravitational pull, with $\boldsymbol{g}$, the vector of the gravitational acceleration, toward the surface of the earth — downwards, that is. After removing the common factor of m, we have

$$\frac{\mathrm{d}}{\mathrm{d}t}\boldsymbol{v} = -\gamma\boldsymbol{v} + \boldsymbol{g} \tag{2.2.12}$$

and notice that

$$\boldsymbol{v}(t) = \frac{1}{\gamma}\boldsymbol{g} = \text{constant} \tag{2.2.13}$$

is one solution of the equation, the solution for which the acceleration vanishes, $\dot{\boldsymbol{v}} = 0$. In this situation, we have motion with constant velocity because the frictional force compensates exactly for the gravitational pull and no net force remains. We expect that, given enough time, all point masses will reach the terminal velocity $\boldsymbol{v}_\infty = \boldsymbol{g}/\gamma$, and this is an invitation to rewrite the equation of motion in the form

$$\frac{\mathrm{d}}{\mathrm{d}t}\left(\boldsymbol{v} - \frac{1}{\gamma}\boldsymbol{g}\right) = -\gamma\left(\boldsymbol{v} - \frac{1}{\gamma}\boldsymbol{g}\right) \tag{2.2.14}$$

or

$$\frac{\mathrm{d}}{\mathrm{d}t}(\boldsymbol{v} - \boldsymbol{v}_\infty) = -\gamma(\boldsymbol{v} - \boldsymbol{v}_\infty)\,, \tag{2.2.15}$$

where we remember that $\boldsymbol{v}$ depends on time and $\boldsymbol{v}_\infty$ is a time-independent constant velocity. The solution of this differential equation is immediate,

$$\boldsymbol{v}(t) - \boldsymbol{v}_\infty = (\boldsymbol{v}_0 - \boldsymbol{v}_\infty)\,\mathrm{e}^{-\gamma t}\,, \tag{2.2.16}$$

which incorporates the initial condition $\boldsymbol{v}(t=0) = \boldsymbol{v}_0$. Indeed, the velocity approaches $\boldsymbol{v}_\infty$ for late times,

$$\boldsymbol{v}(t) \longrightarrow \boldsymbol{v}_\infty \quad \text{as } t \to \infty\,, \tag{2.2.17}$$

where the physicist reads the mathematical "$t \to \infty$" statement as meaning $t \gg 1/\gamma$ and so identifies $1/\gamma$ as the time scale set by the frictional force. In passing, let us note that simple estimates can often be had by making use of $\mathrm{e}^3 \cong 20$; for example, we have $\mathrm{e}^{-\gamma t} \cong \frac{1}{8000}$ when $t \cong 9/\gamma$.

Having found the velocity as a function of time, we get the position by an integration

$$\boldsymbol{r}(t) = \boldsymbol{r}_0 + \int_0^t \mathrm{d}t'\,\boldsymbol{v}(t') = \boldsymbol{r}_0 + \int_0^t \mathrm{d}t' \left[\boldsymbol{v}_\infty + (\boldsymbol{v}_0 - \boldsymbol{v}_\infty)\,\mathrm{e}^{-\gamma t'}\right], \tag{2.2.18}$$

so that

$$\boldsymbol{r}(t) = \boldsymbol{r}_0 + \boldsymbol{v}_\infty t + (\boldsymbol{v}_0 - \boldsymbol{v}_\infty)\frac{1 - \mathrm{e}^{-\gamma t}}{\gamma}\,. \tag{2.2.19}$$

At late times, $t \gg 1/\gamma$, this is approximated by

$$\boldsymbol{r}(t) \cong \boldsymbol{r}_0 + \frac{1}{\gamma}(\boldsymbol{v}_0 - \boldsymbol{v}_\infty) + \boldsymbol{v}_\infty t \cong \boldsymbol{v}_\infty t \quad \text{for} \quad \gamma t \gg 1\,, \tag{2.2.20}$$

consistent with $\boldsymbol{v}(t) \cong \boldsymbol{v}_\infty$ for such late times. Accordingly, we have force-free motion with constant velocity at these late times because the gravitational pull and the frictional force compensate for each other, and no net force acts on the point mass.

We should verify as well that we recover the correct $\boldsymbol{r}(t)$ for $\gamma = 0$. Since $\boldsymbol{v}_\infty = \boldsymbol{g}/\gamma$, the limit $\gamma \to 0$ requires that we keep terms of order γ^2 in

$$1 - \mathrm{e}^{-\gamma t} = \gamma t - \frac{1}{2}(\gamma t)^2 + \cdots\,. \tag{2.2.21}$$

Then

$$\begin{aligned}\boldsymbol{r}(t) &= \boldsymbol{r}_0 + \frac{1}{\gamma}\boldsymbol{g}t + \left(\boldsymbol{v}_0 - \frac{1}{\gamma}\boldsymbol{g}\right)\left(t - \frac{1}{2}\gamma t^2 + \cdots\right)\\ &= \boldsymbol{r}_0 + \boldsymbol{v}_0 t + \frac{1}{2}\boldsymbol{g}t^2 + \cdots\,,\end{aligned} \tag{2.2.22}$$

where the ellipsis stands for terms of order γ in the final expression, so that the three displayed terms are the answer for $\gamma = 0$, the familiar and expected answer, indeed, as we confirm by putting $\boldsymbol{F} = m\boldsymbol{g}$ and $t_0 = 0$ in (2.2.9).

The more realistic frictional force proportional to the square of the speed, as it results from air drag,

$$\boldsymbol{F}_{\text{drag}} = -m\kappa|\boldsymbol{v}|\boldsymbol{v} = -m\kappa v\boldsymbol{v}\,, \tag{2.2.23}$$

leads to an equation of motion that is a bit too difficult to handle in the full three-dimensional setting and, therefore, we will be content with studying its consequences in the simpler situation in which the point mass is released from rest — we let a stone drop, rather than throwing it. Choosing the coordinate system such that $\boldsymbol{e}_z$ points downwards and $x = y = z = 0$ is the initial position with $\boldsymbol{v}(t = 0) = 0$, we have

$$\boldsymbol{v}(t) = v(t)\boldsymbol{e}_z\,, \quad \boldsymbol{g} = g\boldsymbol{e}_z\,, \quad v\boldsymbol{v} = v^2\boldsymbol{e}_z\,, \tag{2.2.24}$$

with $v = \dot{z}$, and the equation of motion reads

$$\dot{v} = g - \kappa v^2 \quad \text{with} \quad g,\, \kappa > 0\,, \tag{2.2.25}$$

after removing the common factor of m from $m\dot{\boldsymbol{v}} = m\boldsymbol{g} - m\kappa v\boldsymbol{v}$. The first thing to note is that here, too, we have a terminal velocity v_∞, given by

$$\kappa v_\infty^2 = g\,, \tag{2.2.26}$$

such that $v(t) = v_\infty$ is one solution of the differential equation for $v(t)$. We use this insight to rewrite the equation in the form

$$\frac{\mathrm{d}}{\mathrm{d}t}v = -\kappa(v^2 - v_\infty^2)\,. \tag{2.2.27}$$

A differential equation of this kind — the derivative of the unknown function is a function of the unknown function itself — can be solved by a method known as "separation of variables."

Here is how it works: We separate v and t by solving for $\mathrm{d}t$

$$\kappa\,\mathrm{d}t = \frac{\mathrm{d}v}{v_\infty^2 - v^2}\,, \tag{2.2.28}$$

and then integrate on both sides,

$$\kappa\int_0^t \mathrm{d}t' = \int_0^{v(t)} \frac{\mathrm{d}v'}{v_\infty^2 - v'^2}\,, \tag{2.2.29}$$

where the initial condition $v(t=0)=0$ is incorporated. This takes us to

$$\kappa t = \frac{1}{2v_\infty}\log\frac{v_\infty + v(t)}{v_\infty - v(t)} \tag{2.2.30}$$

where log denotes the natural logarithm. We solve for $v(t)$ and arrive at

$$v(t) = \frac{1-\mathrm{e}^{-2\kappa v_\infty t}}{1+\mathrm{e}^{-2\kappa v_\infty t}}\,v_\infty = v_\infty \tanh(\kappa v_\infty t)\,. \tag{2.2.31}$$

With $v_\infty = \sqrt{g/\kappa}$, this appears as

$$v(t) = \sqrt{\frac{g}{\kappa}}\tanh\bigl(\sqrt{\kappa g}\,t\bigr)\,, \tag{2.2.32}$$

which tells us that the time scale for the (speed)^2 air-drag is set by $1/\sqrt{\kappa g} \equiv \tau$,

$$v(t) = g\tau\tanh\left(\frac{t}{\tau}\right), \tag{2.2.33}$$

with the terminal velocity $v_\infty = g\tau$ being reached after the lapse of a few τs,

$$v(t) \cong g\tau = v_\infty \quad \text{for} \quad t \gg \tau\,. \tag{2.2.34}$$

Having seen how the "separation of variables" works, we can use an obvious short-cut. What matters is that the left-hand side of

$$\frac{1}{v_\infty^2 - v(t)^2}\frac{\mathrm{d}}{\mathrm{d}t}v(t) = \kappa \tag{2.2.35}$$

is a total time derivative,

$$\frac{\mathrm{d}}{\mathrm{d}t}\left[\frac{1}{2v_\infty}\log\frac{v_\infty + v(t)}{v_\infty - v(t)}\right] = \kappa\,. \tag{2.2.36}$$

As soon as we recognize this, the solution is at hand:

$$\frac{1}{2v_\infty}\log\frac{v_\infty + v(t)}{v_\infty - v(t)} = \kappa t\,, \tag{2.2.37}$$

where $v(t=0)=0$ is taken into account.

For the final integration with $z(t=0)=0$ (substitute $u=\kappa v_\infty t'$)

$$z(t)=\int_0^t \mathrm{d}t'\,v(t')=v_\infty\int_0^t \mathrm{d}t'\,\tanh(\kappa v_\infty t')$$
$$=\frac{1}{\kappa}\int_0^{\kappa v_\infty t}\mathrm{d}u\,\tanh u\,, \tag{2.2.38}$$

we recall that

$$\tanh u=\frac{\sinh u}{\cosh u}=\frac{\mathrm{d}}{\mathrm{d}u}\log(\cosh u) \tag{2.2.39}$$

and so find that

$$z(t)=\frac{1}{\kappa}\log\bigl(\cosh(\kappa v_\infty t)\bigr)=\frac{1}{\kappa}\log\bigl(\cosh(\sqrt{\kappa g}\,t)\bigr)\,, \tag{2.2.40}$$

or

$$z(t)=g\tau^2\log\left(\cosh\left(\frac{t}{\tau}\right)\right)=v_\infty\tau\log\left(\cosh\left(\frac{t}{\tau}\right)\right) \tag{2.2.41}$$

if we introduce the friction time scale τ.

The comparison with the corresponding expression for Newtonian friction,

$$z(t)=v_\infty\tau\left(\frac{t}{\tau}-1+\mathrm{e}^{-t/\tau}\right)\qquad\text{(Newtonian friction)} \tag{2.2.42}$$

obtained by having $\boldsymbol{r}(t)=z(t)\boldsymbol{e}_z$, $\boldsymbol{r}_0=0$, $\boldsymbol{v}_0=0$, $\boldsymbol{v}_\infty=v_\infty\boldsymbol{e}_z$ in (2.2.19), shows that the terminal velocity is reached more slowly for Newtonian friction than for (speed)^2 air-drag, as you will confirm in Exercise 22. The physical reason for this is quite clear: During the initial stage of small speeds, the linear Newtonian friction force is stronger than the quadratic air-drag force.

2.2.4 *Linear restoring force: Harmonic oscillations*

The next example is that of a linear restoring force

$$\boldsymbol{F}=-k\boldsymbol{r}\quad\text{with}\quad k>0\,, \tag{2.2.43}$$

as we encounter it when springs are stretched or compressed (not too much) such that their length is increased or decreased (by not too much), as in this one-dimensional picture:

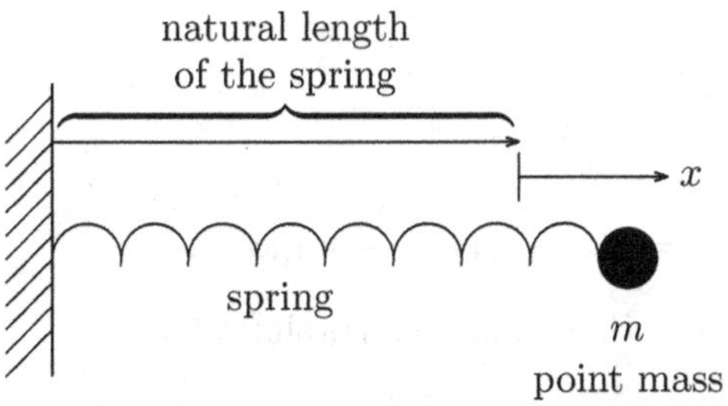

(2.2.44)

The force law above, with the force proportional to the displacement from the equilibrium position is *Hooke's** *Law*: *Ut tensio, sic vis* (As is the extension, so is the force). Hooke's law applies almost always when there is a small deviation from an equilibrium; more about this in Section 3.1.2(a) and much more in Chapter 9.

Since the equation of motion

$$m\frac{\mathrm{d}^2}{\mathrm{d}t^2}\boldsymbol{r} = -k\boldsymbol{r} \tag{2.2.45}$$

is three times the same one-dimensional equation in cartesian coordinates, namely

$$\begin{aligned} m\ddot{x} &= -kx\,, \\ m\ddot{y} &= -ky\,, \\ m\ddot{z} &= -kz\,, \end{aligned} \tag{2.2.46}$$

it is sufficient to deal with one of them, the equation for $x(t)$, say.

We know the answer, of course, because we have encountered this differential equation before, in (1.1.58) and (1.1.59). With $x(0) = x_0$ and $\dot{x}(0) = \dot{x}_0$, the solution is

$$x(t) = x_0 \cos\left(\sqrt{\frac{k}{m}}\,t\right) + \sqrt{\frac{m}{k}}\,\dot{x}_0 \sin\left(\sqrt{\frac{k}{m}}\,t\right), \tag{2.2.47}$$

as one verifies quickly. It is convenient to get rid of the awkward square roots by introducing the circular frequency ω_0 by means of

$$k = m\omega_0^2\,, \quad \omega_0 > 0\,, \tag{2.2.48}$$

*Robert HOOKE (1635–1703)

so that

$$\ddot{x} = -\omega_0^2 x \tag{2.2.49}$$

and then

$$\begin{aligned} x(t) &= x_0 \cos(\omega_0 t) + \frac{\dot{x}_0}{\omega_0} \sin(\omega_0 t)\,, \\ \dot{x}(t) &= \dot{x}_0 \cos(\omega_0 t) - \omega_0 x_0 \sin(\omega_0 t) \end{aligned} \tag{2.2.50}$$

state the position and velocity as a function of time.

It will be useful to know one more method for solving such a linear differential equation with time-independent coefficients, in addition to the three methods in Section 1.1.8. The fourth method is by an *ansatz*: We look for solutions of the kind

$$x(t) = c\,\mathrm{e}^{\lambda t}\,, \tag{2.2.51}$$

where c and λ are constants. An exponential ansatz of this kind is always worth a try for linear differential equations, or systems of them. Clearly, any value of c will do, but for λ we need

$$\lambda^2 = -\omega_0^2\,, \tag{2.2.52}$$

because $\ddot{x} = \lambda^2 x$ for the ansatz. This tells us that $\lambda_1 = \mathrm{i}\omega_0$ and $\lambda_2 = -\mathrm{i}\omega_0$ are possible λ values. Then, linearly combining the two solutions, we have

$$x(t) = c_1\,\mathrm{e}^{\mathrm{i}\omega_0 t} + c_2\,\mathrm{e}^{-\mathrm{i}\omega_0 t} \tag{2.2.53}$$

for the general solution of the differential equation, with the two constants c_1 and c_2 determined by the initial conditions. These are

$$\begin{aligned} x_0 &= x(0) = c_1 + c_2\,, \\ \dot{x}_0 &= \dot{x}(0) = \mathrm{i}\omega_0 c_1 - \mathrm{i}\omega_0 c_2\,, \end{aligned} \tag{2.2.54}$$

so that

$$\begin{aligned} c_1 &= \frac{1}{2} x_0 + \frac{1}{2\mathrm{i}\omega_0} \dot{x}_0\,, \\ c_2 &= \frac{1}{2} x_0 - \frac{1}{2\mathrm{i}\omega_0} \dot{x}_0\,, \end{aligned} \tag{2.2.55}$$

and we have

$$x(t) = \frac{1}{2}x_0 \underbrace{\left(e^{i\omega_0 t} + e^{-i\omega_0 t}\right)}_{=2\cos(\omega_0 t)} + \frac{1}{2i\omega_0}\dot{x}_0 \underbrace{\left(e^{i\omega_0 t} - e^{-i\omega_0 t}\right)}_{=2i\sin(\omega_0 t)}$$

$$= x_0 \cos(\omega_0 t) + \frac{\dot{x}_0}{\omega_0}\sin(\omega_0 t)\,, \tag{2.2.56}$$

hardly a surprise. Any one of the methods gives the same solution, of periodic motion with period $T = 2\pi/\omega_0$,

$$x\left(t + \frac{2\pi}{\omega_0}\right) = x(t) \tag{2.2.57}$$

since $\cos\bigl(\omega_0(t + 2\pi/\omega_0)\bigr) = \cos(\omega_0 t + 2\pi) = \cos(\omega_0 t)$ and likewise for $\sin(\omega_0 t)$. Motion with this simple periodicity is called *harmonic*, and the physical system that obeys the equation of motion

$$\ddot{x} = -\omega_0^2\, x \tag{2.2.58}$$

is a *harmonic oscillator*.

2.2.5 *Damped harmonic oscillations*

Currently, however, we are learning to make good use of the new, fourth, method when considering a harmonic oscillator with Newtonian friction,

$$\ddot{x} = -\gamma\dot{x} - \omega_0^2 x \quad \text{with} \quad \gamma\,, \omega_0 > 0\,, \tag{2.2.59}$$

where $-\gamma\dot{x}$ is the acceleration by the frictional force and $-\omega_0^2 x$ is the acceleration by the linear restoring force. The exponential ansatz $x(t) = c\,e^{\lambda t}$ now gives

$$\lambda^2 = -\gamma\lambda - \omega_0^2\,, \tag{2.2.60}$$

a quadratic equation for λ. We solve it in the usual way, beginning with the completion of the square

$$\left(\lambda + \frac{1}{2}\gamma\right)^2 = \frac{1}{4}\gamma^2 - \omega_0^2\,. \tag{2.2.61}$$

The two solutions

$$\left.\begin{matrix}\lambda_1\\ \lambda_2\end{matrix}\right\} = -\frac{1}{2}\gamma \pm \sqrt{\frac{1}{4}\gamma^2 - \omega_0^2} \tag{2.2.62}$$

are of quite different character, depending in the sign of the discriminant $\frac{1}{4}\gamma^2 - \omega_0^2$. When $\omega_0 > \frac{1}{2}\gamma$ (weak friction) both solutions are complex

$$\frac{1}{2}\gamma < \omega_0 : \qquad \left.\begin{matrix}\lambda_1 \\ \lambda_2\end{matrix}\right\} = -\frac{1}{2}\gamma \pm \mathrm{i}\sqrt{\omega_0^2 - \frac{1}{4}\gamma^2}\,, \tag{2.2.63}$$

but when $\omega_0 < \frac{1}{2}\gamma$ (strong friction) both solutions are real and negative,

$$\frac{1}{2}\gamma > \omega_0 : \qquad \left.\begin{matrix}\lambda_1 \\ \lambda_2\end{matrix}\right\} = -\frac{1}{2}\gamma \pm \sqrt{\frac{1}{4}\gamma^2 - \omega_0^2}\,. \tag{2.2.64}$$

There is also the case of *critical damping* when $\frac{1}{2}\gamma = \omega_0$ and the two λ values coincide,

$$\frac{1}{2}\gamma = \omega_0 : \qquad \lambda_1 = \lambda_2 = -\frac{1}{2}\gamma\,. \tag{2.2.65}$$

We will treat this degenerate case as a limiting situation of $\lambda_1 \neq \lambda_2$, where we have

$$\begin{aligned} x(t) &= c_1\,\mathrm{e}^{\lambda_1 t} + c_2\,\mathrm{e}^{\lambda_2 t}\,, \\ \dot{x}(t) &= c_1\lambda_1\,\mathrm{e}^{\lambda_1 t} + c_2\lambda_2\,\mathrm{e}^{\lambda_2 t}\,, \end{aligned} \tag{2.2.66}$$

so that c_1 and c_2 are determined by

$$x_0 = c_1 + c_2 \quad \text{and} \quad \dot{x}_0 = c_1\lambda_1 + c_2\lambda_2\,. \tag{2.2.67}$$

These give

$$c_1 = \frac{\lambda_2 x_0 - \dot{x}_0}{\lambda_2 - \lambda_1} \quad \text{and} \quad c_2 = \frac{\lambda_1 x_0 - \dot{x}_0}{\lambda_1 - \lambda_2}\,. \tag{2.2.68}$$

We then have

$$\begin{aligned} x(t) &= \frac{1}{\lambda_1 - \lambda_2}\left[x_0\left(\lambda_1\,\mathrm{e}^{\lambda_2 t} - \lambda_2\,\mathrm{e}^{\lambda_1 t}\right) + \dot{x}_0\left(\mathrm{e}^{\lambda_1 t} - \mathrm{e}^{\lambda_2 t}\right)\right], \\ \dot{x}(t) &= \frac{1}{\lambda_1 - \lambda_2}\left[\dot{x}_0\left(\lambda_1\,\mathrm{e}^{\lambda_1 t} - \lambda_2\,\mathrm{e}^{\lambda_2 t}\right) + x_0\lambda_1\lambda_2\left(\mathrm{e}^{\lambda_2 t} - \mathrm{e}^{\lambda_1 t}\right)\right], \end{aligned} \tag{2.2.69}$$

and can now consider the limiting case of $\lambda_1 = \lambda_2$, when the right-hand sides are of the "$\frac{0}{0}$" form. In this limit, we write

$$\left.\begin{matrix}\lambda_1 \\ \lambda_2\end{matrix}\right\} = -\frac{1}{2}\gamma \pm \epsilon\,, \tag{2.2.70}$$

and keep all terms of order ϵ^0 and ϵ^1 in the $[\cdots]$ terms of (2.2.69), finally to be divided by $\lambda_1 - \lambda_2 = 2\epsilon$; there is no need for terms of order ϵ^2, ϵ^3 in our evaluation of the $[\cdots]$ terms because these contributions disappear in the $\epsilon \to 0$ limit.

Accordingly, in $x(t)$ we have

$$\begin{aligned} \mathrm{e}^{\lambda_1 t} - \mathrm{e}^{\lambda_2 t} &= \mathrm{e}^{-\frac{1}{2}\gamma t}\left(\mathrm{e}^{\epsilon t} - \mathrm{e}^{-\epsilon t}\right) \\ &= \mathrm{e}^{-\frac{1}{2}\gamma t}\big[2\epsilon t + \underbrace{O(\epsilon^3)}_{\text{discarded}}\big] = \mathrm{e}^{-\frac{1}{2}\gamma t}\, 2\epsilon t\,, \end{aligned} \tag{2.2.71}$$

and

$$\begin{aligned} \lambda_1\,\mathrm{e}^{\lambda_2 t} - \lambda_2\,\mathrm{e}^{\lambda_1 t} &= -\frac{1}{2}\gamma\,\mathrm{e}^{-\frac{1}{2}\gamma t}\left(\mathrm{e}^{-\epsilon t} - \mathrm{e}^{\epsilon t}\right) + \epsilon\,\mathrm{e}^{-\frac{1}{2}\gamma t}\left(\mathrm{e}^{-\epsilon t} + \mathrm{e}^{\epsilon t}\right) \\ &= \mathrm{e}^{-\frac{1}{2}\gamma t}\Big[\gamma\epsilon t + 2\epsilon + \underbrace{O(\epsilon^3)}_{\text{discarded}}\Big] \\ &= \mathrm{e}^{-\frac{1}{2}\gamma t}\, 2\epsilon\left(1 + \frac{1}{2}\gamma t\right), \end{aligned} \tag{2.2.72}$$

so that

$$\begin{aligned} x(t) &= x_0\,\mathrm{e}^{-\frac{1}{2}\gamma t}\left(1 + \frac{1}{2}\gamma t\right) + \dot{x}_0\,\mathrm{e}^{-\frac{1}{2}\gamma t} t \\ &= x_0\,\mathrm{e}^{-\frac{1}{2}\gamma t} + \left(\dot{x}_0 + \frac{1}{2}\gamma x_0\right) t\,\mathrm{e}^{-\frac{1}{2}\gamma t} \end{aligned} \tag{2.2.73}$$

and, obtained either by differentiation of this $x(t)$ or as the $\epsilon \to 0$ limit of (2.2.69),

$$\dot{x}(t) = \dot{x}_0\,\mathrm{e}^{-\frac{1}{2}\gamma t} - \frac{1}{2}\gamma\left(\dot{x}_0 + \frac{1}{2}\gamma x_0\right) t\,\mathrm{e}^{-\frac{1}{2}\gamma t}\,, \tag{2.2.74}$$

are the position and the velocity at time t in the case of critical damping, $\frac{1}{2}\gamma = \omega_0$. Further differentiation gives

$$\ddot{x}(t) = -\left(\gamma\dot{x}_0 + \frac{1}{4}\gamma^2 x_0\right)\mathrm{e}^{-\frac{1}{2}\gamma t} + \frac{1}{4}\gamma^2\left(\dot{x}_0 + \frac{1}{2}\gamma x_0\right) t\,\mathrm{e}^{-\frac{1}{2}\gamma t}\,, \tag{2.2.75}$$

and then we can verify that these obey the differential equation

$$\ddot{x} + \gamma\dot{x} + \frac{1}{4}\gamma^2 x = 0\,, \tag{2.2.76}$$

which is (2.2.59) for $\omega_0^2 = \frac{1}{4}\gamma^2$.

There is a lesson here about the exponential ansatz $x(t) = c\,\mathrm{e}^{\lambda t}$ and the resulting equation for λ which determines the possible λ values as the zeroes of the polynomial — in the current context the second-order polynomial on the left-hand side of

$$\lambda^2 + \gamma\lambda + \omega_0^2 = 0\,. \tag{2.2.77}$$

The lesson is this: For simple zeroes, we have solutions of the form $c\,\mathrm{e}^{\lambda t}$; for a double zero, we have solutions of the form $(c_0 + c_1 t)\,\mathrm{e}^{\lambda t}$, and we can suspect, what is indeed correct, that a triple zero goes with solutions of the form $(c_0 + c_1 t + c_2 t^2)\,\mathrm{e}^{\lambda t}$ and so forth, in the case of higher-order linear differential equations and their associated higher-order polynomials. More about this in Exercise 27.

Returning to the harmonic oscillator with Newtonian friction, we observe that we have three qualitatively different situations. First, there is the case of weak friction, $\frac{1}{2}\gamma < \omega_0$, when

$$\left.\begin{matrix}\lambda_1\\ \lambda_2\end{matrix}\right\} = -\frac{1}{2}\gamma \pm \mathrm{i}\omega \quad\text{with}\quad \omega = \sqrt{\omega_0^2 - \frac{1}{4}\gamma^2} < \omega_0 \tag{2.2.78}$$

and

$$x(t) = x_0\Bigl[\cos(\omega t) + \frac{\gamma}{2\omega}\sin(\omega t)\Bigr]\,\mathrm{e}^{-\frac{1}{2}\gamma t} + \frac{\dot{x}_0}{\omega}\sin(\omega t)\,\mathrm{e}^{-\frac{1}{2}\gamma t}\,. \tag{2.2.79}$$

This is a damped oscillation: oscillatory sine and cosine terms that are damped by an exponential decrease or, put differently, oscillating terms with exponentially decreasing amplitudes. Of course, we get the undamped motion of the frictionless harmonic oscillator of Section 2.2.4 in the limit of $\gamma \to 0$, when $\omega \to \omega_0$ and the exponential terms disappear.

Quite a different situation is that of a strong frictional force, $\frac{1}{2}\gamma > \omega_0$, when

$$\left.\begin{matrix}\lambda_1\\ \lambda_2\end{matrix}\right\} = -\frac{1}{2}(\gamma \mp \gamma') \quad\text{with}\quad \gamma' = \sqrt{\gamma^2 - 4\omega_0^2} < \gamma \tag{2.2.80}$$

and

$$\begin{aligned} x(t) = {}& \frac{1}{\gamma'}\Bigl[\frac{1}{2}(\gamma + \gamma')x_0 + \dot{x}_0\Bigr]\,\mathrm{e}^{-\frac{1}{2}(\gamma - \gamma')t} \\ & - \frac{1}{\gamma'}\Bigl[\frac{1}{2}(\gamma - \gamma')x_0 + \dot{x}_0\Bigr]\,\mathrm{e}^{-\frac{1}{2}(\gamma + \gamma')t}\,. \end{aligned} \tag{2.2.81}$$

There are no oscillations here, instead we have the sum of two exponentially decaying terms. Alternatively and equivalently, we can write this in terms of hyperbolic functions,

$$\begin{aligned} x(t) = x_0 &\left[\cosh\left(\frac{1}{2}\gamma' t\right) + \frac{\gamma}{\gamma'}\sinh\left(\frac{1}{2}\gamma' t\right)\right] \mathrm{e}^{-\frac{1}{2}\gamma t} \\ &+ \frac{2}{\gamma'}\dot{x}_0 \sinh\left(\frac{1}{2}\gamma' t\right) \mathrm{e}^{-\frac{1}{2}\gamma t}. \end{aligned} \tag{2.2.82}$$

This is also obtained when we put $\omega \to -\mathrm{i}\gamma'/2$ (or $\omega \to \mathrm{i}\gamma'/2$) in the solution for weak friction in (2.2.79) and recall the relations

$$\begin{aligned} \sin(\mathrm{i}x) &= \mathrm{i}\sinh(x)\,, \\ \cos(\mathrm{i}x) &= \cosh(x) \end{aligned} \tag{2.2.83}$$

between trigonometric functions of imaginary arguments and hyperbolic functions of real arguments.

Rather than speaking of weak and strong friction, it is common to speak of the *underdamped* oscillator for $\frac{1}{2}\gamma < \omega_0$ and the *overdamped* oscillator for $\frac{1}{2}\gamma > \omega_0$, whereas $\frac{1}{2}\gamma = \omega_0$ is the case of critical damping. It is available either as the $\omega \to 0$ limit of the underdamped case or as the $\gamma' \to 0$ limit of the overdamped case, and the resulting $x(t)$ is given in (2.2.73). Let us look at the three cases for $\dot{x}_0 = 0$, that is

$$\frac{x(t)}{x_0} = \begin{cases} \left[\cos(\omega t) + \dfrac{\gamma}{2\omega}\sin(\omega t)\right] \mathrm{e}^{-\frac{1}{2}\gamma t} & \text{for } \dfrac{1}{2}\gamma < \omega_0\,, \\ \left(1 + \dfrac{1}{2}\gamma t\right) \mathrm{e}^{-\frac{1}{2}\gamma t} & \text{for } \dfrac{1}{2}\gamma = \omega_0\,, \\ \left[\cosh\left(\frac{1}{2}\gamma' t\right) + \dfrac{\gamma}{\gamma'}\sinh\left(\frac{1}{2}\gamma' t\right)\right] \mathrm{e}^{-\frac{1}{2}\gamma t} & \text{for } \dfrac{1}{2}\gamma > \omega_0\,, \end{cases} \tag{2.2.84}$$

which all have the same behavior for early times

$$x(t) = \left(1 - \frac{1}{2}(\omega_0 t)^2 + \{\text{terms of order } t^3\}\right) x_0\,, \tag{2.2.85}$$

but differ markedly at later times. The three cases are compared in this figure:

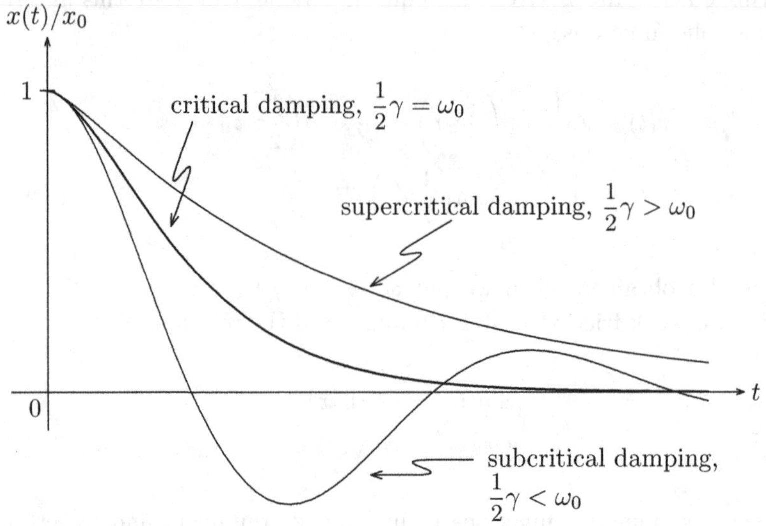

(2.2.86)

The approach to $|x(t)| \ll x_0$ is fastest for critical damping, when we neither have the underdamped oscillations of subcritical damping, nor the slowly decaying exponential factor $e^{-\frac{1}{2}(\gamma - \gamma')t}$ of supercritical damping that characterizes overdamped oscillations.

2.2.6 *Damped and driven harmonic oscillations*

Let us now see what happens if we drive the damped harmonic oscillator with an external force. A real-life situation of this kind is the pushing of a child on a swing, when you apply forces to speed up, or slow down, the swing. The equation of motion is now

$$m\ddot{x} = -m\gamma\dot{x} - m\omega_0^2 x + F(t) \tag{2.2.87}$$

or

$$\ddot{x} + \gamma\dot{x} + \omega_0^2 x = \frac{1}{m}F(t)\,, \tag{2.2.88}$$

where $F(t)$ is the applied external force. This is an *inhomogeneous* linear second-order differential equation. Its general solution is the sum of the general solution of the *homogeneous* equation

$$\ddot{x} + \gamma\dot{x} + \omega_0^2 x = 0 \tag{2.2.89}$$

which we studied in Section 2.2.5, and a special solution of the inhomogeneous equation. This is easy to see: Just imagine that you have found two solutions of the inhomogeneous equation, $x_1(t)$ and $x_2(t)$; their difference obeys the homogeneous equation,

$$\begin{aligned}&\left(\frac{\mathrm{d}}{\mathrm{d}t}\right)^2(x_1-x_2)+\gamma\frac{\mathrm{d}}{\mathrm{d}t}(x_1-x_2)+\omega_0^2(x_1-x_2)\\&=\underbrace{(\ddot{x}_1+\gamma\dot{x}_1+\omega_0^2x_1)}_{=F(t)/m}-\underbrace{(\ddot{x}_2+\gamma\dot{x}_2+\omega_0^2x_2)}_{=F(t)/m}=0\,.\end{aligned}\tag{2.2.90}$$

So, with the general solution of the homogeneous equation at hand (see Section 2.2.5), all we need is one solution of the inhomogeneous equation.

(a) Periodic harmonic force: Resonance

Before we deal with this problem in general, let us consider the particular situation of a periodic external force of the harmonic form

$$F(t)=ma\cos(\Omega t)\,.\tag{2.2.91}$$

Since the solutions of the homogeneous equation decay exponentially, only the special solution of the inhomogeneous equation (or its essential parts) will be relevant after some time has elapsed. Physical intuition tells us that the motion will be periodic with the imposed frequency $\dfrac{\Omega}{2\pi}$ after all traces of the initial conditions have disappeared. This invites the ansatz

$$x(t)=c\cos(\Omega t-\phi)\,,\tag{2.2.92}$$

with an unknown amplitude c and an unknown phase shift ϕ, both to be determined from the inhomogeneous equation of motion such that the ansatz works.

This requires that

$$-c\Omega^2\cos(\Omega t-\phi)-c\gamma\Omega\sin(\Omega t-\phi)+c\omega_0^2\cos(\Omega t-\phi)=a\cos(\Omega t)\tag{2.2.93}$$

holds for all times t. We write the right-hand side as

$$\begin{aligned}a\cos(\Omega t)&=a\cos\big((\Omega t-\phi)+\phi\big)\\&=a\cos(\Omega t-\phi)\cos\phi-a\sin(\Omega t-\phi)\sin\phi\,,\end{aligned}\tag{2.2.94}$$

and then equate the cosine and sine terms on both sides of (2.2.93), which gives the two equations

$$-c\Omega^2+c\omega_0^2=a\cos\phi\,,\quad -c\gamma\Omega=-a\sin\phi\,.\tag{2.2.95}$$

It follows that

$$a^2 = (a\cos\phi)^2 + (a\sin\phi)^2$$
$$= c^2\Big[\left(\omega_0^2 - \Omega^2\right)^2 + (\gamma\Omega)^2\Big] \tag{2.2.96}$$

and, after choosing the root that gives c the same sign as a,

$$c = \frac{a}{\sqrt{\left(\omega_0^2 - \Omega^2\right)^2 + (\gamma\Omega)^2}}, \tag{2.2.97}$$

the phase shift ϕ is determined as the unique solution of

$$\cos\phi = \frac{\omega_0^2 - \Omega^2}{\sqrt{\left(\omega_0^2 - \Omega^2\right)^2 + (\gamma\Omega)^2}},$$
$$\sin\phi = \frac{\gamma\Omega}{\sqrt{\left(\omega_0^2 - \Omega^2\right)^2 + (\gamma\Omega)^2}}. \tag{2.2.98}$$

The implied statement,

$$\tan\phi = \frac{\gamma\Omega}{\omega_0^2 - \Omega^2} \tag{2.2.99}$$

contains less information because it does not distinguish between ϕ and $\phi + \pi$.

The first observation is that the ansatz works: There are special solutions of this kind. Remembering what is said above about the general solution, one such special solution is all we need.

The second observation is that the amplitude c of the resulting oscillation depends crucially on the frequency of the imposed force. The argument of the square root,

$$\left(\omega_0^2 - \Omega^2\right)^2 + (\gamma\Omega)^2$$
$$= \left(\Omega^2 - \omega_0^2 + \frac{1}{2}\gamma^2\right)^2 + \gamma^2\left(\omega_0^2 - \frac{1}{4}\gamma^2\right) \tag{2.2.100}$$

can be quite small if $\Omega^2 = \omega_0^2 - \frac{1}{2}\gamma^2$, which in turn is only possible if the frictional force is not too large. We shall assume $\gamma \ll \omega_0$ from now on, so that $\omega_0^2 - \frac{1}{2}\gamma^2 > 0$ (and $\omega_0^2 - \frac{1}{4}\gamma^2 > 0$), or after recalling $\omega^2 = \omega_0^2 - \frac{1}{4}\gamma^2$ from page (2.2.78),

$$\omega^2 > \frac{1}{4}\gamma^2. \tag{2.2.101}$$

We then have

$$c = \frac{a}{\sqrt{\left(\Omega^2 - \omega^2 + \frac{1}{4}\gamma^2\right)^2 + (\gamma\omega)^2}} \leq \frac{a}{\gamma\omega}, \tag{2.2.102}$$

with the equal sign holding for $\Omega^2 = \omega^2 - \frac{1}{4}\gamma^2$.

When there is little friction, $\gamma \ll \omega_0$, the maximal amplitude can be quite large. We are here encountering the phenomenon of *resonance*: A system driven by a periodic force whose period is close to the natural period of the system tends to exhibit oscillations with large amplitudes. This can have catastrophic consequences. A much cited example is the collapse of the Tacoma Narrows Bridge in 1940, although the actual mechanism that destroyed the bridge is more complicated than a simple forced resonance. And then, of course, there is Oskar Matzerath* and his glass-breaking scream.

For $\Omega > 0$ ($\Omega < 0$ makes no essential difference), we have $\sin\phi > 0$, so that ϕ is in the range $0 \cdots \pi$. For driving frequencies below the natural frequency of the oscillator, $\Omega < \omega_0$, the cosine of ϕ is positive, and the phase shift is in the range $0 < \phi < \frac{1}{2}\pi$. For $\Omega = \omega_0$, we have $\cos\phi = 0$ and $\phi = \frac{1}{2}\pi$, and we get $\frac{1}{2}\pi < \phi < \pi$ for $\Omega > \omega_0$ when $\cos\phi < 0$. The overall picture is this:

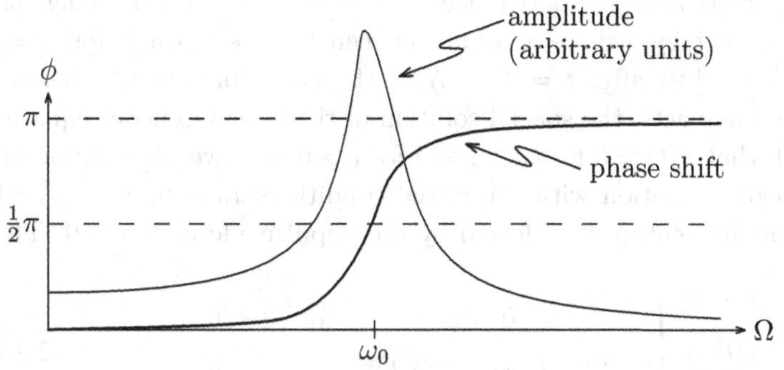

(2.2.103)

which is a plot for $\omega_0 = 10\gamma$. Just for the record, here is a reminder of what a shift by phase ϕ means:

*Oskar MATZERATH (b. 1924)

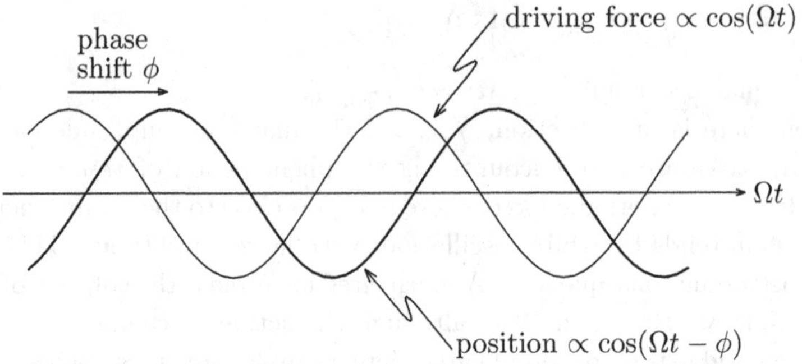

(2.2.104)

The maximal displacement occurs later than the maximum force by a delay of $\Delta t = \phi/\Omega$.

(b) Impulsive force: Heaviside's step function and Dirac's delta function

We move on and turn our attention to another particular case, that of an impulsive force at $t = 0$, meaning that there is a very strong force that acts for a very short time, so short indeed that its net effect is the change of the momentum $m\dot{x}$ by a finite amount between $t = -\epsilon$ ("just before $t = 0$") and $t = +\epsilon$ ("just after $t = 0$"). With the point mass at rest before the impulsive force acts, the special solution of the inhomogeneous equation is then such that $x(t) = 0$ for $t < 0$, and for $t > 0$ we have the solution of the homogeneous equation with the initial conditions $x_0 = 0$, $\dot{x}_0 = v_0$ where mv_0 is the momentum transferred by the impulsive force at $t = 0$. Thus

$$x(t) = \begin{cases} 0 & \text{for} \quad t \leq 0\,, \\ \dfrac{v_0}{\lambda_1 - \lambda_2}\left(e^{\lambda_1 t} - e^{\lambda_2 t}\right) & \text{for} \quad t \geq 0\,, \end{cases} \tag{2.2.105}$$

where λ_1 and λ_2 are the solutions of the quadratic equation that we found in Section 2.2.5. The case $\lambda_1 = \lambda_2 = -\frac{1}{2}\gamma$ for $\frac{1}{2}\gamma = \omega_0$ is, as always, contained as a limit, here

$$x(t) = \begin{cases} 0 & \text{for} \quad t \leq 0\,, \\ v_0 t\, e^{-\frac{1}{2}\gamma t} & \text{for} \quad t \geq 0\,. \end{cases} \tag{2.2.106}$$

We can write these statements more compactly with the aid of *Heaviside's** *unit step function*

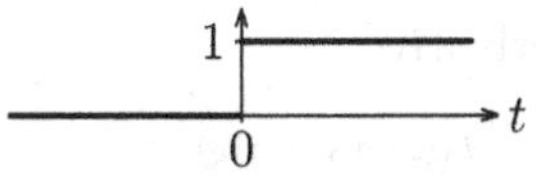

$$\eta(t)=\begin{cases}0 & \text{for}\quad t<0\,,\\ 1 & \text{for}\quad t>0\,,\end{cases} \tag{2.2.107}$$

and we do not need to assign a particular value to $\eta(t=0)$; it is often best to think of $\eta(t=0)$ as a stand-in for all values between 0 and 1. The more compact expression for $x(t)$ is then

$$x(t)=\frac{v_0}{\lambda_1-\lambda_2}\left(\mathrm{e}^{\lambda_1 t}-\mathrm{e}^{\lambda_2 t}\right)\eta(t)\,. \tag{2.2.108}$$

Likewise we have

$$\begin{aligned}\dot{x}(t)&=\left\{\begin{matrix}0 & \text{for}\quad t<0\\ \dfrac{v_0}{\lambda_1-\lambda_2}\left(\lambda_1\,\mathrm{e}^{\lambda_1 t}-\lambda_2\,\mathrm{e}^{\lambda_2 t}\right) & \text{for}\quad t>0\end{matrix}\right\}\\ &=\frac{v_0}{\lambda_1-\lambda_2}\left(\lambda_1\,\mathrm{e}^{\lambda_1 t}-\lambda_2\,\mathrm{e}^{\lambda_2 t}\right)\eta(t)\,.\end{aligned} \tag{2.2.109}$$

The comparison with the derivative of $x(t)$,

$$\begin{aligned}\frac{\mathrm{d}}{\mathrm{d}t}x(t)&=\frac{v_0}{\lambda_1-\lambda_2}\left(\lambda_1\,\mathrm{e}^{\lambda_1 t}-\lambda_2\,\mathrm{e}^{\lambda_2 t}\right)\eta(t)\\ &\quad+\frac{v_0}{\lambda_1-\lambda_2}\left(\mathrm{e}^{\lambda_1 t}-\mathrm{e}^{\lambda_2 t}\right)\frac{\mathrm{d}}{\mathrm{d}t}\eta(t)\,,\end{aligned} \tag{2.2.110}$$

tells us that the second term actually vanishes.

How does this come about? Well, we surely have $\dfrac{\mathrm{d}}{\mathrm{d}t}\eta(t)=0$ for $t<0$ and $t>0$ because the step function is constant in these t ranges; and for $t=0$, the factor $\mathrm{e}^{\lambda_1 t}-\mathrm{e}^{\lambda_2 t}$ equals zero, and then $\dfrac{\mathrm{d}\eta}{\mathrm{d}t}(t=0)$ no longer matters. This reasoning is, however, more than just a bit superficial because if we calculate the derivative of the step function at $t=0$, we need to find

$$\left.\frac{\mathrm{d}}{\mathrm{d}t}\eta(t)\right|_{t\,=\,0}=\left.\frac{\eta(\tau)-\eta(-\tau)}{2\tau}\right|_{0\,<\,\tau\,\to\,0}, \tag{2.2.111}$$

where the symmetric way of taking the limit avoids the problem of stating a specific value of $\eta(t)$ for $t=0$. Since $\tau>0$ here, the right-hand side is

$$\left.\frac{1-0}{2\tau}\right|_{0\,<\,\tau\,\to\,0}=\infty \tag{2.2.112}$$

---

*Oliver HEAVISIDE (1850–1925)

and, for that second term in (2.2.110) we are facing the challenge of giving meaning to a product of the form "0 times ∞" which can have many answers.

Surely, $\eta(t)$ is the anti-derivative of its derivative,

$$\int_{t_2}^{t_1} \mathrm{d}t \, \frac{\mathrm{d}}{\mathrm{d}t}\eta(t) = \eta(t_1) - \eta(t_2) = \begin{cases} 0 & \text{for} \quad t_1 > t_2 > 0\,, \\ 1 & \text{for} \quad t_1 > 0 > t_2\,, \\ 0 & \text{for} \quad 0 > t_1 > t_2\,, \end{cases} \tag{2.2.113}$$

and it follows that $\frac{\mathrm{d}}{\mathrm{d}t}\eta(t)$ is not a function of the usual kind: It vanishes for $t < 0$ and $t > 0$, but in the vicinity of $t = 0$ it is infinite in the sense that the integral from $t_2 < 0$ to $t_1 > 0$ gives unity, irrespective of how close t_2 and t_1 are to $t = 0$. Jointly, these properties characterize *Dirac's* delta function* $\delta(t)$, so that we write

$$\frac{\mathrm{d}}{\mathrm{d}t}\eta(t) = \delta(t)\,. \tag{2.2.114}$$

The delta function is not a function in the usual sense but rather an example of a *distribution*. It defines the mapping of a continuous function $f(t)$ to its value at $t = 0$ by means of the formal integration relation

$$\int_{t_2}^{t_1} \mathrm{d}t \, f(t)\delta(t) = f(0) \quad \text{for} \quad t_2 < 0 < t_1\,. \tag{2.2.115}$$

If the integration interval does not contain $t = 0$, the integral vanishes:

$$\int_{t_2}^{t_1} \mathrm{d}t \, f(t)\delta(t) = 0 \quad \text{for} \quad t_2 < t_1 < 0 \quad \text{or} \quad t_1 > t_2 > 0\,, \tag{2.2.116}$$

and the usual rules for integrals apply, such as

$$\int_{t_2}^{t_1} \mathrm{d}t \, f(t)\delta(t) = -\int_{t_1}^{t_2} \mathrm{d}t \, f(t)\delta(t)\,. \tag{2.2.117}$$

We stumbled upon the delta function in a particular physical context, that of an impulsive force. The appearance of the step function in (2.2.108) results from an over-idealization of the physical situation, namely the description of the strong force of short duration as an instantaneous momentum transfer, as if an infinitely strong force was acting for an infinitesimally

*Paul Adrien Maurice DIRAC (1902–1984)

short time. This over-idealization is very useful, it allows us to focus on the essential effect of the impulse, but it comes with a price, the necessity to deal with the step function and its derivative, the delta function.

We could avoid these mild complications by modeling the impulsive force as a real force of very short duration. Equivalently, we can replace the step function by a normal continuous function that grows from 0 to 1 rapidly near $t = 0$. An example for such a *model* for the step function is

$$\eta_\tau(t) = \frac{1}{2}\left(\frac{t}{\sqrt{t^2+\tau^2}} + 1\right) \tag{2.2.118}$$

where τ is a positive, very short time, which sets the time scale for the switch from $\eta_\tau(t \ll -\tau) = 0$ to $\eta_\tau(t \gg \tau) = 1$:

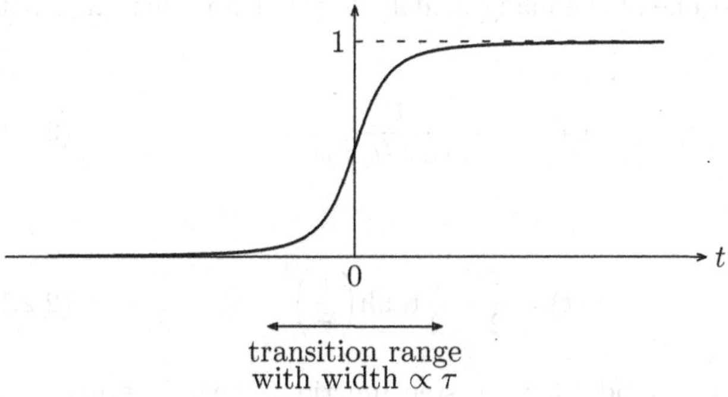

(2.2.119)

In the limit of $\tau \to 0$, $\eta_\tau(\ )$ turns into the step function $\eta(\ )$,

$$\eta_\tau(t) \xrightarrow[\tau\to 0]{} \frac{1}{2}\big(\mathrm{sgn}(t) + 1\big) = \eta(t) \quad \text{for} \quad t \neq 0\,, \tag{2.2.120}$$

and assigns the model-specific value of $\frac{1}{2}$ to $\eta(t = 0)$.

Upon differentiation, the model (2.2.118) for the step function gives

$$\delta_\tau(t) = \frac{1}{2}\frac{\tau^2}{\sqrt{t^2+\tau^2}^{\,3}} \tag{2.2.121}$$

as the corresponding model for the delta function:

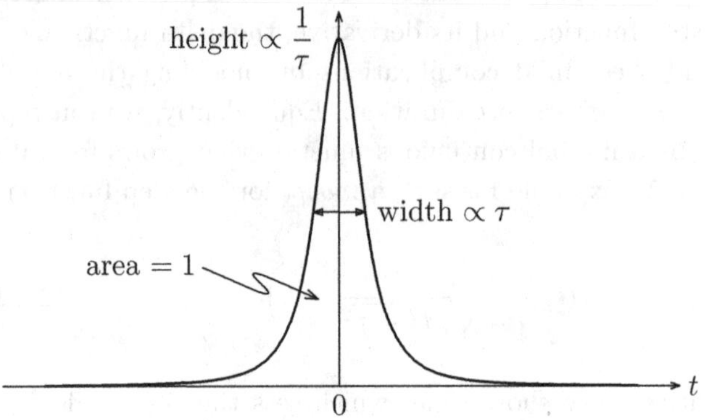

(2.2.122)

a peak of height $\propto 1/\tau$ and width $\propto \tau$ at $t = 0$, with unit area under the curve. This is just one of the many models for the delta function. Another example is

$$\delta_\tau(t) = \frac{1}{2\tau \cosh(t/\tau)^2} \tag{2.2.123}$$

with

$$\eta_\tau(t) = \frac{1}{2} + \frac{1}{2}\tanh\left(\frac{t}{\tau}\right) \tag{2.2.124}$$

as the corresponding model for the step function. There are many more models for the step function and the delta function, and if you want to use one of them at intermediate steps, just choose a model that is convenient for your application and take the $\tau \to 0$ limit at a suitable stage of the calculation.

What is essential though, is that the physics of the impulsive force will not depend on which model we choose to describe it, as long as the model parameter τ is much smaller than all characteristic times of the physical system. For the harmonic oscillator with Newtonian friction, this requires $\omega_0\tau \ll 1$ and $\gamma\tau \ll 1$, and then neither the details of the model nor the actual value of τ matter. It is, then, much more convenient to use the step function and the delta function rather than any model for them with irrelevant details.

Before proceeding with our investigations of the harmonic oscillator with friction and driving force, we need to establish two calculational rules for the delta function. First we note that

$$\int \mathrm{d}t\, f(t)\delta(t-t_0) = f(t_0) \tag{2.2.125}$$

provided that the range of integration contains t_0. Second, in an integral of the form

$$\int \mathrm{d}t\, g(t)f(t)\delta(t-t_0) = g(t_0)f(t_0) \tag{2.2.126}$$

the replacement

$$f(t)\delta(t-t_0) = f(t_0)\delta(t-t_0) \tag{2.2.127}$$

is permissible. Quite generally, we should always think of expressions like $f(t)\delta(t-t_0)$ as parts of an integrand.

Now we take another look at (2.2.110),

$$\begin{aligned}\frac{\mathrm{d}}{\mathrm{d}t}x(t) &= \frac{v_0}{\lambda_1-\lambda_2}\Big(\lambda_1\,\mathrm{e}^{\lambda_1 t} - \lambda_2\,\mathrm{e}^{\lambda_2 t}\Big)\eta(t) \\ &\quad + \underbrace{\frac{v_0}{\lambda_1-\lambda_2}\Big(\mathrm{e}^{\lambda_1 t} - \mathrm{e}^{\lambda_2 t}\Big)}_{=0 \text{ for } t=0}\underbrace{\frac{\mathrm{d}}{\mathrm{d}t}\eta(t)}_{=\delta(t)} \\ &= \frac{v_0}{\lambda_1-\lambda_2}\Big(\lambda_1\,\mathrm{e}^{\lambda_1 t} - \lambda_2\,\mathrm{e}^{\lambda_2 t}\Big)\eta(t)\,,\end{aligned} \tag{2.2.128}$$

where (2.2.127) is used for $t_0 = 0$ and $f(t_0) = 0$. Indeed, we recover (2.2.109), as we should.

What about the acceleration $\ddot{x}(t)$? Let us see,

$$\begin{aligned}\left(\frac{\mathrm{d}}{\mathrm{d}t}\right)^2 x(t) &= \frac{v_0}{\lambda_1-\lambda_2}\Big(\lambda_1^2\,\mathrm{e}^{\lambda_1 t} - \lambda_2^2\,\mathrm{e}^{\lambda_2 t}\Big)\eta(t) \\ &\quad + \frac{v_0}{\lambda_1-\lambda_2}\Big(\lambda_1\,\mathrm{e}^{\lambda_1 t} - \lambda_2\,\mathrm{e}^{\lambda_2 t}\Big)\delta(t)\,,\end{aligned} \tag{2.2.129}$$

where $\lambda_1^2 = -\gamma\lambda_1 - \omega_0^2$ and $\lambda_2^2 = -\gamma\lambda_2 - \omega_0^2$ (recall (2.2.60)) turn the first term into $-\gamma\dot{x}(t) - \omega_0^2\,x(t)$, and we employ $f(t)\delta(t) = f(0)\delta(t)$ in the second term to arrive at

$$\ddot{x}(t) = -\gamma\dot{x}(t) - \omega_0^2 x(t) + v_0\delta(t)\,. \tag{2.2.130}$$

The comparison with (2.2.88),

$$\ddot{x}(t) + \gamma\dot{x}(t) + \omega_0^2 x(t) = \frac{1}{m}F(t)\,, \tag{2.2.131}$$

identifies the impulsive force as

$$F(t) = mv_0\,\delta(t)\,, \tag{2.2.132}$$

and this expression makes perfect sense. For, a force of this kind transfers momentum mv_0 at time $t = 0$,

$$\begin{aligned} m\dot{x}(t=+0) &= m\dot{x}(t=-0) + \int_{-0}^{+0} \mathrm{d}t\,F(t) \\ &= m\dot{x}(t=-0) + mv_0 \end{aligned} \tag{2.2.133}$$

where $t = \pm 0$ are instants just before and after $t = 0$, when the impulsive force is acting.

(c) Arbitrary driving force: Green's function

Upon stripping the $x(t)$ of the impulsive force in (2.2.108) off the velocity parameter v_0, we have the following statement:

$$\left[\left(\frac{\mathrm{d}}{\mathrm{d}t}\right)^2 + \gamma\frac{\mathrm{d}}{\mathrm{d}t} + \omega_0^2\right] G(t) = \delta(t) \tag{2.2.134}$$

with

$$G(t) = \frac{\mathrm{e}^{\lambda_1 t} - \mathrm{e}^{\lambda_2 t}}{\lambda_1 - \lambda_2}\eta(t)\,, \tag{2.2.135}$$

or

$$G(t) = \frac{\sin(\omega t)}{\omega}\,\mathrm{e}^{-\frac{1}{2}\gamma t}\eta(t) \tag{2.2.136}$$

after inserting the λs of the underdamped oscillator in (2.2.78). This $G(t)$ solves the inhomogeneous differential equation for the inhomogeneity $\delta(t)$ and the initial conditions $G(t) = 0$, $\frac{\mathrm{d}G}{\mathrm{d}t}(t) = 0$ for $t < 0$. We call this very special solution the *retarded Green's* function*, where "retarded" refers to the condition that all effects of $\delta(t)$ are after, not before the impulsive force acts. There is also an "advanced Green's function" that is identically zero for $t > 0$, but we have no use for it in the present context.

With the Green's function at hand, we can find the retarded solutions for an arbitrary force $F(t)$ in (2.2.131). This needs two more ingredients. One is the observation that if $x(t)$ is the solution for force $F(t)$, then $x(t-t')$

*George GREEN (1793–1841)

is the solution for the force $F(t-t')$, which is a consequence of the fact that the coefficients of the differential operator in (2.2.131) and (2.2.134),

$$\left(\frac{\mathrm{d}}{\mathrm{d}t}\right)^2 + \gamma\frac{\mathrm{d}}{\mathrm{d}t} + \omega_0^2\,, \tag{2.2.137}$$

do not depend on time t. The other ingredient is the property that if $x_1(t)$ is the solution for the force $F_1(t)$ and $x_2(t)$ is the solution for force $F_2(t)$, then $x_1(t)+x_2(t)$ is the solution for force $F_1(t)+F_2(t)$; this is another consequence of the linearity of the differential equation.

Armed with all these insights, we write the general force $F(t)$ as a sequence of impulsive forces,

$$F(t) = \int \mathrm{d}t'\, F(t')\delta(t-t')\,, \tag{2.2.138}$$

where the product $\mathrm{d}t'\, F(t')$ is the momentum transferred — the value of mv_0, so to say — for the impulsive force acting at time t'. Its solution of the differential equation for $x(t)$ is

$$\frac{1}{m}\mathrm{d}t'\, F(t')\, G(t-t')\,, \tag{2.2.139}$$

and the full solution is found by adding up the contributions for all instants t':

$$x(t) = \int \mathrm{d}t'\, G(t-t')\frac{1}{m}F(t')\,. \tag{2.2.140}$$

We verify that this solves the inhomogeneous differential equation (2.2.131),

$$\begin{aligned}
&\left[\left(\frac{\mathrm{d}}{\mathrm{d}t}\right)^2 + \gamma\frac{\mathrm{d}}{\mathrm{d}t} + \omega_0^2\right] x(t) \\
&= \int \mathrm{d}t' \underbrace{\left[\left(\frac{\mathrm{d}}{\mathrm{d}t}\right)^2 + \gamma\frac{\mathrm{d}}{\mathrm{d}t} + \omega_0^2\right] G(t-t')}_{=\delta(t-t')} \frac{1}{m}F(t') \\
&= \int \mathrm{d}t'\, \delta(t-t')\frac{1}{m}F(t') = \frac{1}{m}F(t)\,.
\end{aligned} \tag{2.2.141}$$

Indeed, it does.

To complete the picture, we need to specify fully the physical situation: At time $t=0$, we have the initial position x_0 and the initial velocity $\dot{x}_0$, and for $t > 0$ the force $F(t)$ is acting. What is the position $x(t)$ for $t > 0$? Answer:

$$x(t) = x_0\Big[\cos(\omega t) + \frac{\gamma}{2\omega}\sin(\omega t)\Big]\,\mathrm{e}^{-\frac{1}{2}\gamma t} + \dot{x}_0\frac{\sin(\omega t)}{\omega}\,\mathrm{e}^{-\frac{1}{2}\gamma t} + \int_0^t \mathrm{d}t'\,\frac{\sin\big(\omega(t-t')\big)}{\omega}\,\mathrm{e}^{-\frac{1}{2}\gamma(t-t')}\,\frac{1}{m}F(t')\,, \tag{2.2.142}$$

where the first two terms make up the solution of the homogeneous equation of motion for the given initial position and initial velocity, see (2.2.79), and the third term is the retarded solution of the inhomogeneous equation for the force acting between $t' = 0$ and $t' = t$. An equivalent expression is

$$x(t) = x_0\Big[\cos(\omega t) + \frac{\gamma}{2\omega}\sin(\omega t)\Big]\,\mathrm{e}^{-\frac{1}{2}\gamma t} + \dot{x}_0\frac{\sin(\omega t)}{\omega}\,\mathrm{e}^{-\frac{1}{2}\gamma t} + \int_0^t \mathrm{d}t'\,\frac{\sin(\omega t')}{\omega}\,\mathrm{e}^{-\frac{1}{2}\gamma t'}\,\frac{1}{m}F(t-t')\,, \tag{2.2.143}$$

which we obtain by the substitution $t' \to t - t'$ in the integral in (2.2.142).

As a simple example, we consider $x_0 = 0$, $\dot{x}_0 = 0$, and $F = ma$ with constant acceleration a, that is: the point mass is at rest before $t = 0$, and then a constant force starts acting. This gives

$$\begin{aligned} x(t) &= \frac{a}{\omega}\int_0^t \mathrm{d}t'\,\sin(\omega t')\,\mathrm{e}^{-\frac{1}{2}\gamma t'} = \frac{a}{\omega}\,\mathrm{Im}\left(\int_0^t \mathrm{d}t'\,\mathrm{e}^{\mathrm{i}\omega t'}\,\mathrm{e}^{-\frac{1}{2}\gamma t'}\right) \\ &= \frac{a}{\omega}\,\mathrm{Im}\left(\frac{1-\mathrm{e}^{\mathrm{i}\omega t}\,\mathrm{e}^{-\frac{1}{2}\gamma t}}{\frac{1}{2}\gamma - \mathrm{i}\omega}\right) \\ &= \frac{a/\omega}{\omega^2 + \left(\frac{1}{2}\gamma\right)^2}\Big[\omega - \omega\cos(\omega t)\,\mathrm{e}^{-\frac{1}{2}\gamma t} - \frac{1}{2}\gamma\sin(\omega t)\,\mathrm{e}^{-\frac{1}{2}\gamma t}\Big] \end{aligned} \tag{2.2.144}$$

or

$$x(t) = \frac{a}{\omega_0^2}\Big[1 - \cos(\omega t)\,\mathrm{e}^{-\frac{1}{2}\gamma t} - \frac{\gamma}{2\omega}\sin(\omega t)\,\mathrm{e}^{-\frac{1}{2}\gamma t}\Big]\,, \tag{2.2.145}$$

where the constant term a/ω_0^2 solves the inhomogeneous differential equation

$$\ddot{x} + \gamma\dot{x} + \omega_0^2 x = a\,, \tag{2.2.146}$$

and the terms $\propto \cos(\omega t)\,\mathrm{e}^{-\frac{1}{2}\gamma t}$, $\sin(\omega t)\,\mathrm{e}^{-\frac{1}{2}\gamma t}$ solve the homogeneous equation. One verifies easily that $x(0) = 0$ and $\dot{x}(0) = 0$.

In Section 2.2.6(a) we dealt with a periodic force of the harmonic form $F(t) = ma\cos(\Omega t)$, and what was found then can also be derived by inserting this $F(t)$ into the expressions in (2.2.142) or (2.2.143). You should regard this an exercise.

(d) Periodic impulsive force: Cyclically steady state

Let us now consider a periodic driving force of another kind: regular impulsive forces. More specifically, an impulsive force is acting with period T, transferring momentum mv_0 whenever it acts. After the initial conditions have become irrelevant because a time much longer than $1/\gamma$ has elapsed, we expect the system to reach a *cyclically steady state* in which the position $x^{(\mathrm{css})}(t)$ is periodic with the imprinted period T,

$$x^{(\mathrm{css})}(t-T) = x^{(\mathrm{css})}(t) = x^{(\mathrm{css})}(t+T)\,, \tag{2.2.147}$$

and we have position x_{css} at the times when the impulsive momentum transfer happens, while the velocity changes from $\dot{x}_{\mathrm{css}}$ to $\dot{x}_{\mathrm{css}} + v_0$ at these moments. The situation is then as pictured here:

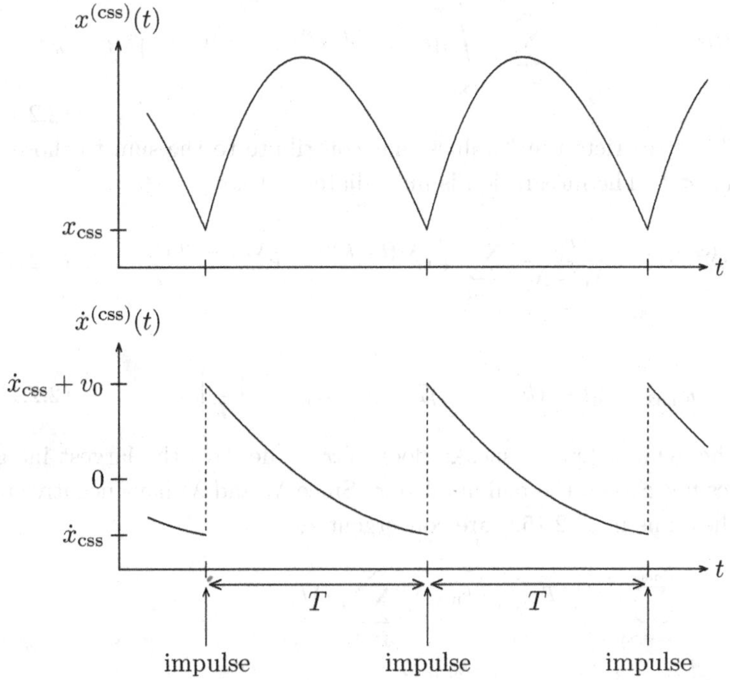

(2.2.148)

We find $x^{(\mathrm{css})}(t)$ by exploiting the power of the Green's function. The periodic impulsive force is

$$F(t) = \sum_{k=-\infty}^{\infty} m v_0 \delta(t - kT) \tag{2.2.149}$$

if the impulses occur at times $t = 0, \pm T, \pm 2T, \pm 3T, \ldots$, and the cyclically steady state is reached after this force has been acting for a long time. Therefore, we have

$$x^{(\mathrm{css})}(t) = \int_{-\infty}^{\infty} \mathrm{d}t'\, G(t - t') \frac{1}{m} F(t'), \tag{2.2.150}$$

where the initial time "$t = -\infty$" simply means that the beginning is so long ago that the contribution of the homogeneous terms in (2.2.142) has become irrelevant, and only the particular solution (2.2.140) of the inhomogeneous equation matters.

With the Green's function of (2.2.135) and the force (2.2.149), we get

$$x^{(\mathrm{css})}(t) = \frac{v_0}{\lambda_1 - \lambda_2} \sum_{k=-\infty}^{\infty} \int_{-\infty}^{t} \mathrm{d}t' \left(\mathrm{e}^{\lambda_1 (t - t')} - \mathrm{e}^{\lambda_2 (t - t')} \right) \delta(t' - kT), \tag{2.2.151}$$

where $t' < t$ restricts the k values that contribute to the sum to those for which $kT < t$. The integration is immediate, so that

$$x^{(\mathrm{css})}(t) = \frac{v_0}{\lambda_1 - \lambda_2} \sum_{k=-\infty}^{k_{\max}} \left(\mathrm{e}^{\lambda_1 (t - kT)} - \mathrm{e}^{\lambda_2 (t - kT)} \right) \tag{2.2.152}$$

with

$$k_{\max} T < t < (k_{\max} + 1) T \quad \text{or} \quad k_{\max} = \left\lfloor \frac{t}{T} \right\rfloor, \tag{2.2.153}$$

where the symbol $\lfloor x \rfloor$ — speak: floor of x — denotes the largest integer that does not exceed the real number x. Since λ_1 and λ_2 have negative real parts, the sums in (2.2.152) are convergent,

$$\begin{aligned} \sum_{k=-\infty}^{k_{\max}} \mathrm{e}^{-k\lambda T} &= \mathrm{e}^{-k_{\max}\lambda T} \sum_{j=0}^{\infty} \mathrm{e}^{j\lambda T} \\ &= \frac{\mathrm{e}^{-k_{\max}\lambda T}}{1 - \mathrm{e}^{\lambda T}} \qquad \text{for } \lambda = \lambda_1 \text{ or } \lambda = \lambda_2, \end{aligned} \tag{2.2.154}$$

and we arrive at

$$x^{(\mathrm{css})}(t) = \frac{v_0}{\lambda_1 - \lambda_2}\left(\frac{\mathrm{e}^{\lambda_1 \bar{t}}}{1 - \mathrm{e}^{\lambda_1 T}} - \frac{\mathrm{e}^{\lambda_2 \bar{t}}}{1 - \mathrm{e}^{\lambda_2 T}}\right), \tag{2.2.155}$$

where

$$\bar{t} = t - k_{\max} T = t - \left\lfloor \frac{t}{T} \right\rfloor T \tag{2.2.156}$$

is the time that has elapsed since the last impulse. Clearly, $x^{(\mathrm{css})}(t)$ is periodic, $x^{(\mathrm{css})}(t+T) = x^{(\mathrm{css})}(t)$, with the period imposed by the periodic force.

We verify that $x^{(\mathrm{css})}(t)$ is continuous at $t = 0, \pm T, \pm 2T, \ldots$, for which purpose we compare the right-hand sides of (2.2.155) for $\bar{t} = 0$ and $\bar{t} = T$, that is

$$\begin{aligned} x_{\mathrm{css}} = x^{(\mathrm{css})}(t)\Big|_{\bar{t}=0} &= \frac{v_0}{\lambda_1 - \lambda_2}\left(\frac{1}{1 - \mathrm{e}^{\lambda_1 T}} - \frac{1}{1 - \mathrm{e}^{\lambda_2 T}}\right) \\ = x^{(\mathrm{css})}(t)\Big|_{\bar{t}=T} &= \frac{v_0}{\lambda_1 - \lambda_2}\left(\frac{\mathrm{e}^{\lambda_1 T}}{1 - \mathrm{e}^{\lambda_1 T}} - \frac{\mathrm{e}^{\lambda_2 T}}{1 - \mathrm{e}^{\lambda_2 T}}\right). \end{aligned} \tag{2.2.157}$$

Both should be equal and give the same value for x_{css} and, indeed, they do, because we have

$$\frac{1}{1 - \mathrm{e}^{\lambda T}} = \frac{\mathrm{e}^{\lambda T}}{1 - \mathrm{e}^{\lambda T}} + 1 \tag{2.2.158}$$

for $\lambda = \lambda_1$ and $\lambda = \lambda_2$. Similarly, we confirm that the velocity

$$\dot{x}^{(\mathrm{css})}(t) = \frac{v_0}{\lambda_1 - \lambda_2}\left(\frac{\lambda_1\,\mathrm{e}^{\lambda_1 \bar{t}}}{1 - \mathrm{e}^{\lambda_1 T}} - \frac{\lambda_2\,\mathrm{e}^{\lambda_2 \bar{t}}}{1 - \mathrm{e}^{\lambda_2 T}}\right), \tag{2.2.159}$$

has a discontinuity of v_0 at the times of instantaneous momentum transfer by the impulsive force,

$$\begin{aligned} \dot{x}^{(\mathrm{css})}(t)\Big|_{\bar{t}=0} &= \frac{v_0}{\lambda_1 - \lambda_2}\left(\frac{\lambda_1}{1 - \mathrm{e}^{\lambda_1 T}} - \frac{\lambda_2}{1 - \mathrm{e}^{\lambda_2 T}}\right) \\ &= \frac{v_0}{\lambda_1 - \lambda_2}\left(\frac{\lambda_1\,\mathrm{e}^{\lambda_1 T}}{1 - \mathrm{e}^{\lambda_1 T}} + \lambda_1 - \frac{\lambda_2\,\mathrm{e}^{\lambda_2 T}}{1 - \mathrm{e}^{\lambda_2 T}} - \lambda_2\right) \\ &= \dot{x}^{(\mathrm{css})}(t)\Big|_{\bar{t}=T} + v_0 = \dot{x}_{\mathrm{css}} + v_0\,, \end{aligned} \tag{2.2.160}$$

with $\dot{x}^{(\mathrm{css})}(t)\big|_{\bar{t}=T}$ equal to the $\dot{x}_{\mathrm{css}}$ indicated in (2.2.148).

For subcritical damping, we have

$$x_{\text{css}} = \frac{v_0}{2\omega} \frac{\sin(\omega T)}{\cosh\left(\frac{1}{2}\gamma T\right) - \cos(\omega T)} . \tag{2.2.161}$$

There is a meaningful $T \to 0$ limit if we keep the average acceleration $a = v_0/T$ fixed,

$$\begin{aligned} x_{\text{css}}\Big|_{T\to 0} &= \frac{aT}{2\omega} \frac{\omega T}{\frac{1}{8}(\gamma T)^2 + \frac{1}{2}(\omega T)^2}\Bigg|_{T\to 0} \\ &= \frac{a}{\omega^2 + \frac{1}{4}\gamma^2} = \frac{a}{\omega_0^2} , \end{aligned} \tag{2.2.162}$$

the constant displacement by the constant force $F = ma$ that we found above in (2.2.145). For $T > 0$, x_{css} is always less than this $T = 0$ value of a/ω_0^2.

The phenomenon of resonance — particularly large amplitudes for specific periods T — that we observed for the periodic harmonic force in Section 2.2.6(a), also occurs for the periodic impulsive force of (2.2.149). Exercise 40 deals with these matters.

Chapter 3

Conservative Forces

3.1 One-dimensional motion

3.1.1 *Kinetic and potential energy*

We now turn to a rather fundamental subject matter, that of conservative forces. For a start, we consider motion in one dimension only, say along the x axis, under the influence of a force that depends on the position of the point mass, but not on its velocity, and also does not have a parametric time dependence. Then Newton's equation of motion has the form

$$m\ddot{x} = F(x) \quad \text{for} \quad x = x(t) \,. \tag{3.1.1}$$

We multiply the equation with the velocity $\dot{x}(t)$ and recognize that the left-hand side is the time derivative of $\frac{1}{2}m\dot{x}^2$,

$$\frac{\mathrm{d}}{\mathrm{d}t}\frac{1}{2}m\dot{x}^2 = m\dot{x}\ddot{x} = F(x)\dot{x} \,. \tag{3.1.2}$$

The right-hand side is also a time derivative, namely of $-V(x)$, which we define by

$$F(x) = -\frac{\mathrm{d}}{\mathrm{d}x}V(x) \tag{3.1.3}$$

or

$$V(x) = V(x_0) - \int_{x_0}^{x} \mathrm{d}x' \, F(x') \,, \tag{3.1.4}$$

where we are free to choose the reference point x_0 and the reference value $V(x_0)$ to our liking. In the equation of motion, only the force, the negative derivative of $V(x)$ matters. Then

$$F(x)\,\dot{x} = -\frac{\mathrm{d}x(t)}{\mathrm{d}t}\frac{\mathrm{d}V}{\mathrm{d}x}(x(t)) = -\frac{\mathrm{d}}{\mathrm{d}t}V(x(t)) \tag{3.1.5}$$

shows an application of the chain rule of differentiation. As a consequence, we have

$$\frac{\mathrm{d}}{\mathrm{d}t}\left(\frac{m}{2}\dot{x}^2 + V(x)\right) = 0 \tag{3.1.6}$$

for each $x(t)$ that solves the equation of motion.

What we have found here is a *conservation law*, that of *energy conservation*. The terms that are summed in

$$E = \frac{m}{2}\dot{x}^2 + V(x) \tag{3.1.7}$$

are the velocity-dependent *kinetic energy* $\frac{m}{2}\dot{x}^2$ and the position-dependent *potential energy* $V(x)$. With $x(t)$ changing in time as required by the equation of motion, both the kinetic energy and the potential energy are time-dependent quantities but their sum E, the total energy or simply *the energy*, is constant.

Forces for which one has energy conservation are termed *conservative forces*. Clearly, in the current one-dimensional situation we can always find a potential energy for the given force $F(x)$, but there are other forces of a nonconservative nature, in particular the velocity-dependent frictional forces.

A typical graph of a potential energy $V(x)$ looks roughly like this:

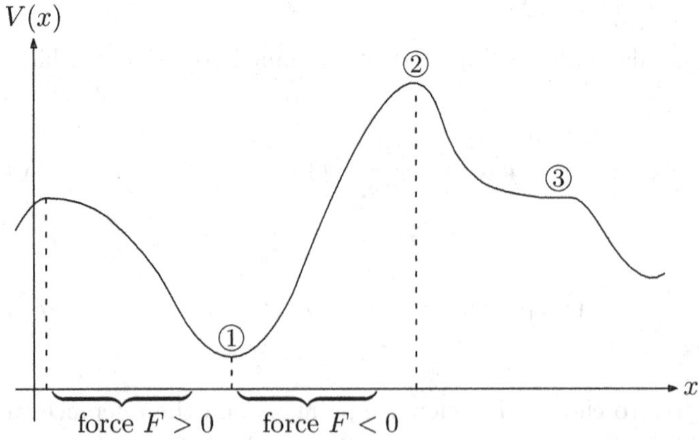

(3.1.8)

There are ranges with $F > 0$ (V decreases with growing x), ranges with $F < 0$ (V increases with growing x), separated by the x values where $F = 0$. These can mark a minimum of the potential energy (point ①), a maximum (point ②), or a saddle point (point ③). If we have the point mass at a potential minimum at rest, it will stay at rest if left undisturbed, or will undergo small oscillations around the position of the minimum. Behavior of this kind identifies the $F = 0$ position as that of a *stable equilibrium.*

By contrast, a point mass at rest at the position of a maximum of the potential energy is in an *unstable equilibrium*: A tiny perturbation will make it move away from the $F = 0$ position for good.

Finally, there are the $F = 0$ points that are neither at a minimum nor a maximum of the potential energy, as exemplified by the point ③ in the sketch of $V(x)$ in (3.1.8). Here the potential increases to one side but decreases to the other. Small perturbations will ultimately make a point mass at rest at such a saddle point move away, so that saddle points are more similar to potential energy maxima in this respect.

3.1.2 *Bounded motion between two turning points*

Consider now the following situation:

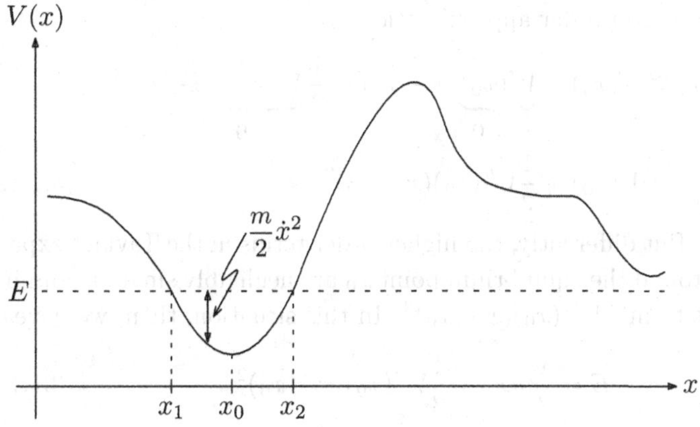

(3.1.9)

The point mass has initial conditions such that it is in the vicinity of the stable equilibrium point x_0 with an initial velocity that gives an energy E somewhat larger than $V(x_0)$ but not much larger. There are neighboring positions x_1 and x_2 where $E = V(x_1) = V(x_2)$, so that the kinetic energy

$$\frac{m}{2}\dot{x}^2 = E - V(x) \tag{3.1.10}$$

vanishes there. We call x_1 and x_2 *turning points*, which refers to their basic geometrical meaning as limits of the x range to which the motion is confined,

$$x_1 < x < x_2 \,. \tag{3.1.11}$$

For, the difference $E - V(x)$ is positive inside this range and we have a permissible value of the kinetic energy, whereas to the left of x_1 and to the right of x_2, $E - V(x)$ is negative and there is no velocity $\dot{x}$ such that the kinetic energy $\frac{1}{2}m\dot{x}^2$ is equal to this negative amount.

The motion thus confined to the stretch between the turning points is periodic, for half a period the point mass moves from x_1 to x_2, fastest at x_0, slowest near the turning points, and for the other half of the period it moves back from x_2 to x_1, then the next period starts — forth and back, forth and back, forever (that is: as long as the physical situation remains as stated).

(a) Small-amplitude oscillations

Let us look at the case of *small oscillations* first, that is: oscillations with small amplitudes. There, all x values are such that $x(t) - x_0$ is small in the sense that the second-order approximation

$$\begin{aligned} V(x) &\cong V(x_0) + \underbrace{V'(x_0)}_{=0}(x - x_0) + \frac{1}{2}\underbrace{V''(x_0)}_{>0}(x - x_0)^2 \\ &= V(x_0) + \frac{1}{2}V''(x_0)(x - x_0)^2 \,, \end{aligned} \tag{3.1.12}$$

is permissible. Put differently, the higher-order terms in the Taylor* expansion of $V(x)$ around the equilibrium point x_0 are negligibly small, beginning with the cubic term $\frac{1}{6}V'''(x_0)(x - x_0)^3$. In this situation, then, we have

$$E = \frac{1}{2}m\dot{x}^2 + \frac{1}{2}V''(x_0)(x - x_0)^2 \tag{3.1.13}$$

for the energy, and the equation of motion that we re-establish by differentiation,

$$0 = \frac{\mathrm{d}}{\mathrm{d}t}E = \big[m\ddot{x} + V''(x_0)(x - x_0)\big]\dot{x} \,, \tag{3.1.14}$$

*Brook TAYLOR (1685–1731)

implying

$$m\ddot{x} = -V''(x_0)(x - x_0)\,, \tag{3.1.15}$$

is that of a harmonic oscillator,

$$\left(\frac{\mathrm{d}}{\mathrm{d}t}\right)^2 (x - x_0) = -\omega_0^2(x - x_0) \tag{3.1.16}$$

with

$$\omega_0^2 = \frac{1}{m}V''(x_0) > 0\,, \tag{3.1.17}$$

where it is crucial that we have a minimum of the potential energy at x_0. In summary, then, for small excursions from a stable equilibrium point, we have a harmonic oscillation with the circular frequency $\omega_0 = \sqrt{V''(x_0)/m}$ and the period $T = 2\pi/\omega_0 = 2\pi\sqrt{m/V''(x_0)}$.

(b) Large-amplitude oscillations

If the energy is larger, the second-order approximation to $V(x)$ is not applicable. But we can still say something rather specific about the properties of the oscillating motion between the turning points. The velocity is

$$\dot{x} = \pm\sqrt{\frac{2}{m}\left[E - V(x)\right]}\,, \tag{3.1.18}$$

with $\dot{x} > 0$ for the half-period of motion during which the point mass moves from x_1 to x_2, and $\dot{x} < 0$ on the way back. The period T is, therefore, equal to twice the time it takes to move from x_1 to x_2,

$$T = 2\int_{x_1}^{x_2} \frac{\mathrm{d}x}{\sqrt{\frac{2}{m}\left[E - V(x)\right]}} \tag{3.1.19}$$

where

$$\frac{\mathrm{d}x}{\sqrt{\frac{2}{m}\left[E - V(x)\right]}} = \frac{\mathrm{d}x}{\dot{x}} = \mathrm{d}t \tag{3.1.20}$$

for the half-period with $\dot{x} > 0$. The integration range covers the stretch between the turning points where the argument of the square root is positive. At the turning points themselves, we have $V(x_1) = V(x_2) = E$, and the velocity vanishes there. Hereby, the turning points and the period are functions of the energy — that is: $x_1(E)$, $x_2(E)$, and $T(E)$ — but we do not want to overload the equations and indicate the energy dependence only where this is necessary or useful.

For any potential $V(x)$, for which we have a stretch $x_1 < x < x_2$ between two turning points, (3.1.19) gives the period of the oscillation as a function of the energy. This covers the situation of small-amplitude oscillations as well, for which we have the potential energy of a harmonic oscillator,

$$V(x) = V(x_0) + \frac{1}{2}m\omega_0^2(x - x_0)^2 ; \tag{3.1.21}$$

this is (3.1.12) with (3.1.17). Then,

$$\begin{aligned} E = V(x_1) &= V(x_0) + \frac{1}{2}m\omega_0^2(x_0 - x_1)^2 \\ = V(x_2) &= V(x_0) + \frac{1}{2}m\omega_0^2(x_2 - x_0)^2 , \end{aligned} \tag{3.1.22}$$

so that

$$\begin{aligned} \frac{2}{m}\left[E - V(x)\right] &= \omega_0^2(x_0 - x_1)^2 - \omega_0^2(x - x_0)^2 \\ &= \omega_0^2(x_2 - x_0)^2 - \omega_0^2(x - x_0)^2 . \end{aligned} \tag{3.1.23}$$

It is convenient to write $x = x_0 + y$ and $x_2 - x_0 = x_0 - x_1 = \bar{y}$. Then $\mathrm{d}x = \mathrm{d}y$; $x = x_1$: $y = -\bar{y}$; $x = x_2$: $y = \bar{y}$, and the period of the oscillation is

$$T = 2\int_{-\bar{y}}^{\bar{y}} \frac{\mathrm{d}y}{\sqrt{\omega_0^2(\bar{y}^2 - y^2)}} = \frac{2}{\omega_0}\int_{-\bar{y}}^{\bar{y}} \frac{\mathrm{d}y}{\sqrt{\bar{y}^2 - y^2}} , \tag{3.1.24}$$

where the energy dependence on the right-hand side is in the value of $\bar{y}$,

$$\bar{y} = \sqrt{\frac{2}{m\omega_0^2}\left[E - V(x_0)\right]} . \tag{3.1.25}$$

The integral, however, does not depend on $\bar{y}$, as wee see when we perform the standard substitution

$$y = \bar{y}\sin\varphi , \ \mathrm{d}y = \bar{y}\,\mathrm{d}\varphi\,\cos\varphi ; \ y = -\bar{y}\colon \varphi = -\frac{\pi}{2} ; \ y = \bar{y}\colon \varphi = \frac{\pi}{2} , \tag{3.1.26}$$

which gives

$$T = \frac{2}{\omega_0}\int_{-\pi/2}^{\pi/2} \frac{\mathrm{d}\varphi\,\cos\varphi}{\sqrt{1 - (\sin\varphi)^2}} = \frac{2\pi}{\omega_0} . \tag{3.1.27}$$

As expected, this is the energy-independent period of a harmonic oscillator.

(c) Potential energy inferred from the energy-dependent period

For a given potential energy $V(x)$, we find the period T for motion bounded by two turning points upon evaluating the integral in (3.1.19). This gives us the period as a function of the energy, $T = T(E)$, because once the energy is specified, the turning points are determined by $E = V(x_1) = V(x_2)$. Let us now reverse the question: If we know the period $T(E)$ for $E > V(x_0)$, what can we then say about the potential energy $V(x)$?

We assume that, in the x range of interest, there are no other extremal points of the potential energy than the minimum at x_0. In other words, the force $F(x) = -\frac{\mathrm{d}}{\mathrm{d}x}V(x)$ is strictly positive for $x < x_0$ and strictly negative for $x > x_0$. This assumption avoids the complications that would arise from discontinuities or singularities in $T(E)$. Accordingly, the situation is as sketched here:

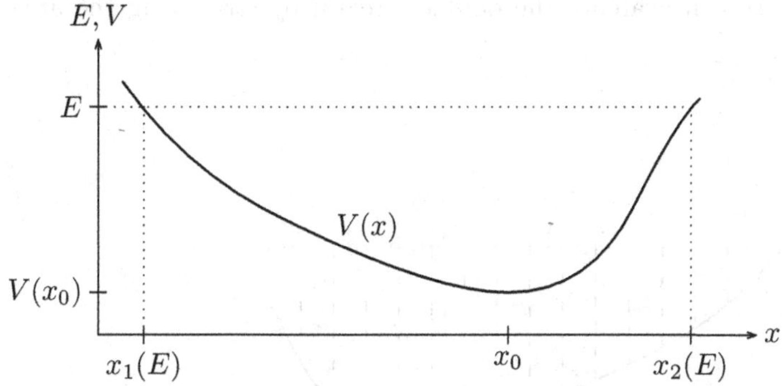

(3.1.28)

The relation (3.1.19) can be regarded as an integral transform that maps the potential energy function $V(x)$ on the period function $T(E)$. The question posed above thus asks whether we can find the inverse of this integral transform.

As a first step toward an answer, we consider the following integral of $T(E)$:

$$\int\limits_{V(x_0)}^{E} \frac{\mathrm{d}E'\, T(E')}{\sqrt{2m(E-E')}} = \int\limits_{V(x_0)}^{E} \mathrm{d}E' \int\limits_{x_1(E')}^{x_2(E')} \frac{\mathrm{d}x}{\sqrt{(E-E')(E'-V(x))}}\,, \quad (3.1.29)$$

where we first integrate over x for the given value of E', and then over E'. This corresponds to a "horizontal" stratification of the integration area in

the x, E plane,

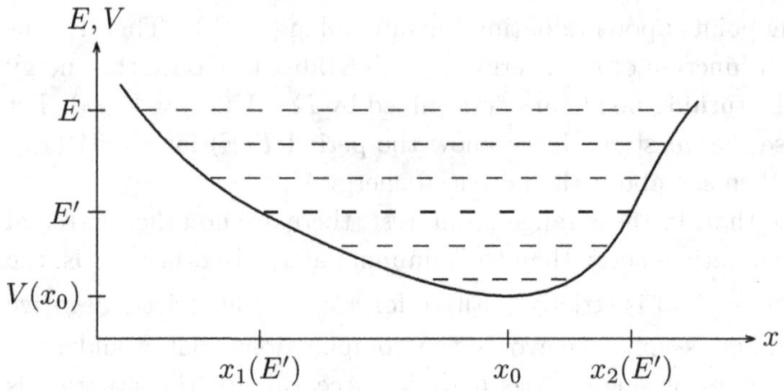

(3.1.30)

Equivalently, we can evaluate the double integral by stratifying the area "vertically,"

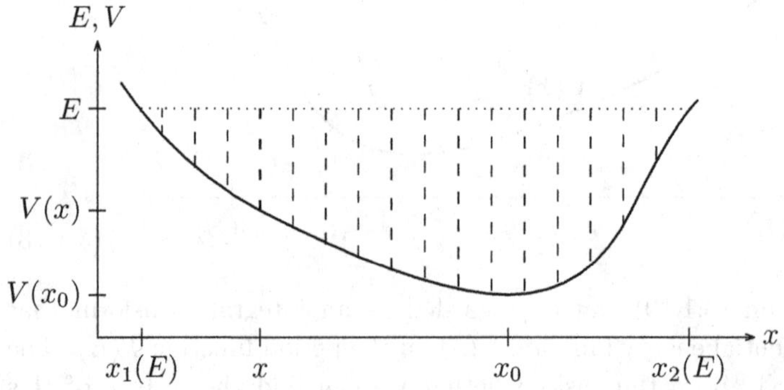

(3.1.31)

so that we now first integrate over E' for the given x value, and then over x,

$$\int_{V(x_0)}^{E} \frac{\mathrm{d}E'\,T(E')}{\sqrt{2m(E-E')}} = \int_{x_1(E)}^{x_2(E)} \mathrm{d}x \underbrace{\int_{V(x)}^{E} \frac{\mathrm{d}E'}{\sqrt{(E-E')(E'-V(x))}}}_{=\pi}$$
$$= \pi\bigl(x_2(E) - x_1(E)\bigr)\,, \qquad (3.1.32)$$

where we make use of the integral in (3.1.40) below. This states that knowledge of $T(E)$ does not tell us $x_1(E)$ and $x_2(E)$, but only their difference

$$x_2(E) - x_1(E) = \frac{1}{\pi} \int_{V(x_0)}^{E} \frac{\mathrm{d}E'\, T(E')}{\sqrt{2m(E-E')}}, \tag{3.1.33}$$

the width, so to say, of the range of allowed x values for energy E. An essentially unique potential can be inferred if we impose additional conditions, such as that the potential energy is symmetric about x_0,

$$x_2(E) - x_0 = x_0 - x_1(E) \quad \text{for all } E > V(x_0)\,. \tag{3.1.34}$$

Then we have an implicit equation for $V(x)$,

$$|x - x_0| = \frac{1}{2\pi} \int_{V(x_0)}^{V(x)} \frac{\mathrm{d}E'\, T(E')}{\sqrt{2m\bigl(V(x)-E'\bigr)}}, \tag{3.1.35}$$

after replacing $x_2 - x_0 = x_0 - x_1$ by $|x - x_0|$ and E by $V(x)$ in (3.1.33). The position x_0 of the potential minimum is, of course, not determined by $T(E)$ and must be chosen appropriately by some convention or on the basis of other information.

As an example consider $T(E) = 2\pi/\omega_0 = \text{constant}$. This gives

$$x_2(E) - x_1(E) = \frac{2}{\omega_0}\sqrt{\frac{2}{m}\bigl[E - V(x_0)\bigr]} \tag{3.1.36}$$

for the general case of (3.1.33) and

$$|x - x_0| = \frac{1}{\omega_0}\sqrt{\frac{2}{m}\bigl[V(x) - V(x_0)\bigr]} \tag{3.1.37}$$

for the symmetric case of (3.1.35). The latter yields, not unexpectedly, the potential energy of a harmonic oscillator,

$$V(x) = V(x_0) + \frac{m}{2}\omega_0^2 (x - x_0)^2\,, \tag{3.1.38}$$

as in (3.1.21), whereas there are many potential energy functions $V(x)$ that are consistent with (3.1.36), such as

$$V(x) = V(x_0) + \begin{cases} \frac{1}{2}k_1(x_0 - x)^2 \text{ for } x < x_0 \\ \frac{1}{2}k_2(x - x_0)^2 \text{ for } x > x_0 \end{cases}$$

$$\text{with} \quad \sqrt{\frac{m}{k_1}} + \sqrt{\frac{m}{k_2}} = \frac{2}{\omega_0}, \tag{3.1.39}$$

for example, where we have different spring constants to the left and to the right of x_0.

Finally, we note that the E' integral in (3.1.32) is a version of

$$\int_b^a \frac{dx}{\sqrt{(a-x)(x-b)}} = \pi \quad \text{for} \quad a > b. \tag{3.1.40}$$

The substitution $x = \frac{a+b}{2} + \frac{a-b}{2}\sin\alpha$ turns it into an elementary integral over α. Note that the integral in (3.1.24) is a special case of (3.1.40).

3.1.3 *Unbounded motion with a single turning point*

Different from the motion between two turning points, depicted in (3.1.9), is the situation with a single turning point:

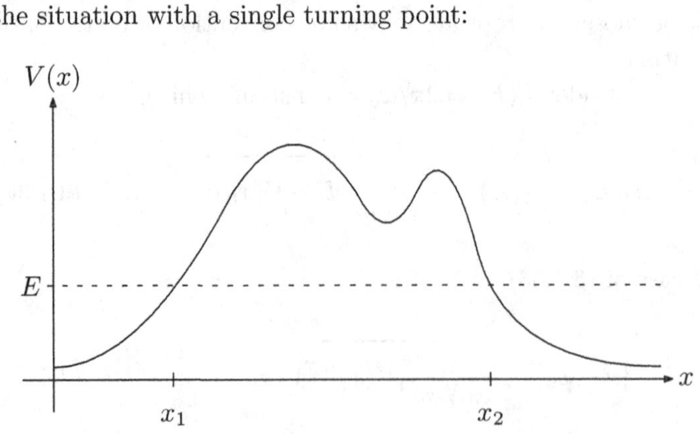

(3.1.41)

Here, depending on the initial position and velocity, the motion is restricted to the range $x < x_1$, or the range $x > x_2$. Suppose, for example, that $x(t=0) < x_1$ and $\dot{x}(t=0) > 0$, that is: The point mass is initially to the left of x_1 and approaches this turning point. Then it will continue to move to the right, eventually reaching x_1 with decreasing velocity, having zero velocity at x_1, then it will get accelerated to the left and finally move even further away.

In this context of unbounded motion, there is no period, simply because the motion is not periodic. But we can ask a more appropriate question, such as: How long does it take to reach the turning point? In the particular scenario specified above, we have

$$
\begin{aligned}
x(t=0) &= x_0 < x_1\,,\\
\dot{x}(t=0) &= \sqrt{\frac{2}{m}\left[E - V(x_0)\right]} > 0\,,\\
\text{and}\quad \dot{x}(t>0) &= \sqrt{\frac{2}{m}\left[E - V\big(x(t)\big)\right]} > 0\,,
\end{aligned}
\tag{3.1.42}
$$

while $x_0 < x(t) < x_1$ and $\dot{x}(t) > 0$, as it is the case during the approach. The total time needed to get to the turning point x_1 is then

$$
T = \int_{x_0}^{x_1} \frac{\mathrm{d}x}{\sqrt{\frac{2}{m}\left[E - V(x)\right]}}\,,
\tag{3.1.43}
$$

which — not by accident — has an obvious resemblance with the expression for the period T in (3.1.19), but we need to remember that the two expressions refer to quite different situations and that the meaning of T is not the same.

There is a particular limiting situation, in which the energy is equal to the value of a (local) maximum of the potential energy:

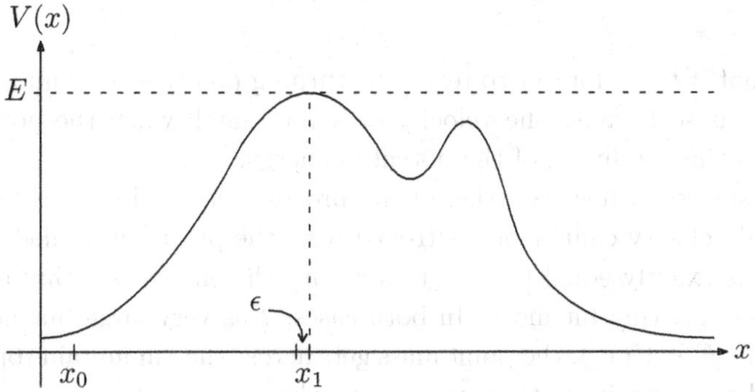

(3.1.44)

Near x_1, the potential energy is well approximated by

$$V(x) = E + \frac{1}{2}V''(x_1)(x-x_1)^2 + \underbrace{\cdots}_{\text{discarded}} \tag{3.1.45}$$

where the higher-order terms of the Taylor expansion are ignored, much like we did it above in the vicinity of the local minimum, see (3.1.12). We break the integral up into the ranges $x_0 < x < x_1 - \epsilon$ and $x_1 - \epsilon < x < x_1$, with ϵ small enough to justify the quadratic approximation for $V(x)$. This gives

$$T = \int\limits_{x_0}^{x_1-\epsilon} \frac{\mathrm{d}x}{\sqrt{\frac{2}{m}[E - V(x)]}} + \int\limits_{x_1-\epsilon}^{x_1} \frac{\mathrm{d}x}{\sqrt{\frac{2}{m}\left[-\frac{1}{2}V''(x_1)(x-x_1)^2\right]}} \tag{3.1.46}$$

with $V''(x_1) < 0$ at this local maximum of $V(x)$. In the second integral we substitute $x = x_1 - x'$ and so turn it into

$$\int\limits_{x_1-\epsilon}^{x_1} \frac{\mathrm{d}x}{\sqrt{\frac{2}{m}\left[-\frac{1}{2}V''(x_1)(x-x_1)^2\right]}} = \sqrt{\frac{m}{-V''(x_1)}} \int\limits_0^{\epsilon} \frac{\mathrm{d}x'}{x'}\,, \tag{3.1.47}$$

which is logarithmically divergent,

$$\int\limits_0^{\epsilon} \frac{\mathrm{d}x'}{x'} = \log\left(\frac{x'}{\epsilon}\right)\Bigg|_{x'=0}^{\epsilon} = \log\left(\frac{\epsilon}{x'}\right)\Bigg|_{0<x'\to 0} = \infty\,. \tag{3.1.48}$$

This tells us that it takes forever to reach the turning point, and we understand that this is so because the velocity gets very small when the point mass is close to the maximum of the potential energy.

Physically speaking, however, there is no process that really takes forever, the initial velocity cannot be controlled with the precision needed to ensure that E is exactly equal to $V(x_1)$, in reality the energy is a tiny bit less than $V(x_1)$ or a tiny bit more. In both cases, T is very large but not infinite. And for $E \gtrsim V(x_1)$, the point mass gets across the "mountain top" at $x = x_1$, and then continues to move to the right.

3.1.4 *Unbounded motion without a turning point*

This brings us to the third kind of motion, that with no turning points:

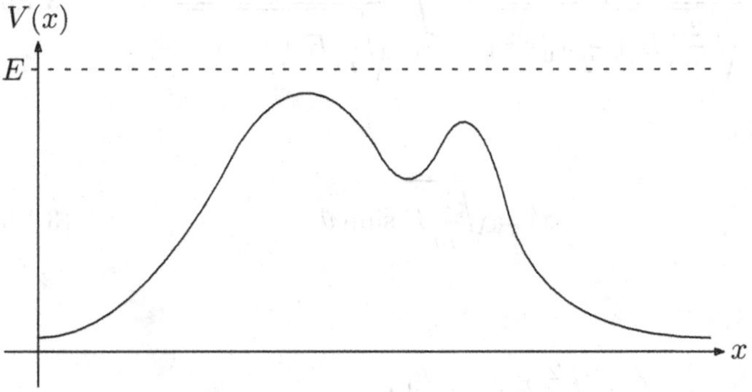

$$(3.1.49)$$

Here, the energy is so large that $\frac{1}{2}m\dot{x}^2 = E - V(x)$ is positive for all positions x. Depending on the sign of the initial velocity we then either have $\dot{x}(t) > 0$ or $\dot{x}(t) < 0$ for all t,

$$\dot{x}(t) = \pm\sqrt{\frac{2}{m}[E - V(x)]} \quad \text{for} \quad \begin{cases} \dot{x}(0) > 0\,, \\ \dot{x}(0) < 0\,. \end{cases} \tag{3.1.50}$$

Upon exploiting

$$\mathrm{d}t = \pm\frac{\mathrm{d}x}{\sqrt{\frac{2}{m}[E - V(x)]}} \tag{3.1.51}$$

once more, we have

$$t = \pm\int_{x_0}^{x(t)} \frac{\mathrm{d}x'}{\sqrt{\frac{2}{m}[E - V(x')]}} \tag{3.1.52}$$

as an implicit solution of the equation of motion — implicit because we need to solve this for $x(t)$ in terms of time t eventually. But for many purposes, such as that of plotting $x(t)$ as a function of t, the implicit solution is fine.

Here is a simple example, the "inverted oscillator" with the potential energy

$$V(x) = -\frac{m}{2}\omega_0^2 x^2\,, \tag{3.1.53}$$

so that, for $E > 0$ and $\dot{x}_0 > 0$,

$$t = \int_{x_0}^{x(t)} \frac{dx'}{\sqrt{\frac{2}{m}\left(E + \frac{m}{2}\omega_0^2 x'^2\right)}} = \int_{x_0}^{x(t)} \frac{dx'}{\sqrt{\frac{2}{m}E + (\omega_0 x')^2}}. \tag{3.1.54}$$

The substitution

$$\omega_0 x' = \sqrt{\frac{2}{m}E}\,\sinh\vartheta \tag{3.1.55}$$

with

$$\omega_0 dx' = \sqrt{\frac{2}{m}E}\,\cosh\vartheta\, d\vartheta\,,$$

$$\sqrt{\frac{2}{m}E + (\omega_0\, x')^2} = \sqrt{\frac{2}{m}E}\,\sqrt{1 + (\sinh\vartheta)^2} = \sqrt{\frac{2}{m}E}\,\cosh\vartheta \tag{3.1.56}$$

and

$$\begin{aligned} \vartheta = \vartheta_0 \quad &\text{for} \quad x' = x_0\,, \\ \vartheta = \vartheta(t) \quad &\text{for} \quad x' = x(t)\,, \end{aligned}$$

gives

$$t = \int_{\vartheta_0}^{\vartheta(t)} \frac{d\vartheta}{\omega_0} = \frac{1}{\omega_0}(\vartheta(t) - \vartheta_0) \tag{3.1.57}$$

or

$$\vartheta(t) = \vartheta_0 + \omega_0 t\,. \tag{3.1.58}$$

Then

$$x(t) = \frac{1}{\omega_0}\sqrt{\frac{2}{m}E}\,\sinh(\vartheta(t)) = \frac{1}{\omega_0}\sqrt{\frac{2}{m}E}\,\sinh(\vartheta_0 + \omega_0 t) \tag{3.1.59}$$

and

$$\dot{x} = \sqrt{\frac{2}{m}E}\,\cosh(\vartheta_0 + \omega_0 t)\,, \tag{3.1.60}$$

which tell us that

$$x(0) = x_0 = \frac{1}{\omega_0}\sqrt{\frac{2}{m}E}\,\sinh\vartheta_0\,,$$
$$\dot{x}(0) = \dot{x}_0 = \sqrt{\frac{2}{m}E}\,\cosh\vartheta_0\,, \tag{3.1.61}$$

and so we arrive at

$$\begin{aligned} x(t) &= \frac{1}{\omega_0}\sqrt{\frac{2}{m}E}\big[\sinh\vartheta_0\cosh(\omega_0 t) + \cosh\vartheta_0\sinh(\omega_0 t)\big] \\ &= x_0\cosh(\omega_0 t) + \frac{\dot{x}_0}{\omega_0}\sinh(\omega_0 t) \end{aligned} \tag{3.1.62}$$

and

$$\dot{x}(t) = \dot{x}_0\cosh(\omega_0 t) + \omega_0 x_0 \sinh(\omega_0 t)\,. \tag{3.1.63}$$

These expressions for $x(t)$ and $\dot{x}(t)$, here derived for $\dot{x}_0 > 0$, are equally valid for $\dot{x}_0 < 0$ and even apply when x_0 and $\dot{x}_0$ are such that the energy $E = \frac{1}{2}m\dot{x}_0^2 - \frac{1}{2}m\omega_0^2 x_0^2$ is negative; just verify that they solve the equation of motion for the potential energy in (3.1.53). We note that $\omega_0 \to \mathrm{i}\omega_0$ in (2.2.50) gives (3.1.62) and (3.1.63), consistent with the change from ω_0^2 to $-\omega_0^2$ in the potential energies of (3.1.21) and (3.1.53).

3.2 Three-dimensional motion

3.2.1 *Kinetic energy, potential energy*

In three dimensions, we look for energy conservation by the same approach that succeeded so easily in one dimension, that is we consider Newton's equation of motion with a force that depends only on the position of the point mass,

$$m\frac{\mathrm{d}^2}{\mathrm{d}t^2}\boldsymbol{r}(t) = \boldsymbol{F}\big(\boldsymbol{r}(t)\big)\,, \tag{3.2.1}$$

and multiply by the velocity $\boldsymbol{v}(t) = \dfrac{\mathrm{d}}{\mathrm{d}t}\boldsymbol{r}(t)$, taking the dot product,

$$m\frac{\mathrm{d}\boldsymbol{r}}{\mathrm{d}t}\cdot\frac{\mathrm{d}^2\boldsymbol{r}}{\mathrm{d}t^2} = \frac{\mathrm{d}\boldsymbol{r}}{\mathrm{d}t}\cdot\boldsymbol{F}(\boldsymbol{r})\,. \tag{3.2.2}$$

On the left-hand side we have the time derivative of the kinetic energy,

$$\frac{\mathrm{d}}{\mathrm{d}t}\left[\frac{m}{2}\left(\frac{\mathrm{d}\boldsymbol{r}}{\mathrm{d}t}\right)^2\right] = \frac{\mathrm{d}\boldsymbol{r}}{\mathrm{d}t}\cdot\boldsymbol{F}(\boldsymbol{r}), \tag{3.2.3}$$

and we would like to have the negative derivative of a potential energy on the right-hand side,

$$\frac{\mathrm{d}\boldsymbol{r}}{\mathrm{d}t}\cdot\boldsymbol{F}(\boldsymbol{r}) \stackrel{?}{=} -\frac{\mathrm{d}}{\mathrm{d}t}V(\boldsymbol{r}). \tag{3.2.4}$$

Since we know from (1.2.22) that an application of the chain rule establishes

$$\frac{\mathrm{d}}{\mathrm{d}t}V(\boldsymbol{r}) = \frac{\mathrm{d}\boldsymbol{r}}{\mathrm{d}t}\cdot\boldsymbol{\nabla}V(\boldsymbol{r}), \tag{3.2.5}$$

this requires

$$\boldsymbol{F}(\boldsymbol{r}) = -\boldsymbol{\nabla}V(\boldsymbol{r}), \tag{3.2.6}$$

the force must be equal to the negative gradient of the potential energy $V(\boldsymbol{r})$.

Forces with this property are called *conservative forces* because for them we have

$$\frac{\mathrm{d}}{\mathrm{d}t}\left[\frac{m}{2}\left(\frac{\mathrm{d}\boldsymbol{r}}{\mathrm{d}t}\right)^2 + V(\boldsymbol{r})\right] = 0, \tag{3.2.7}$$

stating that the energy

$$E = \frac{m}{2}\left(\frac{\mathrm{d}\boldsymbol{r}}{\mathrm{d}t}\right)^2 + V(\boldsymbol{r}) \tag{3.2.8}$$

it conserved — it is constant in time, $\frac{\mathrm{d}}{\mathrm{d}t}E = 0$, although neither the kinetic energy

$$E_{\mathrm{kin}} = \frac{m}{2}\left(\frac{\mathrm{d}\boldsymbol{r}}{\mathrm{d}t}\right)^2 = \frac{m}{2}\boldsymbol{v}^2 \tag{3.2.9}$$

is constant itself, nor the potential energy $V(\boldsymbol{r})$. In passing, we note that the actual value of $V(\boldsymbol{r})$ is not essential, we can add any constant energy to $V(\boldsymbol{r})$ without changing the force $\boldsymbol{F} = -\boldsymbol{\nabla}V(\boldsymbol{r})$.

3.2.2 *Conservative force fields*

(a) Necessity of vanishing curl

Not all forces are conservative. For example,

$$\boldsymbol{F}(\boldsymbol{r}) = \kappa\bigl(y\boldsymbol{e}_x - x\boldsymbol{e}_y\bigr) \quad \text{with} \quad \kappa = \text{constant}, \tag{3.2.10}$$

cannot be written as the gradient of a potential energy. To see this, let us try to find $V(\boldsymbol{r})$ such that

$$\kappa\bigl(y\boldsymbol{e}_x - x\boldsymbol{e}_y\bigr) \stackrel{?}{=} -\boldsymbol{\nabla}V(\boldsymbol{r}) = -\left(\boldsymbol{e}_x\frac{\partial}{\partial x} + \boldsymbol{e}_y\frac{\partial}{\partial y} + \boldsymbol{e}_z\frac{\partial}{\partial z}\right)V(\boldsymbol{r}), \tag{3.2.11}$$

where we recall the cartesian-coordinate version of the gradient in (1.2.8). This requires

$$\frac{\partial}{\partial x}V(\boldsymbol{r}) = -\kappa y\,, \quad \frac{\partial}{\partial y}V(\boldsymbol{r}) = \kappa x\,, \quad \frac{\partial}{\partial z}V(\boldsymbol{r}) = 0\,. \tag{3.2.12}$$

There is no such $V(\boldsymbol{r})$ because we cannot have

$$\frac{\partial}{\partial x}\frac{\partial}{\partial y}V(\boldsymbol{r}) = \frac{\partial}{\partial x}(\kappa x) = \kappa \tag{3.2.13}$$

and also

$$\frac{\partial}{\partial y}\frac{\partial}{\partial x}V(\boldsymbol{r}) = \frac{\partial}{\partial y}(-\kappa y) = -\kappa\,, \tag{3.2.14}$$

as it does not matter whether we evaluate $\dfrac{\partial}{\partial x}\dfrac{\partial}{\partial y}V(\boldsymbol{r})$ by first differentiating with respect to y and then with respect to x, or in the reverse order. It follows that the force in (3.2.10) is not conservative.

More generally, a conservative force

$$\begin{aligned}\boldsymbol{F} &= \boldsymbol{e}_xF_x(\boldsymbol{r}) + \boldsymbol{e}_yF_y(\boldsymbol{r}) + \boldsymbol{e}_zF_z(\boldsymbol{r})\\ &= -\left(\boldsymbol{e}_x\frac{\partial}{\partial x} + \boldsymbol{e}_y\frac{\partial}{\partial y} + \boldsymbol{e}_z\frac{\partial}{\partial z}\right)V(\boldsymbol{r})\end{aligned} \tag{3.2.15}$$

must have cartesian components that obey

$$\begin{aligned}\frac{\partial}{\partial y}F_x &= \frac{\partial}{\partial x}F_y\,,\\ \frac{\partial}{\partial z}F_y &= \frac{\partial}{\partial y}F_z\,,\\ \frac{\partial}{\partial x}F_z &= \frac{\partial}{\partial z}F_x\,.\end{aligned} \tag{3.2.16}$$

The difference of the two expressions in the first of these equations is

$$\begin{aligned} 0 &= \frac{\partial}{\partial x} F_y - \frac{\partial}{\partial y} F_x \\ &= \left(\frac{\partial}{\partial x} \boldsymbol{e}_y - \frac{\partial}{\partial y} \boldsymbol{e}_x \right) \cdot \boldsymbol{F} \\ &= \left(\boldsymbol{e}_z \times \boldsymbol{e}_x \frac{\partial}{\partial x} + \boldsymbol{e}_z \times \boldsymbol{e}_y \frac{\partial}{\partial y} \right) \cdot \boldsymbol{F} \end{aligned} \tag{3.2.17}$$

or

$$0 = (\boldsymbol{e}_z \times \boldsymbol{\nabla}) \cdot \boldsymbol{F} = \boldsymbol{e}_z \cdot (\boldsymbol{\nabla} \times \boldsymbol{F}) \,, \tag{3.2.18}$$

where we exploited the fact that the cartesian unit vectors do not depend on the position vector $\boldsymbol{r}$.

For a conservative force, then, the z component of $\boldsymbol{\nabla} \times \boldsymbol{F}$ vanishes and since there is nothing particular about the z direction (we can choose *any* direction to be the z direction of our coordinate system), we conclude that

$$\boldsymbol{\nabla} \times \boldsymbol{F}(\boldsymbol{r}) = 0 \tag{3.2.19}$$

holds for a conservative force $\boldsymbol{F}(\boldsymbol{r})$. One calls

$$\boldsymbol{\nabla} \times \boldsymbol{A} = \boldsymbol{e}_x \left(\frac{\partial A_z}{\partial y} - \frac{\partial A_y}{\partial z} \right) + \boldsymbol{e}_y \left(\frac{\partial A_x}{\partial z} - \frac{\partial A_z}{\partial x} \right) + \boldsymbol{e}_z \left(\frac{\partial A_y}{\partial x} - \frac{\partial A_x}{\partial y} \right) \tag{3.2.20}$$

the "curl of vector field $\boldsymbol{A}$", or simply the "curl of $\boldsymbol{A}$" for any $\boldsymbol{A}(\boldsymbol{r})$. With this terminology we state (3.2.19) as: *The curl of a conservative force field vanishes.*

With $\boldsymbol{F} = -\boldsymbol{\nabla} V(\boldsymbol{r})$, the statement reads

$$\boldsymbol{\nabla} \times \boldsymbol{\nabla} V(\boldsymbol{r}) = 0 \,, \tag{3.2.21}$$

and in this very suggestive form it looks self-evident, inasmuch as $\boldsymbol{a} \times \boldsymbol{a} = 0$ for any vector $\boldsymbol{a}$. Indeed, $\boldsymbol{\nabla} \times \boldsymbol{\nabla} = 0$ is a correct statement about the gradient differential operator vector.

(b) Sufficiency of vanishing curl — Stokes's theorem

A conservative force field, as we have just seen, is necessarily such that its curl vanishes. The reverse is also true: If the curl of $\boldsymbol{F}(\boldsymbol{r})$ is zero, we can find a potential $V(\boldsymbol{r})$.

Here is how this works. Assume that the point mass is at $\boldsymbol{r}_0$ and then we move it along a certain path to a new position $\boldsymbol{r}_1$:

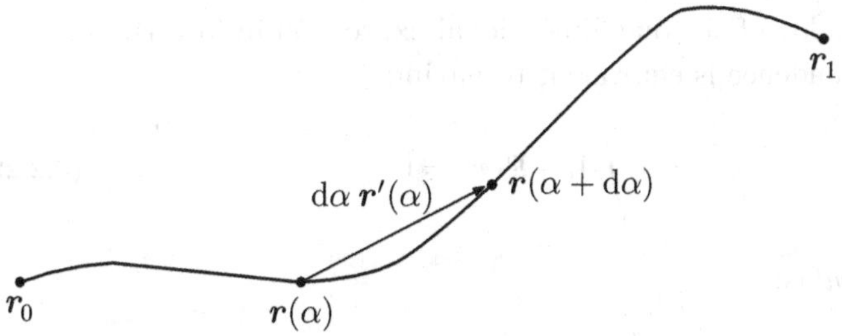

(3.2.22)

The path is parameterized by a parameter α, such that

$$\boldsymbol{r}(\alpha = 0) = \boldsymbol{r}_0 \quad \text{and} \quad \boldsymbol{r}(\alpha = 1) = \boldsymbol{r}_1 \,. \tag{3.2.23}$$

How much work do we have to do *against* the force field? The path element between α and $\alpha + \mathrm{d}\alpha$ has the net displacement

$$\boldsymbol{r}(\alpha + \mathrm{d}\alpha) - \boldsymbol{r}(\alpha) = \mathrm{d}\alpha \frac{\mathrm{d}}{\mathrm{d}\alpha} \boldsymbol{r}(\alpha) = \mathrm{d}\alpha\, \boldsymbol{r}'(\alpha) \,, \tag{3.2.24}$$

and the work for this path element is

$$-\mathrm{d}\boldsymbol{r} \cdot \boldsymbol{F}(\boldsymbol{r}) = -\mathrm{d}\alpha\, \boldsymbol{r}'(\alpha) \cdot \boldsymbol{F}\big(\boldsymbol{r}(\alpha)\big) \,. \tag{3.2.25}$$

The total work along the path is then

$$-\int_0^1 \mathrm{d}\alpha\, \boldsymbol{r}'(\alpha) \cdot \boldsymbol{F}\big(\boldsymbol{r}(\alpha)\big) = -\int_\Gamma \mathrm{d}\boldsymbol{r} \cdot \boldsymbol{F}(\boldsymbol{r}) \,, \tag{3.2.26}$$

where Γ symbolizes the path and the latter version recognizes that it does not matter how the path is parameterized, all that matters are the dot products $\mathrm{d}\boldsymbol{r} \cdot \boldsymbol{F}(\boldsymbol{r})$ along the path.

For a conservative force, we have

$$-\mathrm{d}\boldsymbol{r} \cdot \boldsymbol{F}(\boldsymbol{r}) = \mathrm{d}\boldsymbol{r} \cdot \boldsymbol{\nabla} V(\boldsymbol{r}) = V(\boldsymbol{r} + \mathrm{d}\boldsymbol{r}) - V(\boldsymbol{r}) \,, \tag{3.2.27}$$

that is: each $-\mathrm{d}\boldsymbol{r} \cdot \boldsymbol{F}(\boldsymbol{r})$ contribution adds the potential-energy difference between the neighboring points. The sum of these incremental differences is then the difference in potential energy between the end points of the path,

$$-\int_\Gamma \mathrm{d}\boldsymbol{r} \cdot \boldsymbol{F}(\boldsymbol{r}) = V(\boldsymbol{r}_1) - V(\boldsymbol{r}_0) \,, \tag{3.2.28}$$

and the actual path Γ connecting $\boldsymbol{r}_0$ and $\boldsymbol{r}_1$ does not matter: The value of the integral is *path-independent.* Clearly, we have here the three-dimensional analog of the one-dimensional expression in (3.1.4).

Path independence is equivalent to having

$$\int_{\Gamma} \mathrm{d}\boldsymbol{r} \cdot \boldsymbol{F}(\boldsymbol{r}) = 0 \qquad (3.2.29)$$

for all *closed paths*:

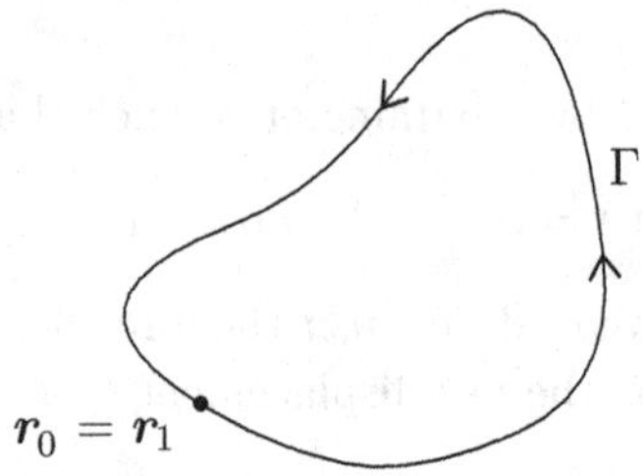

(3.2.30)

because we would have $V(\boldsymbol{r}_1) - V(\boldsymbol{r}_0) = 0$ for such a situation with coinciding end points. What we have to show, then, is that the integral of $\mathrm{d}\boldsymbol{r} \cdot \boldsymbol{F}(\boldsymbol{r})$ along *any* closed path vanishes if $\boldsymbol{\nabla} \times \boldsymbol{F} = 0$. But we do not really have to deal with arbitrarily shaped closed paths because we can break up a path into smaller pieces:

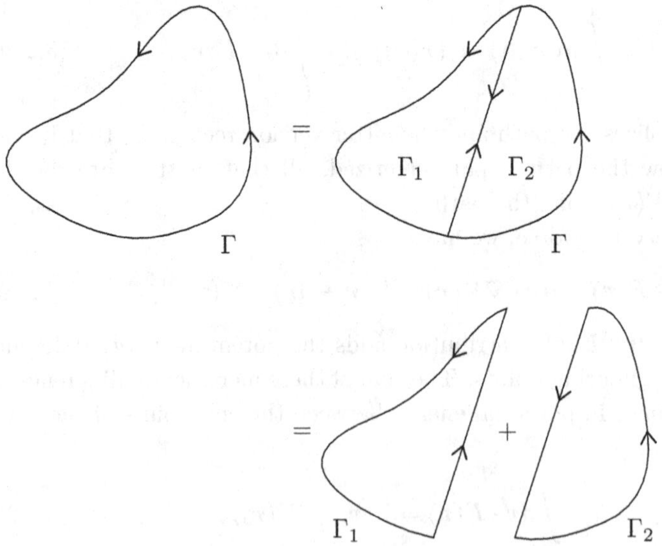

(3.2.31)

The intermediate line that is common to paths Γ_1 and Γ_2 cancels in the sum

$$\int_{\Gamma_1} \mathrm{d}\boldsymbol{r} \cdot \boldsymbol{F} + \int_{\Gamma_2} \mathrm{d}\boldsymbol{r} \cdot \boldsymbol{F} = \int_{\Gamma} \mathrm{d}\boldsymbol{r} \cdot \boldsymbol{F}, \tag{3.2.32}$$

because positive work for one of the paths is fully compensated for by corresponding negative work for the other path.

We can continue this division process until we have covered the surface bounded by path Γ (think of a membrane supported by a wire loop in the shape of Γ) by very small loops. If we can now show that

$$\int_{\Gamma} \mathrm{d}\boldsymbol{r} \cdot \boldsymbol{F}(\boldsymbol{r}) = 0 \tag{3.2.33}$$

for any tiny closed loop Γ, we know that the closed-loop integral vanishes for all loops. The tiny loop can be of triangular shape because we can triangulate the surface bounded by the loop in the way surveyors triangulate a landscape.

For such a tiny triangle, then, we choose coordinates that put the triangle into the xy plane

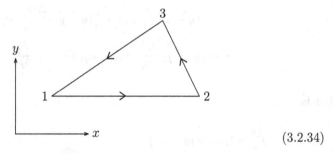

(3.2.34)

We integrate from (x_1, y_1) to $(x_2, y_2) = (x_1 + a, y_1)$, then to $(x_3, y_3) = (x_1 + b, y_1 + c)$, then back to (x_1, y_1). The triangle is small enough to allow the replacement of $F_x(x, y) \equiv F_x(x, y, z = 0)$ and $F_y(x, y) \equiv F_y(x, y, z = 0)$ by

$$F_x(x, y) = F_x(x_1, y_1) + (x - x_1)\frac{\partial F_x}{\partial x}(x_1, y_1) + (y - y_1)\frac{\partial F_x}{\partial y}(x_1, y_1) \tag{3.2.35}$$

and

$$F_y(x, y) = F_y(x_1, y_1) + (x - x_1)\frac{\partial F_y}{\partial x}(x_1, y_1) + (y - y_1)\frac{\partial F_y}{\partial y}(x_1, y_1)\,. \tag{3.2.36}$$

The three sides of the triangle contribute the following amounts to the closed-loop integral:

$$\begin{aligned}
\int\limits_{1\to 2} \mathrm{d}\boldsymbol{r}\cdot\boldsymbol{F}(\boldsymbol{r}) &= \int\limits_0^1 \mathrm{d}\alpha\, a\boldsymbol{e}_x\cdot\boldsymbol{F}(\boldsymbol{r}_1+\alpha a\boldsymbol{e}_x)\,,\\
\int\limits_{2\to 3} \mathrm{d}\boldsymbol{r}\cdot\boldsymbol{F}(\boldsymbol{r}) &= \int\limits_0^1 \mathrm{d}\alpha\,\bigl[(b-a)\boldsymbol{e}_x+c\boldsymbol{e}_y\bigr]\cdot\boldsymbol{F}\bigl(\boldsymbol{r}_2+\alpha(b-a)\boldsymbol{e}_x+\alpha c\boldsymbol{e}_y\bigr)\,,\\
\int\limits_{3\to 1} \mathrm{d}\boldsymbol{r}\cdot\boldsymbol{F}(\boldsymbol{r}) &= \int\limits_0^1 \mathrm{d}\alpha\,(-b\boldsymbol{e}_x-c\boldsymbol{e}_y)\cdot\boldsymbol{F}(\boldsymbol{r}_3-\alpha b\boldsymbol{e}_x-\alpha c\boldsymbol{e}_y)\,.
\end{aligned}\tag{3.2.37}$$

But before we insert the above expressions for the x and y components of $\boldsymbol{F}$, let us note that we can write them in the form

$$\begin{aligned}
\boldsymbol{e}_xF_x+\boldsymbol{e}_yF_y = \boldsymbol{\nabla}\biggl[&(x-x_1)F_x(\boldsymbol{r}_1)+\frac{1}{2}(x-x_1)^2\frac{\partial F_x}{\partial x}(\boldsymbol{r}_1)\\
&+(y-y_1)F_y(\boldsymbol{r}_1)+\frac{1}{2}(y-y_1)^2\frac{\partial F_y}{\partial y}(\boldsymbol{r}_1)\biggr]\\
&+\boldsymbol{e}_x(y-y_1)\frac{\partial F_x}{\partial y}(\boldsymbol{r}_1)+\boldsymbol{e}_y(x-x_1)\frac{\partial F_y}{\partial x}(\boldsymbol{r}_1)\,,
\end{aligned}\tag{3.2.38}$$

that is

$$\begin{aligned}
\boldsymbol{e}_xF_x+\boldsymbol{e}_yF_y &= \{\text{gradient part}\}\\
&\quad+\boldsymbol{e}_x(y-y_1)\frac{\partial F_x}{\partial y}(\boldsymbol{r}_1)+\boldsymbol{e}_y(x-x_1)\frac{\partial F_y}{\partial x}(\boldsymbol{r}_1)\,.
\end{aligned}\tag{3.2.39}$$

As we know, the gradient part does not contribute to the closed-loop integral and, therefore, we only need to take the contributions of the other two terms into detailed account. We have, then,

$$\int\limits_{1\to 2}\mathrm{d}\boldsymbol{r}\cdot\boldsymbol{F}(\boldsymbol{r}) = \{\text{gradient contribution}\}+0 \tag{3.2.40}$$

as well as

$$\int\limits_{2\to 3} \mathrm{d}\boldsymbol{r}\cdot\boldsymbol{F}(\boldsymbol{r}) = \{\text{gradient contribution}\} + \int\limits_0^1 \mathrm{d}\alpha \left[(b-a)\alpha c\frac{\partial F_x}{\partial y}(\boldsymbol{r}_1) + \big(ac + c\alpha(b-a)\big)\frac{\partial F_y}{\partial x}(\boldsymbol{r}_1)\right]$$
$$= \{\text{gradient contribution}\} + \frac{1}{2}bc\left(\frac{\partial F_y}{\partial x} + \frac{\partial F_x}{\partial y}\right)(\boldsymbol{r}_1) + \frac{1}{2}ac\left(\frac{\partial F_y}{\partial x} - \frac{\partial F_x}{\partial y}\right)(\boldsymbol{r}_1) \tag{3.2.41}$$

and

$$\int\limits_{3\to 1} \mathrm{d}\boldsymbol{r}\cdot\boldsymbol{F}(\boldsymbol{r}) = \{\text{gradient contribution}\} + \int\limits_0^1 \mathrm{d}\alpha \left[-b(1-\alpha)c\frac{\partial F_x}{\partial y}(\boldsymbol{r}_1) - c(1-\alpha)b\frac{\partial F_y}{\partial x}(\boldsymbol{r}_1)\right]$$
$$= \{\text{gradient contribution}\} - \frac{1}{2}bc\left(\frac{\partial F_x}{\partial y} + \frac{\partial F_y}{\partial x}\right)(\boldsymbol{r}_1)\,. \tag{3.2.42}$$

Their sum is the value of the closed-loop integral for the tiny triangle,

$$\int\limits_{\Delta} \mathrm{d}\boldsymbol{r}\cdot\boldsymbol{F} = \frac{1}{2}ac\left(\frac{\partial F_y}{\partial x} - \frac{\partial F_x}{\partial y}\right)(\boldsymbol{r}_1)$$
$$= \frac{1}{2}ac\boldsymbol{e}_z\cdot\big(\boldsymbol{\nabla}\times\boldsymbol{F}\big)(\boldsymbol{r}_1)$$
$$= \mathrm{d}\boldsymbol{S}\cdot\big(\boldsymbol{\nabla}\times\boldsymbol{F}\big)(\boldsymbol{r}_1)\,, \tag{3.2.43}$$

where $\mathrm{d}\boldsymbol{S} = \frac{1}{2}ac\boldsymbol{e}_z$ is the vectorial surface element for the triangle, a vector normal to the surface, with the right-hand rule applying to the sense of the loop, and the length equal to the area of the triangle.

There are two observations here. First, if we have a curl-free force field, $\boldsymbol{\nabla}\times\boldsymbol{F} = 0$ everywhere, the closed-loop integral for each tiny triangle is zero, and it follows by triangulation that the integral for any closed loops vanishes. In conjunction with (3.2.28), this implies that

$$V(\boldsymbol{r}) = V(\boldsymbol{r}_0) - \int_{\boldsymbol{r}_0}^{\boldsymbol{r}} \mathrm{d}\boldsymbol{r}' \cdot \boldsymbol{F}(\boldsymbol{r}') \tag{3.2.44}$$

expresses the potential energy as a line integral of the force whereby the choice of reference point $\boldsymbol{r}_0$ and the value that we assign to the potential energy at the reference point can be chosen to our liking. The path from $\boldsymbol{r}_0$ to $\boldsymbol{r}_1$ that we need for the integration in (3.2.44) is at our discretion as well, except that it should stay away from singularities of the force field if there are any.

Second, if we have a vector field that has a non-vanishing curl we can add up the contributions of all tiny triangles to establish the identity

$$\int_\Gamma \mathrm{d}\boldsymbol{r} \cdot \boldsymbol{F} = \int_S \mathrm{d}\boldsymbol{S} \cdot (\boldsymbol{\nabla} \times \boldsymbol{F}) \,, \tag{3.2.45}$$

known as *Stokes's** *Theorem*. It equates the line integral along the boundary of the surface S to the surface integral of the curl. The surface is thereby oriented in accordance with the right-handed rule:

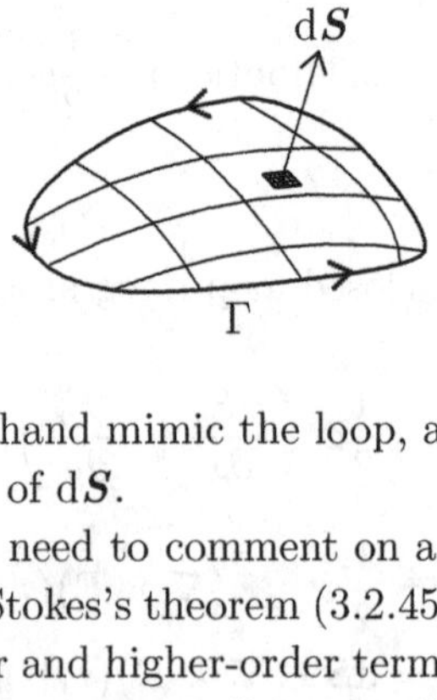

(3.2.46)

The fingers of your right hand mimic the loop, and the thumb of the right hand shows the direction of $\mathrm{d}\boldsymbol{S}$.

Before moving on, we need to comment on a technical detail in the argument that established Stokes's theorem (3.2.45). In (3.2.35) and (3.2.36), we neglected second-order and higher-order terms in the Taylor expansions and, as a consequence, the corresponding contributions to the one-triangle integral in (3.2.43) are not taken into account. This is a negligible correction for each tiny triangle, but we have very many triangles. Could it not be that the very many very small corrections add up to a sizeable contribution that modifies the right-hand side of Stokes's theorem?

No, this is not happening for the following reason. Suppose the whole surface, of area A, is broken up in N tiny triangles of about the same

*Sir George Gabriel STOKES (1819–1903)

area A/N. Then distances within one of the tiny triangles are of the order $\sqrt{A/N}$, which sets the length scale. Now, the second-order terms in (3.2.35) and (3.2.36) involve $(x-x_1)^2$ or $(y-y_1)^2$ or $(x-x_1)(y-y_1)$, each a squared length, and they contribute amounts proportional to a cubed length to (3.2.43). The single-triangle corrections are thus proportional to $\sqrt{A^3/N^3}$ and all N tiny triangles together contribute an amount proportional to $\sqrt{A^3/N}$ to the right-hand side in (3.2.45). For large enough N, that is: for a fine enough triangulation, this contribution is as small as we like, and vanishes in the $N \to \infty$ limit.

In summary, then, the condition

$$\boldsymbol{\nabla} \times \boldsymbol{F}(\boldsymbol{r}) = 0 \tag{3.2.47}$$

is both necessary and sufficient for a conservative force; we can find the potential energy as a line integral of the force, and we get the force as the negative gradient of the potential energy.

There is one caveat, however. In must be possible to attach a surface to the loop in Stokes's theorem — with the loop as the boundary of the surface. Therefore, the force must not be singular along a line, or a pipe, that goes through the loop. Finite, bounded volumes of singular regions in space do not matter because we can choose the surface such that it avoids them. In particular, isolated points of singularity are of no concern. If there are such infinite lines, or pipes, where the force field is undefined, it may or may not be possible to write the force as a negative gradient of a potential energy. This must be investigated case by case.

One example that illustrates this remark is provided by

$$\boldsymbol{F}(\boldsymbol{r}) = \frac{\kappa}{s}\boldsymbol{e}_\varphi \mathrel{\widehat{=}} \frac{\kappa}{x^2+y^2}\begin{pmatrix} -y \\ x \\ 0 \end{pmatrix} \quad \text{for} \quad s > 0 \quad \text{with} \quad \kappa = \text{constant}\,, \tag{3.2.48}$$

which has a singularity on the z axis. As indicated, cylindrical coordinates are convenient, for which the gradient is given in (1.2.11). We note that the curl of this force field vanishes away from the z axis,

$$\begin{aligned} \boldsymbol{\nabla} \times \boldsymbol{F}(\boldsymbol{r}) &= -\boldsymbol{e}_\varphi \times \boldsymbol{\nabla}\frac{\kappa}{s} + \frac{\kappa}{s}\boldsymbol{\nabla} \times \boldsymbol{e}_\varphi \\ &= \boldsymbol{e}_\varphi \times \frac{\kappa}{s^2}\boldsymbol{e}_s + \frac{\kappa}{s^2}\boldsymbol{e}_\varphi \times \underbrace{\frac{\partial}{\partial\varphi}\boldsymbol{e}_\varphi}_{=-\boldsymbol{e}_s} = 0 \quad \text{for} \quad s > 0\,. \end{aligned} \tag{3.2.49}$$

It follows that the line integral of the force (3.2.48) around a closed loop vanishes if the loop does not encircle the z axis. And Stokes's theorem says nothing about loops around the z axis. In fact, for a loop that goes around the z axis once in the counterclockwise sense (that is: following the direction of $\boldsymbol{e}_\varphi$), the line integral of the force (3.2.48) is $2\pi\kappa$; regard this integration as an exercise. Since there is no unique value for the line integral of the force field (3.2.48) between any two points off the z axis, this force is not the negative gradient of a potential energy.

3.2.3 *Extremal points of the potential energy: Maxima, minima, saddle points*

If we now have a conservative force $\boldsymbol{F}(\boldsymbol{r}) = -\boldsymbol{\nabla}V(\boldsymbol{r})$, indeed, the sum of the kinetic energy and the potential energy — that is: the energy — does not change in time, it is conserved,

$$\frac{\mathrm{d}}{\mathrm{d}t}E = 0 \quad \text{for} \quad E = \frac{m}{2}\left(\frac{\mathrm{d}\boldsymbol{r}}{\mathrm{d}t}\right)^2 + V(\boldsymbol{r}) \tag{3.2.50}$$

with $\boldsymbol{r} = \boldsymbol{r}(t)$ the position of the point mass at time t. For the given value of E, determined by the initial position $\boldsymbol{r}(0) = \boldsymbol{r}_0$ and velocity $\dot{\boldsymbol{r}}(0) = \boldsymbol{v}_0$,

$$E = \frac{m}{2}\boldsymbol{v}_0^2 + V(\boldsymbol{r}_0)\,, \tag{3.2.51}$$

the motion of the point mass is confined to those regions of space where $E \geq V(\boldsymbol{r})$ because the kinetic energy cannot be negative. This is, of course, fully analogous to the one-dimensional situation of Sections 3.1.2–3.1.4, only that in three dimensions there are no simple turning points, rather we have two-dimensional surfaces of constant potential energy — equipotential surfaces — as the boundaries of the accessible regions. The analog of the one-dimensional motion between two turning points is three-dimensional motion bounded by a closed surface (motion inside a sphere, for instance), and the analog of the one-dimensional motion with one or no turning point is three-dimensional motion bounded by an open surface (motion above the xy plane, for instance) or not bounded at all.

Intermediate is two-dimensional motion, such as motion confined to the xy plane, where we can visualize the potential energy $V(x, y)$ by a contour plot, composed of curved lines along which the potential energy is constant — equipotential lines — as exemplified by this sketch:

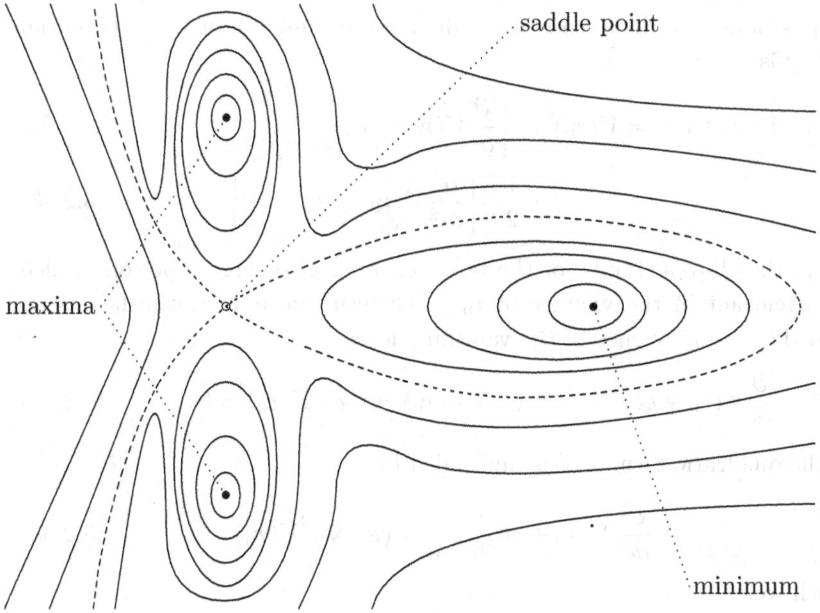

(3.2.52)

Here we have a potential minimum in the vicinity of two maxima (like a mountain lake near two peaks). On the way from one maximum to the other we have a saddle point (a mountain pass, if you like), where the force vanishes and the potential energy increases in both directions along one straight line (roughly that from maximum to maximum), but decreases in both directions along the orthogonal straight line. At a maximum, of course, the potential energy decreases in all directions, and it increases in all directions at a potential minimum. When passing through the saddle point on the way from maximum to maximum, the saddle point appears as the location of a potential-energy minimum, but when passing through the saddle point in the perpendicular direction, it appears as a potential energy maximum (your experience when crossing a mountain pass). The dashed equipotential line that separates the maxima from the minimum self-intersects at the saddle point.

3.2.4 *Potential energy in the vicinity of an extremal point*

In a manner that is fully analogous to what we did for one-dimensional motion, we can classify the points of vanishing force, $\boldsymbol{\nabla} V = 0$, as maxima, minima, or saddle points of the potential energy in three dimensions. Say the force is zero at $\boldsymbol{r} = \boldsymbol{r}_0$, then we have $\boldsymbol{F}(\boldsymbol{r}_0) = -\boldsymbol{\nabla}V(\boldsymbol{r}_0) = 0$, and for positions at a distance ϵ in the direction of unit vector $\boldsymbol{e}$ the potential energy is

$$V(\boldsymbol{r}_0+\epsilon\boldsymbol{e}) = V(\boldsymbol{r}_0) + \epsilon\left[\frac{\partial}{\partial\epsilon}V(\boldsymbol{r}_0+\epsilon\boldsymbol{e})\Big|_{\epsilon=0}\right] + \frac{1}{2}\epsilon^2\left[\frac{\partial^2}{\partial\epsilon^2}V(\boldsymbol{r}_0+\epsilon\boldsymbol{e})\Big|_{\epsilon=0}\right] + \cdots, \tag{3.2.53}$$

where the ellipsis stands for the terms of order ϵ^3 or higher powers, which are irrelevant in the vicinity of $\boldsymbol{r}_0$. The term linear in ϵ vanishes as an immediate consequence of the vanishing force,

$$\frac{\partial}{\partial\epsilon}V(\boldsymbol{r}_0+\epsilon\boldsymbol{e})\Big|_{\epsilon=0} = \boldsymbol{e}\cdot\boldsymbol{\nabla}V(\boldsymbol{r}_0) = -\boldsymbol{e}\cdot\boldsymbol{F}(\boldsymbol{r}_0) = 0\,. \tag{3.2.54}$$

In the quadratic term, $\frac{1}{2}\epsilon^2$ is multiplied by

$$\frac{\partial^2}{\partial\epsilon^2}V(\boldsymbol{r}_0+\epsilon\boldsymbol{e})\Big|_{\epsilon=0} = (\boldsymbol{e}\cdot\boldsymbol{\nabla})^2V(\boldsymbol{r}_0)\,, \tag{3.2.55}$$

which we read as

$$(\boldsymbol{e}\cdot\boldsymbol{\nabla})^2V(\boldsymbol{r}_0) = \begin{pmatrix} e_x & e_y & e_z \end{pmatrix} \begin{pmatrix} \frac{\partial^2}{\partial x^2}V & \frac{\partial}{\partial x}\frac{\partial}{\partial y}V & \frac{\partial}{\partial x}\frac{\partial}{\partial z}V \\ \frac{\partial}{\partial y}\frac{\partial}{\partial x}V & \frac{\partial^2}{\partial y^2}V & \frac{\partial}{\partial y}\frac{\partial}{\partial z}V \\ \frac{\partial}{\partial z}\frac{\partial}{\partial x}V & \frac{\partial}{\partial z}\frac{\partial}{\partial y}V & \frac{\partial^2}{\partial z^2}V \end{pmatrix}\Bigg|_{\boldsymbol{r}_0} \begin{pmatrix} e_x \\ e_y \\ e_z \end{pmatrix}. \tag{3.2.56}$$

This comes about upon using both versions of

$$\boldsymbol{e}\cdot\boldsymbol{\nabla} = \begin{pmatrix} e_x & e_y & e_z \end{pmatrix} \begin{pmatrix} \frac{\partial}{\partial x} \\ \frac{\partial}{\partial y} \\ \frac{\partial}{\partial z} \end{pmatrix} = \begin{pmatrix} \frac{\partial}{\partial x} & \frac{\partial}{\partial y} & \frac{\partial}{\partial z} \end{pmatrix} \begin{pmatrix} e_x \\ e_y \\ e_z \end{pmatrix} \tag{3.2.57}$$

once in $(\boldsymbol{e}\cdot\boldsymbol{\nabla})^2$. That is, we represent the vectors $\boldsymbol{e}$ and $\boldsymbol{\nabla}$ alternatively by columns or rows, and realize the dot product $\boldsymbol{e}\cdot\boldsymbol{\nabla}$ as a row-times-column product of matrix multiplication.

We can do this quite systematically if we regard $\boldsymbol{\nabla}$, $\boldsymbol{r}$, $\boldsymbol{F}$, and so forth, as vectors of column type and introduce their transposes $\boldsymbol{\nabla}^{\mathrm{T}}$, $\boldsymbol{r}^{\mathrm{T}}$, $\boldsymbol{F}^{\mathrm{T}}$, and so forth, as vectors of row type. The scalar product of two column-type vectors is then equal to the simple product of a row-type vector with a column-type vector,

$$\boldsymbol{a}\cdot\boldsymbol{b}=\boldsymbol{a}^{\mathrm{T}}\boldsymbol{b}=\boldsymbol{b}^{\mathrm{T}}\boldsymbol{a}\,. \tag{3.2.58}$$

Then

$$(\boldsymbol{e}\cdot\boldsymbol{\nabla})^2V=\boldsymbol{e}^{\mathrm{T}}\boldsymbol{\nabla}\boldsymbol{\nabla}^{\mathrm{T}}\boldsymbol{e}V=\boldsymbol{e}^{\mathrm{T}}\Bigl(\boldsymbol{\nabla}\boldsymbol{\nabla}^{\mathrm{T}}V\Bigr)\boldsymbol{e} \tag{3.2.59}$$

recognizes the *dyadic* $\boldsymbol{\nabla}\boldsymbol{\nabla}^{\mathrm{T}}V$ — the double gradient of the potential energy — that is represented by its 3×3 matrix of cartesian coordinates

$$\boldsymbol{\nabla}\boldsymbol{\nabla}^{\mathrm{T}}V\,\widehat{=}\begin{pmatrix}\dfrac{\partial}{\partial x}\\[1ex]\dfrac{\partial}{\partial y}\\[1ex]\dfrac{\partial}{\partial z}\end{pmatrix}\begin{pmatrix}\dfrac{\partial}{\partial x}&\dfrac{\partial}{\partial y}&\dfrac{\partial}{\partial z}\end{pmatrix}V\,, \tag{3.2.60}$$

where we meet the product of column-times-row type — the dyadic product of two vectors. So, $\boldsymbol{\nabla}\boldsymbol{\nabla}^{\mathrm{T}}V$ is a column vector on the left and a row vector on the right, if we wish to be pedantic about the row and column character.

It is good and important to keep this detail in mind, but the notation gets awkwardly overloaded if we are keeping track of the rows and columns explicitly. Instead, it is more convenient and more common to simply write

$$(\boldsymbol{e}\cdot\boldsymbol{\nabla})^2V=\boldsymbol{e}\cdot\bigl(\boldsymbol{\nabla}\boldsymbol{\nabla}V\bigr)\cdot\boldsymbol{e}\,, \tag{3.2.61}$$

and we just have to remember that the $\boldsymbol{e}$ on the left is understood as a row, the $\boldsymbol{e}$ on the right as a column, and $\boldsymbol{\nabla}\boldsymbol{\nabla}$ is the dyadic product of the gradient vector with itself — the dyadic square of the gradient, if you like — with column on the left and row on the right, if we need to be specific.

Another popular terminology calls dyadics "tensors", more precisely: tensors of second rank. And then vectors, such as $\boldsymbol{\nabla}V$, are tensors of first rank, and the potential energy itself would be a tensor of zeroth rank. And guess what? Yes, correct, there are also tensors of third, fourth, ... rank. All of them are having their show in expressions like

$$V(\boldsymbol{r}+\epsilon\boldsymbol{e})=\sum_{n=0}^{\infty}\frac{\epsilon^n}{n!}(\boldsymbol{e}\cdot\boldsymbol{\nabla})^nV(\boldsymbol{r})=\mathrm{e}^{\epsilon\boldsymbol{e}\cdot\boldsymbol{\nabla}}V(\boldsymbol{r})\,, \tag{3.2.62}$$

where you recognize the Taylor expansion in its compact exponential form. We shall, however, not have much use of these tensors of higher rank and stick to the terminology of calling $\boldsymbol{\nabla}\boldsymbol{\nabla}V$ a dyadic, $\boldsymbol{\nabla}V$ a vector, and V

itself a scalar — or, more precisely, a dyadic field, a vector field, and a scalar field, respectively.

In passing, we note that dyadics have appeared earlier without receiving attention then. For example, if we combine (1.1.3) and (1.1.7) into one statement,

$$\begin{aligned} \boldsymbol{r} &= \boldsymbol{e}_x\boldsymbol{e}_x \cdot \boldsymbol{r} + \boldsymbol{e}_y\boldsymbol{e}_y \cdot \boldsymbol{r} + \boldsymbol{e}_z\boldsymbol{e}_z \cdot \boldsymbol{r} \\ &= \left(\boldsymbol{e}_x\boldsymbol{e}_x + \boldsymbol{e}_y\boldsymbol{e}_y + \boldsymbol{e}_z\boldsymbol{e}_z\right) \cdot \boldsymbol{r} = \mathbf{1} \cdot \boldsymbol{r}, \end{aligned} \tag{3.2.63}$$

we meet the decomposition of the *unit dyadic* **1** into the sum of the dyadic squares of the cartesian unit vectors,

$$\mathbf{1} = \boldsymbol{e}_x\boldsymbol{e}_x + \boldsymbol{e}_y\boldsymbol{e}_y + \boldsymbol{e}_z\boldsymbol{e}_z\,. \tag{3.2.64}$$

In terms of the 3×3 matrix representation,

$$\begin{pmatrix} 1\,0\,0 \\ 0\,1\,0 \\ 0\,0\,1 \end{pmatrix} = \begin{pmatrix} 1 \\ 0 \\ 0 \end{pmatrix} (1\,0\,0) + \begin{pmatrix} 0 \\ 1 \\ 0 \end{pmatrix} (0\,1\,0) + \begin{pmatrix} 0 \\ 0 \\ 1 \end{pmatrix} (0\,0\,1)\,, \tag{3.2.65}$$

this is quite an obvious matter. For more properties and examples of dyadics, see Exercises 64–74.

Returning now to the potential energy near a point of vanishing force,

$$V(\boldsymbol{r}_0 + \epsilon\boldsymbol{e}) = V(\boldsymbol{r}_0) + \frac{1}{2}\epsilon^2 \boldsymbol{e} \cdot \boldsymbol{\nabla\nabla} V(\boldsymbol{r}_0) \cdot \boldsymbol{e} \tag{3.2.66}$$

or, with $\boldsymbol{\epsilon} = \epsilon\boldsymbol{e}$,

$$V(\boldsymbol{r}_0 + \boldsymbol{\epsilon}) = V(\boldsymbol{r}_0) + \frac{1}{2}\boldsymbol{\epsilon} \cdot \boldsymbol{\nabla\nabla} V(\boldsymbol{r}_0) \cdot \boldsymbol{\epsilon}\,, \tag{3.2.67}$$

the properties of the $\boldsymbol{\nabla\nabla} V$ dyadic at $\boldsymbol{r} = \boldsymbol{r}_0$ distinguish potential energy maxima from minima and saddle points. In the case of a maximum, we have

$$\boldsymbol{\epsilon} \cdot \boldsymbol{\nabla\nabla} V(\boldsymbol{r}_0) \cdot \boldsymbol{\epsilon} < 0 \quad \text{for all } \boldsymbol{\epsilon} \neq 0 \tag{3.2.68}$$

because the potential energy must be less at all neighboring points. Likewise

$$\boldsymbol{\epsilon} \cdot \boldsymbol{\nabla\nabla} V(\boldsymbol{r}_0) \cdot \boldsymbol{\epsilon} > 0 \quad \text{for all } \boldsymbol{\epsilon} \neq 0 \tag{3.2.69}$$

characterizes a minimum. At a saddle point, we have

$$\begin{aligned} &\boldsymbol{\epsilon} \cdot \boldsymbol{\nabla\nabla} V(\boldsymbol{r}_0) \cdot \boldsymbol{\epsilon} > 0 \quad \text{for some } \boldsymbol{\epsilon} \\ \text{and} \quad &\boldsymbol{\epsilon} \cdot \boldsymbol{\nabla\nabla} V(\boldsymbol{r}_0) \cdot \boldsymbol{\epsilon} < 0 \quad \text{for some other } \boldsymbol{\epsilon}\,. \end{aligned} \tag{3.2.70}$$

Since the 3×3 matrix of cartesian coefficients that represents $\nabla\nabla V(\boldsymbol{r}_0)$ is symmetric — $\frac{\partial}{\partial x}\frac{\partial}{\partial y}V = \frac{\partial}{\partial y}\frac{\partial}{\partial x}V$, for example — we know that it has three real eigenvalues and different eigenvalues have orthogonal eigencolumns. Translated into a geometrical statement about $\nabla\nabla V(\boldsymbol{r}_0)$ this says that we can always find a local cartesian coordinate system, such that the 3×3 matrix of $\nabla\nabla V(\boldsymbol{r}_0)$ is diagonal,

$$\nabla\nabla V(\boldsymbol{r}_0)\cdot\boldsymbol{e}_j = \lambda_j\boldsymbol{e}_j\,, \tag{3.2.71}$$

where $\boldsymbol{e}_1$, $\boldsymbol{e}_2$, $\boldsymbol{e}_3$ are the unit vectors of the privileged local coordinate system, and λ_1, λ_2, λ_3 are the eigenvalues of $\nabla\nabla V(\boldsymbol{r}_0)$. Then

$$\nabla\nabla V(\boldsymbol{r}_0) = \sum_{j=1}^{3}\boldsymbol{e}_j\lambda_j\boldsymbol{e}_j\,, \tag{3.2.72}$$

where we recall that $\boldsymbol{e}_j\boldsymbol{e}_j$ is a dyadic product, with the left $\boldsymbol{e}_j$ of column type and the right $\boldsymbol{e}_j$ of row type. It follows from (3.2.68) and (3.2.69) that

$$\begin{aligned}&\text{all } \lambda_j > 0 \text{ at a potential energy minimum,}\\&\text{all } \lambda_j < 0 \text{ at a potential energy maximum,}\end{aligned} \tag{3.2.73}$$

and we have a saddle point if the eigenvalues do not have a common sign. These conditions are clearly the three-dimensional analogs of $V''(x_0) > 0$ and $V''(x_0) < 0$ for a minimum or maximum, respectively, in one dimension. Indeed, one writes

$$\nabla\nabla V(\boldsymbol{r}_0) > 0 \tag{3.2.74}$$

if $\boldsymbol{\epsilon}\cdot\nabla\nabla V(\boldsymbol{r}_0)\cdot\boldsymbol{\epsilon} > 0$ for all $\boldsymbol{\epsilon}\neq 0$ and then calls $\nabla\nabla V(\boldsymbol{r}_0)$ a positive dyadic; and likewise $\nabla\nabla V(\boldsymbol{r}_0) < 0$ for a negative dyadic.

The unit vectors of the privileged local coordinate system have a geometrical-physical significance. Consider a point mass with an energy that is slightly in excess of the potential energy at a minimum of $V(\boldsymbol{r})$, say, at $\boldsymbol{r}_0$. Then

$$E = \frac{m}{2}\left(\frac{\mathrm{d}\boldsymbol{r}}{\mathrm{d}t}\right)^2 + V(\boldsymbol{r}_0) + \frac{1}{2}(\boldsymbol{r}-\boldsymbol{r}_0)\cdot\nabla\nabla V(\boldsymbol{r}_0)\cdot(\boldsymbol{r}-\boldsymbol{r}_0) \tag{3.2.75}$$

applies because the motion will be confined to a small region around the location of the minimum at $\boldsymbol{r}_0$. The resulting equation of motion is

$$m\frac{\mathrm{d}^2}{\mathrm{d}t^2}\boldsymbol{r} = \boldsymbol{F} = -\boldsymbol{\nabla}\left[\frac{1}{2}(\boldsymbol{r}-\boldsymbol{r}_0)\cdot\boldsymbol{\nabla}\boldsymbol{\nabla}V(\boldsymbol{r}_0)\cdot(\boldsymbol{r}-\boldsymbol{r}_0)\right]$$
$$= -\boldsymbol{\nabla}\boldsymbol{\nabla}V(\boldsymbol{r}_0)\cdot(\boldsymbol{r}-\boldsymbol{r}_0)\,, \qquad (3.2.76)$$

where we now employ the privileged local coordinates for a parameterization of $\boldsymbol{r}(t)$,

$$\boldsymbol{r}(t) = \boldsymbol{r}_0 + x_1(t)\boldsymbol{e}_1 + x_2(t)\boldsymbol{e}_2 + x_3(t)\boldsymbol{e}_3\,, \qquad (3.2.77)$$

and so get

$$m(\ddot{x}_1\boldsymbol{e}_1 + \ddot{x}_2\boldsymbol{e}_2 + \ddot{x}_3\boldsymbol{e}_3) = -\lambda_1 x_1\boldsymbol{e}_1 - \lambda_2 x_2\boldsymbol{e}_2 - \lambda_3 x_3\boldsymbol{e}_3 \qquad (3.2.78)$$

or, since all λ_js are positive at the minimum,

$$\ddot{x}_j = -\omega_j^2 x_j \quad \text{with} \quad \omega_j^2 = \lambda_j/m \quad \text{for} \quad j = 1,2,3\,. \qquad (3.2.79)$$

Accordingly, the motion is composed of three independent harmonic oscillations along the axes of the privileged local coordinate system, with respective periods $T_j = 2\pi/\omega_j = 2\pi\sqrt{m/\lambda_j}$.

If the point mass is initially at rest near a maximum of the potential energy, you get quite analogously three independent inverted oscillators of the kind discussed in Section 3.1.4 as long as the point mass has not yet moved away too far. Near a saddle point, finally, we have harmonic oscillation for each direction with $\lambda_j > 0$ and the hyperbolic motion of an inverted oscillator for each direction with $\lambda_j < 0$.

It can also happen that one or more of the eigenvalues of $\boldsymbol{\nabla}\boldsymbol{\nabla}V(\boldsymbol{r}_0)$ vanish. In this situation, which is not typical but does occur occasionally, we must go beyond the second order of the Taylor expansion in (3.2.53) in order to decide whether there is a maximum, or a minimum, or a saddle point. Exercise 42 deals with such higher-order minima (for $\nu > 2$) in a one-dimensional situation.

3.2.5 *Example: Electrostatic potentials have no maxima or minima*

Here is an application of quite some importance for contemporary experiments with trapped ions, atoms with a net charge, usually the result of having one electron removed. A point mass m that carries electric charge q moves under the influence of the force

$$\boldsymbol{F}(\boldsymbol{r}) = q\boldsymbol{E}(\boldsymbol{r}) \tag{3.2.80}$$

in a static electric field $\boldsymbol{E}(\boldsymbol{r})$ that originates in distant charges. Maxwell's* equations tell us that

$$\boldsymbol{\nabla}\times\boldsymbol{E}(\boldsymbol{r}) = 0 \quad\text{and}\quad \boldsymbol{\nabla}\cdot\boldsymbol{E}(\boldsymbol{r}) = 0\,. \tag{3.2.81}$$

The first equation states that the curl of the electric field vanishes and thus implies that the force is conservative

$$\boldsymbol{F}(\boldsymbol{r}) = -q\boldsymbol{\nabla}\Phi(\boldsymbol{r}) = -\boldsymbol{\nabla}\bigl(q\Phi(\boldsymbol{r})\bigr)\,, \tag{3.2.82}$$

where $\Phi(\boldsymbol{r})$ is the electrostatic potential, and $q\Phi(\boldsymbol{r})$ is the potential energy of the charged point mass in the static electric field. The second equation states that the so-called *divergence* of the electrostatic field vanishes; it follows that

$$\boldsymbol{\nabla}\cdot\boldsymbol{\nabla}V(\boldsymbol{r}) = 0 \quad\text{for}\quad V(\boldsymbol{r}) = q\Phi(\boldsymbol{r})\,, \tag{3.2.83}$$

or

$$\boldsymbol{\nabla}^2V(\boldsymbol{r}) = 0 \tag{3.2.84}$$

with the *Laplacian*† *differential operator*

$$\boldsymbol{\nabla}^2 = \boldsymbol{\nabla}\cdot\boldsymbol{\nabla} = \left(\frac{\partial}{\partial x}\right)^2 + \left(\frac{\partial}{\partial y}\right)^2 + \left(\frac{\partial}{\partial z}\right)^2, \tag{3.2.85}$$

here expressed in cartesian coordinates. But $\boldsymbol{\nabla}^2V$ is simply the trace of the dyadic $\boldsymbol{\nabla}\boldsymbol{\nabla}V$, immediately confirmed by a look at the cartesian 3×3 matrix in (3.2.56) or (3.2.60). Since the trace of a dyadic (or of a square matrix) is equal to the sum of its eigenvalues, we have

$$\lambda_1 + \lambda_2 + \lambda_3 = 0 \tag{3.2.86}$$

for each point of vanishing force in the electrostatic field; actually, this is true everywhere, not just at the points with $\boldsymbol{E}(\boldsymbol{r}_0) = 0$. Now, at a minimum of the potential energy we have three positive eigenvalues, so that

$$\boldsymbol{\nabla}^2V(\boldsymbol{r}_0) > 0 \text{ at a minimum,} \tag{3.2.87}$$

*James Clerk MAXWELL (1831–1879) †Marquis de Pierre Simon LAPLACE (1749–1827) and likewise

$$\nabla^2 V(\boldsymbol{r}_0) < 0 \text{ at a maximum,} \tag{3.2.88}$$

and it follows that

$$\nabla^2 V(\boldsymbol{r}_0) = 0 \tag{3.2.89}$$

is only possible at a saddle point.

We conclude, therefore, that one cannot trap an ion in an electrostatic field, irrespective of how ingeniously the field is configured. The laws of physics — the combination of Maxwell's electrodynamics with Newton's mechanics — simply do not allow it. If we manage to put the ion at such a saddle point, the tiniest perturbation will set it into motion to where the potential energy $q\Phi(\boldsymbol{r})$ is lower. But there is a way out: It is possible to confine a charged particle to a small region of space by forces that result from a time-dependent electric field, an insight that eventually was rewarded with a Nobel Prize for the inventor of the so-called Paul* trap. More about this in Exercise 75.

If we have a collection of point charges at rest, then each individual charge finds itself in the joint electrostatic field of all other charges. It follows that none of the charges is at a minimum of the potential energy, at best they can all be at saddle points. Therefore, there is no stable configuration of point charges, which fact is known as *Earnshaw's*† *Theorem*.

Now, matter is composed of point-like charge carriers — electrons and atomic nuclei — and, in defiance of Earnshaw's theorem, matter *is* stable. Perhaps these charges are not at rest but in bound motion, somewhat similar to the sun and the planets? That is not possible either, because the charges would be in accelerated motion, and accelerated charges radiate electromagnetic waves, which carry energy away. The solution to this "stability-of-matter puzzle" involves the laws of quantum physics that apply on the atomic scale, where the above classical-mechanical considerations are not valid.

*Wolfgang PAUL (1913–1993) †Samuel EARNSHAW (1805–1888)

Chapter 4

Pair Forces

4.1 Reciprocal forces: Conservation of momentum

In a system composed of several, perhaps many, point masses, there will be forces that the constituents exert on each other. It could be that they are electrically charged or attracted by gravitational forces or simply linked by an elastic string. Whatever the physical mechanism at work, the forces between the point masses are made up of *pair forces.* For each pair of point masses — the jth mass and the kth mass, say — we have two forces: the force $\boldsymbol{F}_{jk}$ on the jth mass exerted by the kth mass, and the force $\boldsymbol{F}_{kj}$ that the jth mass exerts on the kth mass. In this example of a system of four point masses:

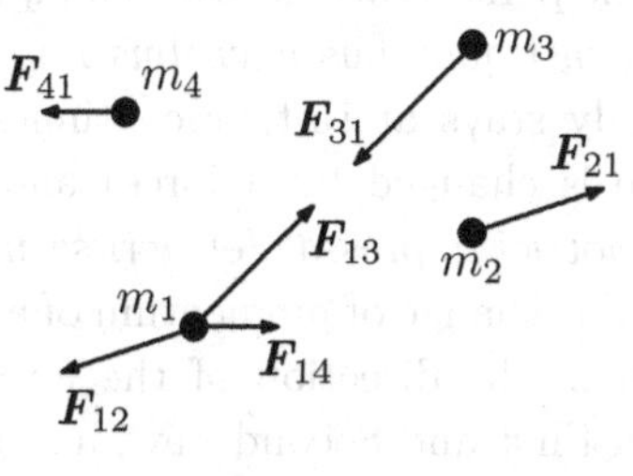

(4.1.1)

where we indicate all forces related to mass m_1, the net force $\boldsymbol{F}_1$ on this mass is the sum of the three force contributions from the other masses, $\boldsymbol{F}_1 = \boldsymbol{F}_{12} + \boldsymbol{F}_{13} + \boldsymbol{F}_{14}$. More generally, the force on the jth mass is the sum over k of all $\boldsymbol{F}_{jk}$s,

$$\boldsymbol{F}_j = \sum_{k(\neq j)} \boldsymbol{F}_{jk} \,. \tag{4.1.2}$$

We do not include a term with $k = j$ in the summation because the jth mass does not exert a force on itself.

This decomposition into pair forces is an insight that we owe to Newton. He recognized that there is a fundamental symmetry,

$$\boldsymbol{F}_{jk} = -\boldsymbol{F}_{kj}\,, \tag{4.1.3}$$

the forces that the jth and kth mass exert on each other are of equal strength and opposite direction, famously known as *Newton's Third Law*, the *law of reciprocal forces*: *Actioni contrariam semper et aequalem esse reactionem* (To every action there is always opposed an equal reaction). This is often paraphrased as "action equals reaction" — "force equals counterforce" is more to the point, however. The forces indicated in (4.1.1) take this into account. We should not miss, however, that in addition to the explicit requirement that the paired forces are balanced, $\boldsymbol{F}_{jk} + \boldsymbol{F}_{kj} = 0$, the law of reciprocal forces also states implicitly that the forces between constituents are pair forces in the first place.

Since (4.1.3) is a property of the pair (jk), it is understood that $j \neq k$. But if we like we can also apply it to $j = k$ and infer what is stated above: $\boldsymbol{F}_{jj} = 0$, no force of the jth mass on itself. We may, therefore, lift the restriction in (4.1.2) and include the term with $k = j$ in the sum whenever this is convenient.

Incidentally, there are, of course, also *Newton's First Law* and *Newton's Second Law*: *Corpus omne perseverare in statu suo quiescendi vel movendi uniformiter in directum, nisi quatenus a viribus impressis cogitur statum illum mutare* (Every body stays at rest or continues to move uniformly forward, unless its state is changed by a force) and *Mutationem motus proportionalem esse vi motrici impressae, et fieri secundum lineam rectam qua vis illa imprimitur* (The change of momentum of a body is proportional to the force acting, and in the direction of the force). But we are not using this terminology of First and Second Law, and prefer to refer to the fundamental statement of dynamics — in the present context: $\dfrac{\mathrm{d}}{\mathrm{d}t} m_j \boldsymbol{v}_j = \boldsymbol{F}_j$ for the jth point mass — simply as Newton's equation of motion. It is the joint content of the First Law and the Second Law.

An important consequence of the law of reciprocal forces is the conservation of the *total momentum*

$$\boldsymbol{P}_{\text{tot}} = \sum_j m_j \boldsymbol{v}_j = \sum_j m_j \frac{\mathrm{d}}{\mathrm{d}t} \boldsymbol{r}_j \tag{4.1.4}$$

of a closed system, where "closed" means that all relevant masses are included or, put differently, that there are no forces acting on the constituents of the system from outside — no *external* forces are present. Upon making use of the equation of motion for the jth mass and the decomposition (4.1.2) of $\boldsymbol{F}_j$ into the contributions from the other point masses, we get

$$\frac{\mathrm{d}}{\mathrm{d}t}\boldsymbol{P}_{\mathrm{tot}} = \sum_j m_j \frac{\mathrm{d}}{\mathrm{d}t}\boldsymbol{v}_j = \sum_j \boldsymbol{F}_j = \sum_{j\neq k} \boldsymbol{F}_{jk} \tag{4.1.5}$$

for the time derivative of $\boldsymbol{P}_{\mathrm{tot}}$, and since the pair forces are subject to (4.1.3), we have

$$\sum_{j\neq k} \boldsymbol{F}_{jk} = \sum_{k\neq j} \boldsymbol{F}_{kj} = \frac{1}{2}\sum_{j\neq k}\Bigl(\boldsymbol{F}_{jk} + \boldsymbol{F}_{kj}\Bigr) = 0\,, \tag{4.1.6}$$

where the first equality simply recognizes that the dummy indices of the summation can be interchanged without changing the value of the sum. The double sum in (4.1.5) vanishes because for each pair (jk) it contains both forces, $\boldsymbol{F}_{jk}$ and $\boldsymbol{F}_{kj}$. Indeed, the total momentum of a closed system, the sum of the momenta $m_j\boldsymbol{v}_j$ of the individual point masses, is conserved:

$$\frac{\mathrm{d}}{\mathrm{d}t}\boldsymbol{P}_{\mathrm{tot}} = 0\,, \qquad \boldsymbol{P}_{\mathrm{tot}}(t) = \boldsymbol{P}_{\mathrm{tot}}(t_0)\,, \tag{4.1.7}$$

with the initial time t_0 at which the positions and velocities of all masses are specified.

4.2 Conservative pair forces: Conservation of energy

If the pair forces are conservative, if they derive from potential energies, we have

$$\begin{aligned} \boldsymbol{F}_{jk} &= -\boldsymbol{\nabla}_j V_{(jk)}\bigl(\boldsymbol{r}_j, \boldsymbol{r}_k\bigr) \\ \text{and}\quad \boldsymbol{F}_{kj} &= -\boldsymbol{\nabla}_k V_{(jk)}\bigl(\boldsymbol{r}_j, \boldsymbol{r}_k\bigr) \end{aligned} \tag{4.2.1}$$

for the pair (jk) and the potential energy $V_{(jk)}\bigl(\boldsymbol{r}_j, \boldsymbol{r}_k\bigr)$ for this pair. In view of the law of reciprocal forces, $V_{(jk)}$ cannot be an arbitrary function of $\boldsymbol{r}_j$ and $\boldsymbol{r}_k$, rather it has to obey

$$(\boldsymbol{\nabla}_j + \boldsymbol{\nabla}_k)V_{(jk)}\bigl(\boldsymbol{r}_j, \boldsymbol{r}_k\bigr) = 0\,. \tag{4.2.2}$$

Here and in (4.2.1), $\boldsymbol{\nabla}_j$ is the gradient with respect to $\boldsymbol{r}_j$, and $\boldsymbol{\nabla}_k$ differentiates $\boldsymbol{r}_k$. The two position vectors are, of course, independent variables, which is compactly stated by

$$\boldsymbol{\nabla}_j \boldsymbol{r}_k = \delta_{jk}\mathbf{1}\,, \tag{4.2.3}$$

where the unit dyadic is multiplied by Kronecker's delta symbol.

We take the scalar product of (4.2.2) with an infinitesimal vector $\delta\boldsymbol{\epsilon}$ and recognize that

$$\begin{aligned} 0 &= (\delta\boldsymbol{\epsilon}\cdot\boldsymbol{\nabla}_j + \delta\boldsymbol{\epsilon}\cdot\boldsymbol{\nabla}_k)V_{(jk)}\big(\boldsymbol{r}_j,\boldsymbol{r}_k\big) \\ &= V_{(jk)}\big(\boldsymbol{r}_j+\delta\boldsymbol{\epsilon},\boldsymbol{r}_k+\delta\boldsymbol{\epsilon}\big) - V_{(jk)}\big(\boldsymbol{r}_j,\boldsymbol{r}_k\big)\,, \end{aligned} \tag{4.2.4}$$

a two-fold application of the defining property (1.2.2) of the gradient. Since finite displacements are made up of many infinitesimal ones, this states that the potential energy $V_{(jk)}\big(\boldsymbol{r}_j,\boldsymbol{r}_k\big)$ does not change if both point masses of the pair are displaced by the same vector $\boldsymbol{\epsilon}$, $\boldsymbol{r}_j \to \boldsymbol{r}_j+\boldsymbol{\epsilon}$ and $\boldsymbol{r}_k \to \boldsymbol{r}_k+\boldsymbol{\epsilon}$. If follows that $V_{(jk)}\big(\boldsymbol{r}_j,\boldsymbol{r}_k\big)$ is a function of the difference $\boldsymbol{r}_j-\boldsymbol{r}_k$, rather than depending on the two vectors separately. For, this difference does not change under the joint displacements whereas $2\boldsymbol{\epsilon}$ is added to the sum $\boldsymbol{r}_j+\boldsymbol{r}_k$, and we can think of $V_{(jk)}\big(\boldsymbol{r}_j,\boldsymbol{r}_k\big)$ as a function of $\boldsymbol{r}_j-\boldsymbol{r}_k$ and $\boldsymbol{r}_j+\boldsymbol{r}_k$, which are linearly independent.

Therefore, conservative pair forces have potential energies that are functions of the relative positions $\boldsymbol{r}_j-\boldsymbol{r}_k$ of the two masses of the pair, not of their absolute positions relative to a reference point, such as the origin of the coordinate system. Upon writing $V_{(jk)}\big(\boldsymbol{r}_j-\boldsymbol{r}_k\big)$ for the potential energy of the pair (jk), the paired forces are

$$\boldsymbol{F}_{jk} = -\boldsymbol{\nabla}V_{(jk)}(\boldsymbol{r})\Big|_{\boldsymbol{r}\,=\,\boldsymbol{r}_j\,-\,\boldsymbol{r}_k} = -\boldsymbol{F}_{kj}\,, \tag{4.2.5}$$

where for each pair we agree on an order in the labeling of the two point masses. The convention $j<k$ is convenient for this purpose, and we shall adopt it.

In the *total energy*

$$E_{\text{tot}} = \sum_j \frac{1}{2}m_j v_j^2 + \sum_{j<k} V_{(jk)}\big(\boldsymbol{r}_j-\boldsymbol{r}_k\big)\,, \tag{4.2.6}$$

we are summing the kinetic energies of all masses and the potential energies of all pairs, counting each pair only once. We differentiate E_{tot} with respect to time t and find

$$\begin{aligned}\frac{\mathrm{d}}{\mathrm{d}t}E_{\mathrm{tot}} &= \sum_j m_j \boldsymbol{v}_j \cdot \frac{\mathrm{d}}{\mathrm{d}t}\boldsymbol{v}_j + \sum_{j<k} \boldsymbol{\nabla} V_{(jk)}(\boldsymbol{r})\Big|_{\boldsymbol{r}\,=\,\boldsymbol{r}_j\,-\,\boldsymbol{r}_k} \cdot (\boldsymbol{v}_j - \boldsymbol{v}_k) \\ &= \sum_j \boldsymbol{v}_j \cdot \boldsymbol{F}_j - \sum_{j<k} \boldsymbol{F}_{jk} \cdot (\boldsymbol{v}_j - \boldsymbol{v}_k) \end{aligned} \tag{4.2.7}$$

after making use of (4.2.5) and the equation of motion for the jth mass. Now we observe that the double summation can be written in a more symmetric way,

$$\begin{aligned}\sum_{j<k} \boldsymbol{F}_{jk} \cdot (\boldsymbol{v}_j - \boldsymbol{v}_k) &= \sum_j \left(\sum_{k(>j)} \boldsymbol{F}_{jk} \cdot \boldsymbol{v}_j + \sum_{k(>j)} \boldsymbol{F}_{kj} \cdot \boldsymbol{v}_k \right) \\ &= \sum_j \sum_{k(>j)} \boldsymbol{F}_{jk} \cdot \boldsymbol{v}_j + \sum_k \sum_{j(>k)} \boldsymbol{F}_{jk} \cdot \boldsymbol{v}_j \\ &= \sum_j \sum_{k(\neq j)} \boldsymbol{F}_{jk} \cdot \boldsymbol{v}_j = \sum_j \boldsymbol{F}_j \cdot \boldsymbol{v}_j \,. \end{aligned} \tag{4.2.8}$$

Therefore, the two terms on the right-hand side of (4.2.7) compensate for each other exactly, and

$$\frac{\mathrm{d}}{\mathrm{d}t}E_{\mathrm{tot}} = 0\,, \qquad E_{\mathrm{tot}}(t) = E_{\mathrm{tot}}(t_0) \tag{4.2.9}$$

follows. The total energy of a closed system with conservative pair forces is conserved, it does not change in time.

4.3 Line-of-sight forces: Conservation of angular momentum

Another situation is that of forces that are along the line-of-sight between the point masses of the pair,

$$\boldsymbol{F}_{jk} \parallel \boldsymbol{r}_j - \boldsymbol{r}_k \,. \tag{4.3.1}$$

In (4.1.1), this is the case for the pairs (12) and (13), with repulsive forces for (12) and attractive forces for (13), whereas the pair (14) does not have line-of-sight forces.

If all pairs have line-of-sight forces, the *total angular momentum*

$$\boldsymbol{L}_{\mathrm{tot}} = \sum_j \boldsymbol{r}_j \times m_j \boldsymbol{v}_j \,, \tag{4.3.2}$$

the sum of the angular momenta of the individual masses, is conserved. We first note that the time derivative of the angular momentum of the jth mass is the torque $\boldsymbol{\tau}_j = \boldsymbol{r}_j \times \boldsymbol{F}_j$ that results from the force $\boldsymbol{F}_j$ acting on the jth mass,

$$\frac{\mathrm{d}}{\mathrm{d}t}(\boldsymbol{r}_j \times m_j\boldsymbol{v}_j) = \underbrace{\frac{\mathrm{d}\boldsymbol{r}_j}{\mathrm{d}t}}_{=\boldsymbol{v}_j} \times m_j\boldsymbol{v}_j + \boldsymbol{r}_j \times \underbrace{\frac{\mathrm{d}}{\mathrm{d}t}(m_j\boldsymbol{v}_j)}_{=\boldsymbol{F}_j} = \boldsymbol{r}_j \times \boldsymbol{F}_j = \boldsymbol{\tau}_j\,. \tag{4.3.3}$$

For the time derivative of $\boldsymbol{L}_{\mathrm{tot}}$ this then gives

$$\frac{\mathrm{d}}{\mathrm{d}t}\boldsymbol{L}_{\mathrm{tot}} = \sum_j \boldsymbol{\tau}_j = \sum_j \boldsymbol{r}_j \times \boldsymbol{F}_j = \sum_{j\neq k} \boldsymbol{r}_j \times \boldsymbol{F}_{jk} = \sum_{j\neq k} \boldsymbol{r}_k \times \boldsymbol{F}_{kj} \tag{4.3.4}$$

or, if we take half the sum of the two latter expressions and invoke the law of reciprocal forces,

$$\begin{aligned}\frac{\mathrm{d}}{\mathrm{d}t}\boldsymbol{L}_{\mathrm{tot}} &= \frac{1}{2}\sum_{j\neq k}\Big(\boldsymbol{r}_j \times \boldsymbol{F}_{jk} + \boldsymbol{r}_k \times \boldsymbol{F}_{kj}\Big)\\ &= \frac{1}{2}\sum_{j\neq k}(\boldsymbol{r}_j - \boldsymbol{r}_k) \times \boldsymbol{F}_{jk} = 0\,,\end{aligned} \tag{4.3.5}$$

because $(\boldsymbol{r}_j - \boldsymbol{r}_k) \times \boldsymbol{F}_{jk} = 0$ for each pair for which we have line-of-sight forces with the defining property (4.3.1) as, by assumption, is the case for all pairs. Accordingly,

$$\frac{\mathrm{d}}{\mathrm{d}t}\boldsymbol{L}_{\mathrm{tot}} = 0\,, \qquad \boldsymbol{L}_{\mathrm{tot}}(t) = \boldsymbol{L}_{\mathrm{tot}}(t_0) \tag{4.3.6}$$

for closed systems with pair forces of the line-of-sight kind. Indeed, the total angular momentum is conserved.

4.4 Conservative line-of-sight forces

The combination of (4.2.5) and (4.3.1) requires

$$\boldsymbol{\nabla}V_{(jk)}(\boldsymbol{r}) \parallel \boldsymbol{r} \tag{4.4.1}$$

if the pair forces are conservative and along the line-of-sight. We take the scalar product of $\boldsymbol{\nabla}V_{(jk)}(\boldsymbol{r})$ and $\boldsymbol{\delta\phi}\times\boldsymbol{r}$, where $\boldsymbol{\delta\phi}$ describes an infinitesimal rotation (see Section 1.1.7), and recognize that

$$\begin{aligned} 0 &= (\delta\boldsymbol{\phi} \times \boldsymbol{r}) \cdot \boldsymbol{\nabla} V_{(jk)}(\boldsymbol{r}) \\ &= V_{(jk)}(\boldsymbol{r} + \delta\boldsymbol{\phi} \times \boldsymbol{r}) - V_{(jk)}(\boldsymbol{r}) . \end{aligned} \tag{4.4.2}$$

This states that the potential energy $V_{(jk)}(\boldsymbol{r})$ does not change if the vector $\boldsymbol{r}$ is rotated. Accordingly, $V_{(jk)}(\boldsymbol{r})$ must be a function of the length $r = |\boldsymbol{r}|$ of $\boldsymbol{r}$ only and cannot depend on the direction of $\boldsymbol{r}$ and, therefore, the potential energy of the pair (jk) is a function of the distance $|\boldsymbol{r}_j - \boldsymbol{r}_k|$ between the jth and kth masses if we have conservative line-of-sight forces between them.

In summary, then, for potential energies $V_{(jk)}(|\boldsymbol{r}_j - \boldsymbol{r}_k|)$ for the pairs of point masses, the total energy

$$E_{\text{tot}} = \sum_j \frac{1}{2} m_j \boldsymbol{v}_j^2 + \sum_{j<k} V_{(jk)}(|\boldsymbol{r}_j - \boldsymbol{r}_k|) \tag{4.4.3}$$

is conserved, and so are the total momentum (4.1.4) and the total angular momentum (4.3.2). This situation is actually quite typical. For example, if the point masses carry electric charges and the only relevant forces are the electrostatic forces that the charges exert on each other, we have

$$E_{\text{tot}} = \sum_j \frac{1}{2} m_j \boldsymbol{v}_j^2 + \frac{1}{2} \sum_{j \neq k} \frac{q_j q_k}{|\boldsymbol{r}_j - \boldsymbol{r}_k|} \tag{4.4.4}$$

for the total energy, where q_j is the electric charge of the jth mass (we are using cgs units for the q_js). We have here replaced the constraint $j < k$ in the double summation by $j \neq k$, and the factor of $\frac{1}{2}$ ensures that the potential energy for each pair is counted only once.

4.5 Additional external forces

4.5.1 *Transfer of momentum, energy, and angular momentum*

In addition to the internal pair forces between the constituents of the system, there can be external forces. For instance, we could have an extended object consisting of a large number of point masses exposed to the gravitational pull of the earth, in which case we usually do not regard the earth as part of the system. The total force on the jth point mass is then

$$\boldsymbol{F}_j = \underbrace{\sum_{k(\neq j)} \boldsymbol{F}_{jk}}_{\substack{\text{internal}\\ \text{force}}} + \underbrace{\boldsymbol{F}_j^{(\text{ext})}}_{\substack{\text{external}\\ \text{force}}} \tag{4.5.1}$$

and the law of reciprocal forces (4.1.3) continues to apply to the internal pair forces.

The total momentum now changes if the net external force is nonzero,

$$\frac{\mathrm{d}}{\mathrm{d}t}\boldsymbol{P}_{\text{tot}} = \sum_j \boldsymbol{F}_j = \underbrace{\sum_{j\neq k} \boldsymbol{F}_{jk}}_{=0} + \sum_j \boldsymbol{F}_j^{(\text{ext})} = \boldsymbol{F}^{(\text{ext})}\,. \tag{4.5.2}$$

Likewise, if the internal forces are conservative, we have an additional term in (4.2.7),

$$\frac{\mathrm{d}}{\mathrm{d}t}E_{\text{tot}} = \sum_j \boldsymbol{v}_j \cdot \boldsymbol{F}_j^{(\text{ext})} = P^{(\text{ext})}\,, \tag{4.5.3}$$

where $P^{(\text{ext})}$ is the total power (work per unit time) associated with the external forces. The energy of the system increases if $P^{(\text{ext})} > 0$ but decreases when $P^{(\text{ext})} < 0$. There can very well be an increase at one time and a decrease at another.

Thus, external forces transfer momentum and energy to the system in accordance with (4.5.2) and (4.5.3), respectively. They also transfer angular momentum. For internal line-of-sight forces, we have

$$\frac{\mathrm{d}}{\mathrm{d}t}\boldsymbol{L}_{\text{tot}} = \sum_j \boldsymbol{r}_j \times \boldsymbol{F}_j^{(\text{ext})} = \boldsymbol{\tau}^{(\text{ext})}\,. \tag{4.5.4}$$

This sum over the torques $\boldsymbol{\tau}_j^{(\text{ext})} = \boldsymbol{r}_j \times \boldsymbol{F}_j^{(\text{ext})}$ that the external forces exert on the individual point masses is, of course, just the total external torque that gives rise to a time-dependence of the total angular momentum.

4.5.2 *Center-of-mass motion*

For a given system composed of several point masses, there is the total mass

$$M = \sum_j m_j \tag{4.5.5}$$

and the mass-weighted average position

$$\boldsymbol{R} = \frac{1}{M}\sum_j m_j \boldsymbol{r}_j\,, \tag{4.5.6}$$

also known as the *center-of-mass position*. Its time derivative

$$\boldsymbol{V} = \frac{\mathrm{d}}{\mathrm{d}t}\boldsymbol{R} = \frac{1}{M}\sum_j m_j \boldsymbol{v}_j \tag{4.5.7}$$

is the *center-of-mass velocity*, the mass-weighted average of the velocities of the constituent point masses. We recognize that the total momentum of (4.1.4) is the product of the total mass and the center-of-mass velocity,

$$\boldsymbol{P}_{\mathrm{tot}} = M\,\boldsymbol{V}\,. \tag{4.5.8}$$

Then (4.5.2) has the appearance of an equation of motion for a fictitious point mass M located at $\boldsymbol{R}$, on which the total external force is acting,

$$M\frac{\mathrm{d}}{\mathrm{d}t}\boldsymbol{V} = \boldsymbol{F}^{(\mathrm{ext})}\,. \tag{4.5.9}$$

Upon associating the angular momentum

$$\boldsymbol{L}_{\mathrm{CM}} = \boldsymbol{R}\times M\,\boldsymbol{V} \tag{4.5.10}$$

with the center-of-mass motion, we have

$$\frac{\mathrm{d}}{\mathrm{d}t}\boldsymbol{L}_{\mathrm{CM}} = \boldsymbol{R}\times\boldsymbol{F}^{(\mathrm{ext})}\,. \tag{4.5.11}$$

This, too, looks as if the force $\boldsymbol{F}^{(\mathrm{ext})}$ were acting on a point mass at $\boldsymbol{R}$. Further, we can assign the kinetic energy $\frac{1}{2}M\,\boldsymbol{V}^2$ to the center-of-mass motion and find

$$\frac{\mathrm{d}}{\mathrm{d}t}\left(\frac{1}{2}M\,\boldsymbol{V}^2\right) = \boldsymbol{V}\cdot\boldsymbol{F}^{(\mathrm{ext})} \tag{4.5.12}$$

for its time derivative, where the right-hand side is the power for a force on a point mass that is moving with the center-of-mass velocity $\boldsymbol{V}$.

In summary, the motion of the system as a whole, in its response to external forces, is *as if* the total mass were located at the center-of-mass position and responding to the external forces like a point mass that is accelerated by the total external force. The "as if" nature of the center-of-mass deserves emphasis: The center-of-mass position is not the location of a material object; it is quite possible, and often the case, that none of the constituent masses is at the center-of-mass position.

The observations in (4.5.9), (4.5.11), and (4.5.12) are some of the reasons why we can treat extended objects as point-like masses if their internal structure and internal dynamics are not relevant. There is also a certain separation of the center-of-mass dynamics and the internal dynamics that we will discuss now.

With the center-of-mass position $\boldsymbol{R}$ available as a natural reference, we are invited to refer the positions of the physical point masses that make up the system to $\boldsymbol{R}$,

$$\boldsymbol{r}_j = \boldsymbol{R} + \boldsymbol{r}'_j\,. \tag{4.5.13}$$

The *internal positions* $\boldsymbol{r}'_j$ are linearly dependent because the definition of $\boldsymbol{R}$ in (4.5.6) implies

$$\sum_j m_j \boldsymbol{r}'_j = 0\,. \tag{4.5.14}$$

As an immediate consequence, the *internal velocities* $\boldsymbol{v}'_j = \dot{\boldsymbol{r}}'_j$ are linearly dependent as well,

$$\sum_j m_j \boldsymbol{v}'_j = 0\,. \tag{4.5.15}$$

We express the total energy of (4.2.6) in terms of the center-of-mass velocity, the internal velocities, and the internal positions,

$$\begin{aligned} E_{\mathrm{tot}} &= \sum_j \frac{1}{2} m_j \bigl(\boldsymbol{V} + \boldsymbol{v}'_j\bigr)^2 + \sum_{j<k} V_{(jk)}(\boldsymbol{r}'_j - \boldsymbol{r}'_k) \\ &= \frac{1}{2}\sum_j m_j\, \boldsymbol{V}^2 + \underbrace{\boldsymbol{V}\cdot\sum_j m_j \boldsymbol{v}'_j}_{=0} + \sum_j \frac{1}{2} m_j \boldsymbol{v}'^2_j + \sum_{j<k} V_{(jk)}(\boldsymbol{r}'_j - \boldsymbol{r}'_k) \\ &= \frac{1}{2} M\,\boldsymbol{V}^2 + \underbrace{\sum_j \frac{1}{2} m_j \boldsymbol{v}'^2_j + \sum_{j<k} V_{(jk)}(\boldsymbol{r}'_j - \boldsymbol{r}'_k)}_{\text{internal energy}}\,, \end{aligned} \tag{4.5.16}$$

where (4.5.5) and (4.5.15) enter at the intermediate stage. The final expression states that the total energy is the sum of the internal energy and the kinetic energy of the center-of-mass that is differentiated in (4.5.12).

Quite the same splitting occurs in the total angular momentum of (4.3.2),

$$\begin{aligned}
\boldsymbol{L}_{\text{tot}} &= \sum_j (\boldsymbol{R} + \boldsymbol{r}'_j) \times m_j(\boldsymbol{V} + \boldsymbol{v}'_j) \\
&= \underbrace{\boldsymbol{R} \times \sum_j m_j \boldsymbol{V}}_{= \boldsymbol{L}_{\text{CM}}} + \underbrace{\sum_j m_j \boldsymbol{r}'_j \times \boldsymbol{V} + \boldsymbol{R} \times \sum_j m_j \boldsymbol{v}'_j}_{=0} + \sum_j \boldsymbol{r}'_j \times m_j \boldsymbol{v}'_j \\
&= \boldsymbol{L}_{\text{CM}} + \underbrace{\sum_j \boldsymbol{r}'_j \times m_j \boldsymbol{v}'_j}_{\text{internal angular momentum}} . \qquad (4.5.17)
\end{aligned}$$

Again, we have a sum of a center-of-mass term and a term for the internal dynamics. And the same is true for the total momentum (4.5.8) which has only a center-of-mass term and no contribution from the internal momenta.

Suggestive as it may be, the splitting of $\boldsymbol{P}_{\text{tot}}$, E_{tot}, and $\boldsymbol{L}_{\text{tot}}$ into a center-of-mass part and an internal part does not imply that there is a complete dynamical separation of the center-of-mass motion and the internal motion. Such a separation would require that the total force $\boldsymbol{F}^{(\text{ext})}$ is a function of $\boldsymbol{R}$ and $\boldsymbol{V}$ only and does not depend on the internal positions and velocities. If this is the case, we can solve (4.5.9) without reference to the internal dynamics. Unless $\boldsymbol{F}^{(\text{ext})} = 0$, the center-of-mass position follows an accelerated motion and the internal position vectors $\boldsymbol{r}'_j$ refer to this accelerated reference position; as a consequence, the differential equations obeyed by the $\boldsymbol{r}'_j(t)$s will then contain so-called inertial forces that result from the accelerated center-of-mass motion and add to the physical pair forces. We will return to this matter of an accelerated frame of reference in Chapter 12.

4.5.3 *Conservative external forces*

If the external forces $\boldsymbol{F}_j^{(\text{ext})}$ are conservative forces, so that they are negative gradients of an external potential energy $V^{(\text{ext})}(\boldsymbol{r}_1, \boldsymbol{r}_2, \boldsymbol{r}_3, \dots)$,

$$\boldsymbol{F}_j^{(\text{ext})} = -\boldsymbol{\nabla}_j V^{(\text{ext})}(\boldsymbol{r}_1, \boldsymbol{r}_2, \boldsymbol{r}_3, \dots) , \qquad (4.5.18)$$

then the power $P^{(\text{ext})}$ in (4.5.3) is the negative time derivative of $V^{(\text{ext})}$,

$$P^{(\text{ext})} = -\sum_j \boldsymbol{v}_j \cdot \boldsymbol{\nabla}_j V^{(\text{ext})} = -\frac{\mathrm{d}}{\mathrm{d}t} V^{(\text{ext})} . \qquad (4.5.19)$$

It follows that the sum $E_{\text{tot}} + V^{(\text{ext})}$ is constant in time,

$$\frac{\mathrm{d}}{\mathrm{d}t}\Bigl(E_{\text{tot}} + V^{(\text{ext})}\Bigr) = 0\,. \tag{4.5.20}$$

In this sum we have the kinetic energy of all point masses and the potential energy of their pair forces — these make up E_{tot} — as well as the potential energy associated with the external forces. In this sense, then, the time-independent sum

$$E_{\text{all}} = E_{\text{tot}} + V^{(\text{ext})} \tag{4.5.21}$$

is "all of the energy," and one may be tempted to call it the "total energy," but we have earlier given this name to E_{tot} alone.

In (4.5.18), we wrote the external potential energy as a general function of all positions. The situation is usually somewhat simpler; almost always one has a sum of single-particle potential energies,

$$V^{(\text{ext})}(\boldsymbol{r}_1, \boldsymbol{r}_2, \boldsymbol{r}_3, \ldots) = \sum_j V_j^{(\text{ext})}(\boldsymbol{r}_j)\,, \tag{4.5.22}$$

with one term each for the point masses.

A typical scenario is familiar from school physics: All point masses are subject to the gravitational pull of the earth, which gives all masses the same constant acceleration $\boldsymbol{g}$,

$$\boldsymbol{F}_j^{(\text{ext})} = m_j\boldsymbol{g} = -\boldsymbol{\nabla}_j(-m_j\boldsymbol{g}\cdot\boldsymbol{r}_j)\,. \tag{4.5.23}$$

Then we have

$$V^{(\text{ext})} = -\sum_j m_j\boldsymbol{g}\cdot\boldsymbol{r}_j = -M\boldsymbol{g}\cdot\boldsymbol{R} \tag{4.5.24}$$

for the external potential energy and

$$E_{\text{all}} = E_{\text{tot}} - M\boldsymbol{g}\cdot\boldsymbol{R} \tag{4.5.25}$$

for all of the energy, where the additional term refers solely to the center-of-mass position, not to the individual positions of the constituent point masses. This is a valid description if the distances between the constituent masses are so small that there is no noticeable change in the gravitational acceleration of the earth from one point mass to another. For example, a fist-size stone is small enough that this is the case. By contrast, in a system of satellites in orbit around the earth, the distances can be so large that the direction of the gravitational acceleration differs markedly from satellite to satellite.

In terms of the center-of-mass position $\boldsymbol{R}$ and the internal positions $\boldsymbol{r}'_j$, (4.5.22) reads

$$V^{(\text{ext})} = \sum_j V_j^{(\text{ext})}(\boldsymbol{R} + \boldsymbol{r}'_j)\,. \tag{4.5.26}$$

When we add the external potential energy to E_{tot} of (4.5.16),

$$E_{\text{all}} = \frac{1}{2}M\,\boldsymbol{V}^2 + \sum_j \frac{1}{2}m_j \boldsymbol{v}'^2_j + \sum_{j<k} V_{(jk)}(\boldsymbol{r}'_j - \boldsymbol{r}'_k) + \sum_j V_j^{(\text{ext})}(\boldsymbol{R} + \boldsymbol{r}'_j)\,, \tag{4.5.27}$$

it becomes obvious that the external forces can lead to a coupling between the center-of-mass motion and the internal motion, and then the two motions are not evolving independently, as mentioned above after (4.5.17).

An important exception is the situation of (4.5.25), where the external potential energy of (4.5.24) depends on the center-of-mass position only. As a consequence, the center-of-mass motion of a small object that is falling towards the surface of the earth is that of a falling point mass, and is not affected by the internal motion of its constituent point masses.

Chapter 5

Two-Body Systems

5.1 Center-of-mass motion and relative motion, reduced mass

In particular, there is the important special case of a closed system consisting of just two point masses with conservative line-of-sight forces between them. We have then the two-mass versions of (4.5.5)–(4.5.6) and (4.5.13),

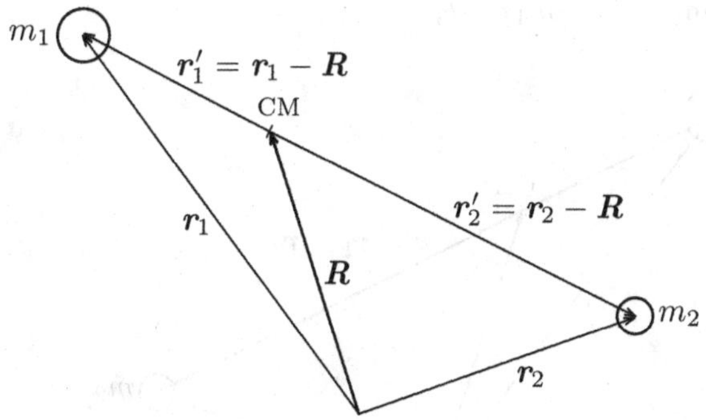

$$M = m_1 + m_2\,, \quad \boldsymbol{R} = \frac{m_1\boldsymbol{r}_1 + m_2\boldsymbol{r}_2}{M}\,, \quad \boldsymbol{V} = \frac{\mathrm{d}\boldsymbol{R}}{\mathrm{d}t} = \frac{m_1\boldsymbol{v}_1 + m_2\boldsymbol{v}_2}{M}\,,$$
$$\boldsymbol{r}_1' = \boldsymbol{r}_1 - \boldsymbol{R} = \frac{m_2}{M}(\boldsymbol{r}_1 - \boldsymbol{r}_2)\,, \quad \boldsymbol{v}_1' = \frac{m_2}{M}(\boldsymbol{v}_1 - \boldsymbol{v}_2)\,,$$
$$\boldsymbol{r}_2' = \boldsymbol{r}_2 - \boldsymbol{R} = \frac{m_1}{M}(\boldsymbol{r}_2 - \boldsymbol{r}_1)\,, \quad \boldsymbol{v}_2' = \frac{m_1}{M}(\boldsymbol{v}_2 - \boldsymbol{v}_1)\,. \tag{5.1.1}$$

Since there is only one pair now, the double sums in (4.4.3) and (4.5.16) reduce to the single $(jk) = (12)$ term and we drop the subscript $_{(12)}$ on the potential energy. Accordingly, we have

$$\begin{aligned} E_{\text{tot}} &= \frac{1}{2}m_1\boldsymbol{v}_1^2 + \frac{1}{2}m_2\boldsymbol{v}_2^2 + V\big(|\boldsymbol{r}_1 - \boldsymbol{r}_2|\big) \\ &= \frac{1}{2}M\boldsymbol{V}^2 + \frac{1}{2}m_1\boldsymbol{v}_1'^2 + \frac{1}{2}m_2\boldsymbol{v}_2'^2 + V\big(|\boldsymbol{r}_1' - \boldsymbol{r}_2'|\big) \end{aligned} \tag{5.1.2}$$

for the total energy.

The linear dependence of the internal positions and velocities, as stated in (4.5.13) and (4.5.14), is immediate in (5.1.1): $m_1\boldsymbol{r}_1' + m_2\boldsymbol{r}_2' = 0$ and $m_1\boldsymbol{v}_1' + m_2\boldsymbol{v}_2' = 0$, and we could eliminate $\boldsymbol{r}_2'$ and $\boldsymbol{v}_2'$ in favor of $\boldsymbol{r}_1'$ and $\boldsymbol{v}_1'$. This would give one of the two point masses a privileged role, although there is no such asymmetry between them. It is, therefore, preferable to use a more symmetric variable for the parameterization of the relative motion of the two constituent masses, namely the *relative position*

$$\boldsymbol{r} = \boldsymbol{r}_1 - \boldsymbol{r}_2 = \boldsymbol{r}_1' - \boldsymbol{r}_2', \tag{5.1.3}$$

the vector pointing from point mass 2 to point mass 1 that appears in $\boldsymbol{r}_1' = m_2\boldsymbol{r}/M$ and $\boldsymbol{r}_2' = -m_1\boldsymbol{r}/M$,

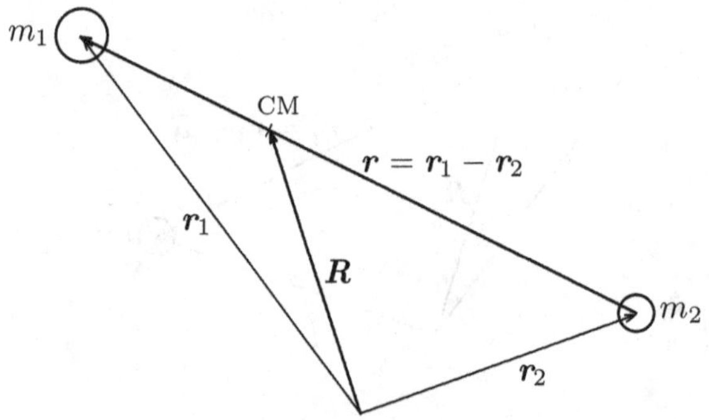

(5.1.4)

and its time derivative, the *relative velocity*

$$\frac{\mathrm{d}\boldsymbol{r}}{\mathrm{d}t} = \boldsymbol{v} = \boldsymbol{v}_1 - \boldsymbol{v}_2 = \boldsymbol{v}_1' - \boldsymbol{v}_2'. \tag{5.1.5}$$

The actual positions $\boldsymbol{r}_1$ and $\boldsymbol{r}_2$ of the point masses are then computed from the center-of-mass position $\boldsymbol{R}$ and the relative position $\boldsymbol{r}$ by means of

$$\boldsymbol{r}_1 = \boldsymbol{R} + \frac{m_2}{M}\boldsymbol{r}, \qquad \boldsymbol{r}_2 = \boldsymbol{R} - \frac{m_1}{M}\boldsymbol{r}, \tag{5.1.6}$$

and likewise

$$\boldsymbol{v}_1 = \boldsymbol{V} + \frac{m_2}{M}\boldsymbol{v}\,, \qquad \boldsymbol{v}_2 = \boldsymbol{V} - \frac{m_1}{M}\boldsymbol{v} \tag{5.1.7}$$

for the velocities.

The kinetic-energy contribution of the internal motion to the total energy in (5.1.2), expressed in terms of the relative velocity $\boldsymbol{v}$, is

$$\frac{m_1}{2}\boldsymbol{v}_1'^2 + \frac{m_2}{2}\boldsymbol{v}_2'^2 = \frac{1}{2}\frac{m_1 m_2^2 + m_2 m_1^2}{M^2}(\boldsymbol{v}_1 - \boldsymbol{v}_2)^2 = \frac{1}{2}m\boldsymbol{v}^2 \tag{5.1.8}$$

with the *reduced mass*

$$m = \frac{m_1 m_2}{M} = \frac{m_1 m_2}{m_1 + m_2} = \left(\frac{1}{m_1} + \frac{1}{m_2}\right)^{-1}, \tag{5.1.9}$$

half the harmonic mean of m_1 and m_2. We note two extreme situations,

$$\begin{aligned} m &= \frac{1}{2}m_1 \ \text{ if } m_1 = m_2, \\ m &\cong m_1 \ \text{ if } m_1 \ll m_2, \end{aligned} \tag{5.1.10}$$

those of two equal masses and of one dominating mass.

The resulting expression for the total energy of the two-particle system is then

$$E_{\text{tot}} = \frac{1}{2}M\boldsymbol{V}^2 + \frac{1}{2}m\boldsymbol{v}^2 + V(r) = E_{\text{CM}} + E_{\text{rel}}\,, \tag{5.1.11}$$

which is the sum of the kinetic energy of the center-of-mass motion,

$$E_{\text{CM}} = \frac{1}{2}M\boldsymbol{V}^2\,, \tag{5.1.12}$$

and the energy of the relative motion,

$$E_{\text{rel}} = \frac{1}{2}m\boldsymbol{v}^2 + V(r)\,. \tag{5.1.13}$$

The corresponding break-up for the angular momentum (4.5.17) is

$$\boldsymbol{L}_{\text{tot}} = \underbrace{\boldsymbol{R}\times M\boldsymbol{V}}_{=\boldsymbol{L}_{\text{CM}}} + \boldsymbol{r}\times m\boldsymbol{v} = \boldsymbol{L}_{\text{CM}} + \boldsymbol{l}\,, \tag{5.1.14}$$

where we write $\boldsymbol{l}$ for the contribution of the relative motion, and as we know from Section 4.5.2, the relative motion does not contribute to the total momentum,

$$\boldsymbol{P}_{\text{tot}} = M\boldsymbol{V} = \boldsymbol{P}_{\text{CM}}\,. \tag{5.1.15}$$

Since there are no external forces, $\boldsymbol{P}_{\text{CM}}$ does not change in time, which says that $\boldsymbol{V}(t) = \boldsymbol{V}(t_0)$ is constant. It follows that the center-of-mass has the constant-velocity trajectory of force-free motion,

$$\boldsymbol{R}(t) = \boldsymbol{R}(t_0) + \boldsymbol{V}(t_0)(t - t_0). \tag{5.1.16}$$

The constancy of $\boldsymbol{P}_{\text{CM}}$ and thus of $\boldsymbol{V}$ implies that the center-of-mass energy E_{CM} is also constant in time, as is the total energy $E_{\text{tot}} = E_{\text{CM}} + E_{\text{rel}}$. Therefore, the energy E_{rel} of the relative motion is constant in time as well. The same reasoning applies to angular momentum: The time derivatives of both $\boldsymbol{L}_{\text{CM}}$ and $\boldsymbol{l}$ vanish.

What has thus been achieved is the complete separation of the center-of-mass motion and the relative motion. For the latter, we have the equation of motion

$$m\frac{\mathrm{d}^2}{\mathrm{d}t^2}\boldsymbol{r} = -\boldsymbol{\nabla}V(r) = -\frac{\partial V(r)}{\partial r}\frac{\boldsymbol{r}}{r}. \tag{5.1.17}$$

Accordingly, the relative motion is that of a fictitious point mass $m = m_1 m_2/(m_1 + m_2)$ in a central-field force. This is a situation of great importance because one encounters such two-body problems in very many contexts.

5.2 Kepler's ellipses and Newton's force law

Arguably, the example of greatest historical importance — and of continuing importance for the writer and the readers of these notes — is the two-body system of sun and planet, for which Kepler* found his three laws by very careful evaluation of the data collected by Brahe† in the decades before the year 1600. We put ourselves in Newton's position and ask which potential energy $V(r)$, or force $F(r) = -\frac{\partial}{\partial r}V(r)$, is needed to be consistent with Kepler's Laws.

As we recall (from school education and first-year general physics modules), *Kepler's First Law* of planetary motion states that the orbits are ellipses with the sun at one focus of the ellipse. It is implicit in this statement that the influence of the other planets is so small that it is not noticeable in the data at Kepler's disposal. We follow this lead and regard each planet as moving independently around the sun. Further, the blunt statement of the sun stationary at the focus implies that the sun is so much more massive that it essentially stands for the total mass. That is: we have

*Johannes KEPLER (1571–1630) †Tycho BRAHE (1546–1601)

the m_1(planet) $\ll m_2$(sun) situation of (5.1.10), so that $\boldsymbol{R} \cong \boldsymbol{r}_2$ as well as $m \cong m_1$ and $\boldsymbol{r} \cong \boldsymbol{r}'_1$ are very good approximations. Under these circumstances, then, the fictitious point mass of the relative motion is essentially the planet.

So, in the equation of motion (5.1.17), the reduced mass m is simply the mass of the planet and $\boldsymbol{r}(t)$ is the planet in orbit about the sun, and since Kepler tells us that the orbit is in a plane, we use cylindrical coordinates to parameterize $\boldsymbol{r}(t)$ in the xy plane,

$$\boldsymbol{r} \mathrel{\widehat{=}} \begin{pmatrix} s\cos\varphi \\ s\sin\varphi \\ 0 \end{pmatrix}, \qquad \frac{\boldsymbol{r}}{r} = \boldsymbol{e}_s \mathrel{\widehat{=}} \begin{pmatrix} \cos\varphi \\ \sin\varphi \\ 0 \end{pmatrix}. \tag{5.2.18}$$

Kepler's ellipse is then parameterized as depicted here:

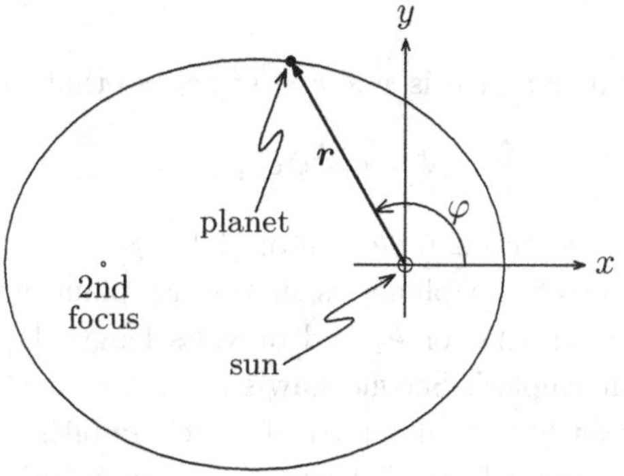

(5.2.19)

We have the sun at $\boldsymbol{r} = 0$ and the second focus of the ellipse on the negative x axis, and we choose the xy plane such that the planet is orbiting the sun counter-clockwise, that is: $\dot{\varphi} > 0$.

The message of *Kepler's Second Law* is illustrated in this figure:

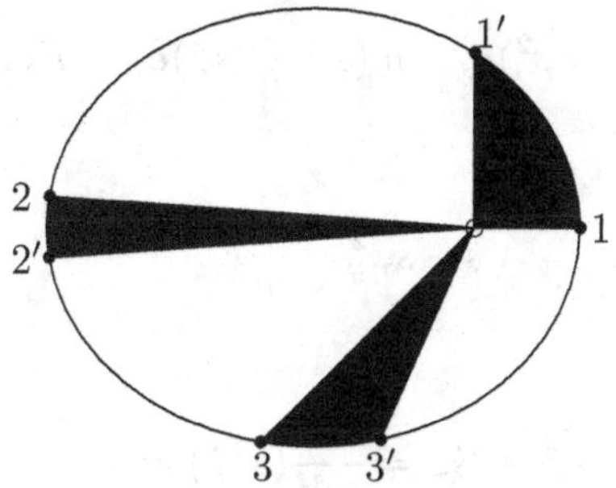

(5.2.20)

Kepler recognized that the area covered by the line from the sun to the planet is equal for equal lapses of time. In (5.2.20), it takes the planet the same time to travel from point 1 to point 1′ as from 2 to 2′ or from 3 to 3′ because these three segments of the ellipse have the same area.

Upon recalling the geometrical meaning of the vector product, see (1.1.16), we note that the area of the infinitesimal segment for positions $\boldsymbol{r}(t)$ and $\boldsymbol{r}(t+\mathrm{d}t) = \boldsymbol{r}(t) + \boldsymbol{v}(t)\mathrm{d}t$ is

$$\frac{1}{2}\big|\boldsymbol{r}(t) \times \boldsymbol{r}(t+\mathrm{d}t)\big| = \frac{1}{2}\big|\boldsymbol{r}(t) \times \boldsymbol{v}(t)\mathrm{d}t\big| = \frac{\mathrm{d}t}{2m}|\boldsymbol{l}|\,, \tag{5.2.21}$$

where $\boldsymbol{l} = \boldsymbol{r} \times m\boldsymbol{v}$ is the angular momentum of the planet. With

$$\boldsymbol{r} = s\boldsymbol{e}_s\,, \quad \boldsymbol{v} = \dot{s}\boldsymbol{e}_s + s\dot{\varphi}\boldsymbol{e}_\varphi \mathrel{\widehat{=}} \begin{pmatrix} \dot{s}\cos\varphi - s\dot{\varphi}\sin\varphi \\ \dot{s}\sin\varphi + s\dot{\varphi}\cos\varphi \\ 0 \end{pmatrix}, \tag{5.2.22}$$

the vector product of $\boldsymbol{r}$ and $\boldsymbol{v}$ is $\boldsymbol{r} \times \boldsymbol{v} = s^2\dot{\varphi}\boldsymbol{e}_z$, so that

$$\boldsymbol{l} = ms^2\dot{\varphi}\boldsymbol{e}_z\,, \tag{5.2.23}$$

where we recognize the $z(t) \equiv 0$ version of (1.1.118).

Since the motion of the planet is in the xy plane with $\dot{\varphi} > 0$, the direction of $\boldsymbol{l}$ is always that of $\boldsymbol{e}_z$ and only its length $|\boldsymbol{l}| = ms^2\dot{\varphi}$ could change in time. But Kepler's Second Law says it does not: $|\boldsymbol{l}| = \text{constant}$, because "equal areas for equal lapses of time" implies "equal areas for equal *infinitesimal* lapses of time." Kepler had thus found that the angular momentum of the planet in orbit is conserved, which tells us that there is a line-of-sight force attracting the planet to the sun: $m\ddot{\boldsymbol{r}} = \boldsymbol{F} = F\frac{\boldsymbol{r}}{r} = F\boldsymbol{e}_s$ with $F < 0$.

With the expression for the acceleration in cylindrical coordinates in (1.1.114), the equation of motion now reads

$$m(\ddot{s} - s\dot{\varphi}^2)\boldsymbol{e}_s + m(s\ddot{\varphi} + 2\dot{s}\dot{\varphi})\boldsymbol{e}_\varphi = F\boldsymbol{e}_s \tag{5.2.24}$$

or rather

$$\ddot{s} - s\dot{\varphi}^2 = \frac{1}{m}F \tag{5.2.25}$$

after we note that

$$s\ddot{\varphi} + 2\dot{s}\dot{\varphi} = \frac{1}{s}\frac{\mathrm{d}}{\mathrm{d}t}(s^2\dot{\varphi}) = 0 \tag{5.2.26}$$

since $s^2\dot{\varphi} = |\boldsymbol{l}|/m$ is constant. The value κ of this constant,

$$\kappa = s^2\dot{\varphi}\,, \tag{5.2.27}$$

is determined by the initial conditions.

We use the conservation of angular momentum to relate $\varphi(t)$ to $s(t)$ by

$$\dot{\varphi}(t) = \frac{\kappa}{s(t)^2}\,, \tag{5.2.28}$$

and can so reduce the problem to finding s as a function of φ,

$$\begin{aligned} s(t) &= s(\varphi(t))\,, \\ \dot{s}(t) &= \frac{\mathrm{d}}{\mathrm{d}t}s(\varphi(t)) = \frac{\mathrm{d}\varphi(t)}{\mathrm{d}t}\frac{\mathrm{d}s(\varphi)}{\mathrm{d}\varphi}\bigg|_{\varphi(t)} = \dot{\varphi}s'(\varphi)\,, \end{aligned} \tag{5.2.29}$$

with the prime denoting differentiation with respect to φ. Now,

$$\dot{s} = \dot{\varphi}s' = \frac{\kappa}{s^2}s' = -\kappa\frac{\mathrm{d}}{\mathrm{d}\varphi}\frac{1}{s} \tag{5.2.30}$$

and

$$\ddot{s} = \dot{\varphi}\frac{\mathrm{d}}{\mathrm{d}\varphi}\dot{s} = \frac{\kappa}{s^2}\left[-\kappa\left(\frac{\mathrm{d}}{\mathrm{d}\varphi}\right)^2\frac{1}{s}\right] = -\frac{\kappa^2}{s^2}\left(\frac{\mathrm{d}}{\mathrm{d}\varphi}\right)^2\frac{1}{s}\,. \tag{5.2.31}$$

Together with $s\dot{\varphi}^2 = \frac{\kappa^2}{s^2}\frac{1}{s}$ this turns (5.2.25) into

$$\left(\frac{\mathrm{d}}{\mathrm{d}\varphi}\right)^2\frac{1}{s} + \frac{1}{s} = -\frac{s^2F}{m\kappa^2}\,. \tag{5.2.32}$$

We are now ready to exploit Kepler's finding that the orbits are ellipses with the force center at a focus, for which purpose we need $s(\varphi)$ for such an elliptical orbit. We recall the basic definition of the ellipse as the points for which the sum of the distances from the two foci is constant:

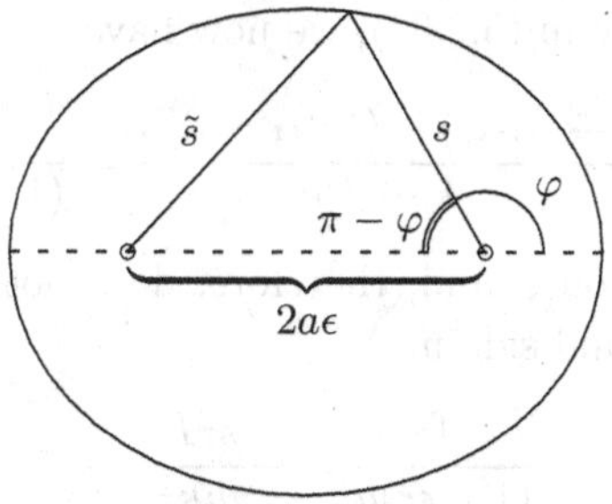

(5.2.33)

meaning

$$s + \tilde{s} = 2a\,, \tag{5.2.34}$$

where a is the major half-axis of the ellipse and $2a\epsilon$ is the distance between the foci, with the numerical eccentricity ϵ from the range $0 \le \epsilon < 1$. We continue to denote by $s(\varphi)$ the distance from one focus, then

$$\begin{aligned}\tilde{s} &= \sqrt{(2a\epsilon)^2 + s^2 - 2(2a\epsilon)s\cos(\pi - \varphi)}\\ &= \sqrt{(2a\epsilon)^2 + s^2 + 4a\epsilon s\cos\varphi}\end{aligned} \tag{5.2.35}$$

is the distance from the second focus and $s + \tilde{s} = 2a$ requires

$$s(\varphi) = \frac{1-\epsilon^2}{1+\epsilon\cos\varphi}a\,; \tag{5.2.36}$$

see also (A.18) in Appendix A. We have chosen the x axis ($\varphi = 0$ or $\varphi = \pi$) as the major axis of the ellipse; more generally we could have

$$s(\varphi) = \frac{1-\epsilon^2}{1+\epsilon\cos(\varphi - \varphi_0)}a\,, \tag{5.2.37}$$

if φ_0 specifies the direction for the major axis, but in the context of the Kepler problem at hand, this generality is of no use. So we stick to $\varphi_0 = 0$, and then

$$s(\varphi = 0) = \frac{1-\epsilon^2}{1+\epsilon}a = (1-\epsilon)a \tag{5.2.38}$$

is the smallest distance on the orbit (terminology: the planet is at its *perihelion* for $\varphi = 0$), and

$$s(\varphi = \pi) = \frac{1-\epsilon^2}{1-\epsilon}a = (1+\epsilon)a \tag{5.2.39}$$

is the largest distance (planet at its *aphelion*).

For the left-hand side in (5.2.32), we now have

$$\left(\frac{\mathrm{d}}{\mathrm{d}\varphi}\right)^2\frac{1}{s} + \frac{1}{s} = \frac{(-\epsilon\cos\varphi) + (1+\epsilon\cos\varphi)}{(1-\epsilon^2)a} = \frac{1}{(1-\epsilon^2)a}\,, \tag{5.2.40}$$

which does not depend on φ and, therefore, does not depend on $s(\varphi)$. It follows that the right-hand side in

$$\frac{1}{(1-\epsilon^2)a} = -\frac{s^2F}{m\kappa^2} \tag{5.2.41}$$

does not depend on s or φ as well, which in turn tells us that the force

$$F(s) = -\frac{m\kappa^2/a}{1-\epsilon^2}\frac{1}{s^2} \tag{5.2.42}$$

is proportional to $\frac{1}{s^2}$ with a negative factor, as we anticipated for an attractive force. This is Newton's famous *inverse-square law* for the gravitational attraction between sun and planets — and then, more generally, between all massive bodies. Newton communicated this truly seminal result in 1684 in a manuscript he sent to Halley* and published his findings in 1687 in his *Philosophiae Naturalis Principia Mathematica* (Mathematical Principles of Natural Philosophy). Newton acknowledged earlier work by Boulliau,† Hooke, Wren,‡ and Halley, who had conjectured the inverse-square law but lacked the mathematical skills to connect it to the empirical evidence summarized by Kepler's Laws. There is quite a big difference between their conjectures and Newton's clear-cut demonstration of the case. As Clairaut§ put it: *quelle distance il y a entre une vérité entrevue & une vérité démontrée* (what a distance there is between a truth that is glimpsed and a truth that is demonstrated).

If all of this has any significance beyond a statement about one class of orbits, the coefficient

$$\frac{m\kappa^2/a}{1-\epsilon^2}, \tag{5.2.43}$$

which is partly determined by the mass of the planet and partly by the initial conditions that select the actual elliptical orbit, must have a universal meaning. The proportionality to the planet's mass suggests by symmetry that we should also have proportionality to the mass of the sun $m_\odot$, and so we arrive at

$$\frac{\kappa^2/a}{1-\epsilon^2} = Gm_\odot \tag{5.2.44}$$

and

$$F(s) = -\frac{Gm_\odot m}{s^2} \tag{5.2.45}$$

with a universal constant G. This is, of course, the gravitational constant,

$$G = 6.67 \times 10^{-11}\mathrm{Nm^2/kg^2}\,, \tag{5.2.46}$$

---

*Edmond HALLEY (1656–1742) †Ismael BOULLIAU (1605–1694)
‡Sir Christopher WREN (1632–1723) §Alexis CLAIRAUT (1713–1765)

and the solar mass is about $m_\odot = 2 \times 10^{30}$ kg, 330 000 times the mass of the earth.

We exploited Kepler's First and Second Laws (of 1606) and arrived at Newton's inverse-square law for the force between the sun and the planets. But how can we be sure that the left-hand side in (5.2.44), which is entirely determined by the parameters of the elliptical orbit of the planet considered, has the same value for all planets? This assurance is given by *Kepler's Third Law* (of 1619) which links the period T of the elliptical orbit to the major half-axis by stating that the ratio a^3/T^2 is the same for all planets, that is: the same for all elliptical orbits if we continue, as we do, to insist that these statements have a universal character. Kepler's Second Law implies that

$$\text{area of the orbit ellipse} = \frac{1}{2}\kappa T\,, \tag{5.2.47}$$

which is just the full-orbit integral of (5.2.21) with $|\boldsymbol{l}|/m = \kappa$, and since the minor half-axis of the ellipse is $a\sqrt{1-\epsilon^2}$, we have

$$\frac{1}{2}\kappa T = \pi a^2\sqrt{1-\epsilon^2} \tag{5.2.48}$$

for this area. Therefore,

$$\left(\frac{2\pi}{T}\right)^2 a^3 = \left(\frac{\kappa/a^2}{\sqrt{1-\epsilon^2}}\right)^2 a^3 = \frac{\kappa^2/a}{1-\epsilon^2}\,, \tag{5.2.49}$$

which is exactly the combination of orbit parameters in (5.2.44). The constant ratio a^3/T^2 of Kepler's Third Law is essentially the product $Gm_\odot$ of the gravitational constant and the mass of the sun,

$$\frac{a^3}{T^2} = \frac{Gm_\odot}{(2\pi)^2} \cong 3.4 \times 10^{18}\,\frac{\mathrm{m}^3}{\mathrm{s}^2}\,, \tag{5.2.50}$$

quite consistent with the orbit parameters of the earth: $T = 1\,\text{year}$ and $a = 1.5 \times 10^8$ km.

In summary, Newton could derive the inverse-square force law (5.2.45) for the gravitational attraction between point masses from Kepler's First and Second Laws and conduct a check of consistency offered by the Third Law. But, of course, this is only the starting point. The correctness of the inverse-square law cannot be established by a single class of observations; we cannot verify the inverse-square law by deriving Kepler's Laws from it — the data that suggest a hypothesis cannot be used to test it. What is required is the study of other phenomena.

For example, there is the earth-moon system in orbit around the sun. From Section 4.5.2, we know that the center-of-mass of the earth-moon system follows an equation of motion of a point mass under the influence of the total external force, that is: the sum of the gravitational attraction of the earth and the moon by the sun. Roughly, then, this center-of-mass propagates along a Kepler ellipse, but not precisely because the internal configuration of earth and moon determines the total external force. This coupling between the internal dynamics and the center-of-mass motion makes the three-body problem much more complicated and more interesting. Many details of the effects that the forces between the planets, and between the planets and their moons, have on the evolution of the solar system as a whole have been worked out diligently by ingenious methods developed for this purpose in the time span between Newton's days and the mid-19th century.

But even the rather simple system of sun and one single planet is not completely cut and dry, as one may perhaps conclude after deriving the inverse-square law from Kepler's Laws. For, with the ellipse equation for $s(\varphi)$ in (5.2.36), we get the differential equation (5.2.28) for $\varphi(t)$,

$$\frac{\mathrm{d}}{\mathrm{d}t}\varphi(t) = \frac{\kappa}{s\big(\varphi(t)\big)^2} = \frac{\kappa}{a^2}\left(\frac{1+\epsilon\cos\big(\varphi(t)\big)}{1-\epsilon^2}\right)^2 , \tag{5.2.51}$$

which we solve implicitly by

$$\kappa t = \int\limits_{\varphi(0)}^{\varphi(t)} \mathrm{d}\varphi'\, s(\varphi')^2 = \int\limits_{\varphi(0)}^{\varphi(t)} \mathrm{d}\varphi'\, \frac{(1-\epsilon^2)^2 a^2}{(1+\epsilon\cos\varphi')^2} . \tag{5.2.52}$$

Integrals of this kind — rational functions of trigonometric functions — can be converted by a standard substitution,

$$u = \tan\frac{\varphi'}{2}, \quad \mathrm{d}u = \frac{\mathrm{d}\varphi'}{2\cos(\varphi'/2)^2} \quad \text{or} \quad \mathrm{d}\varphi' = \mathrm{d}u\frac{2}{1+u^2},$$
$$f(\sin\varphi', \cos\varphi') = f\left(\frac{2u}{1+u^2}, \frac{1-u^2}{1+u^2}\right), \tag{5.2.53}$$

into integrals of rational functions, which in turn can be solved by partial fractions or other familiar methods. Eventually, this results in an implicit expression for $\varphi(t)$. We leave it to the reader to work out the details as an exercise (see Appendix B).

5.3 Motion in a central-force field

5.3.1 *Bounded motion*

For motion in an arbitrary central-force field $\boldsymbol{F} = -\boldsymbol{\nabla}V(r)$, with the conservative line-of-sight force deriving from the potential energy $V(r)$ that depends only on the distance $r = |\boldsymbol{r}|$ from the center (center-of-mass, that is, for the closed two-body system), we have the equation of motion

$$m\dot{\boldsymbol{v}} = -\frac{\partial V(r)}{\partial r}\frac{\boldsymbol{r}}{r}. \tag{5.3.1}$$

The energy

$$E = \frac{1}{2}mv^2 + V(r) \tag{5.3.2}$$

and the angular momentum

$$\boldsymbol{l} = \boldsymbol{r} \times m\boldsymbol{v} \tag{5.3.3}$$

are conserved quantities, with their values determined by the initial position $\boldsymbol{r}(t=0) = \boldsymbol{r}_0$ and the initial velocity $\boldsymbol{v}(t=0) = \boldsymbol{v}_0$,

$$E = \frac{1}{2}mv_0^2 + V(r_0), \qquad \boldsymbol{l} = \boldsymbol{r}_0 \times m\boldsymbol{v}_0. \tag{5.3.4}$$

Since $\boldsymbol{l}$ is constant in time, both $\boldsymbol{r}(t)$ and $\boldsymbol{v}(t)$ are in the plane perpendicular to $\boldsymbol{l}$ at all times, the plane spanned by $\boldsymbol{r}_0$ and $\boldsymbol{v}_0$. As in (5.2.22), we choose cylindrical coordinates such that $\boldsymbol{l} = m\kappa\boldsymbol{e}_z$ with $\kappa = s^2\dot{\varphi} > 0$. Then

$$\varphi(t) = \varphi(0) + \int_0^t \mathrm{d}t'\, \frac{\kappa}{s(t')^2} \tag{5.3.5}$$

determines the azimuth as a function of time as soon as the distance $s(t)$ from the z axis is known.

For $s(t)$, we have the equation of motion (5.2.25) with $F = -\frac{\partial}{\partial s}V(s)$,

$$m\left(\ddot{s} - s\dot{\varphi}^2\right) = -\frac{\partial}{\partial s}V(s), \tag{5.3.6}$$

or

$$m\ddot{s} = \frac{m\kappa^2}{s^3} - \frac{\partial V(s)}{\partial s} = -\frac{\partial}{\partial s}\left(V(s) + \frac{m\kappa^2}{2s^2}\right) \tag{5.3.7}$$

after using $\dot{\varphi} = \kappa/s^2$ for the elimination of $\dot{\varphi}$. This has the mathematical structure of one-dimensional motion in the effective potential energy

$$V_{\text{eff}}(s) = V(s) + \frac{m\kappa^2}{2s^2}, \tag{5.3.8}$$

for which we have the conserved energy

$$E = \frac{m}{2}\dot{s}^2 + V_{\text{eff}}(s). \tag{5.3.9}$$

The extra term in the effective potential energy is actually part of the kinetic energy, inasmuch as

$$\begin{aligned} E_{\text{kin}} &= \frac{m}{2}\boldsymbol{v}^2 = \frac{m}{2}\left(\dot{s}\boldsymbol{e}_s + s\dot{\varphi}\boldsymbol{e}_\varphi\right)^2 \\ &= \frac{m}{2}\dot{s}^2 + \frac{m}{2}(s\dot{\varphi})^2 = \frac{m}{2}\dot{s}^2 + \frac{m}{2}\left(\frac{\kappa}{s}\right)^2. \end{aligned} \tag{5.3.10}$$

But in the specific context of the pseudo–one-dimensional s motion it enters the formalism as if it were a contribution to the potential energy.

As an example, let us reconsider Newton's inverse-square force law (5.2.45),

$$F(s) = \frac{\partial}{\partial s}\frac{Gm_\odot m}{s}, \tag{5.3.11}$$

for which we have the potential energy

$$V(s) = -\frac{Gm_\odot m}{s} \tag{5.3.12}$$

up to an additive constant. Adopting the usual convention, we choose that constant such that $V(s \to \infty) = 0$. For this potential energy, then, the effective potential is

$$V_{\text{eff}}(s) = -\frac{Gm_\odot m}{s} + \frac{m\kappa^2}{2s^2}, \tag{5.3.13}$$

where the physical potential energy is dominating for large distances whereas the contribution from the kinetic energy is dominating at small distances:

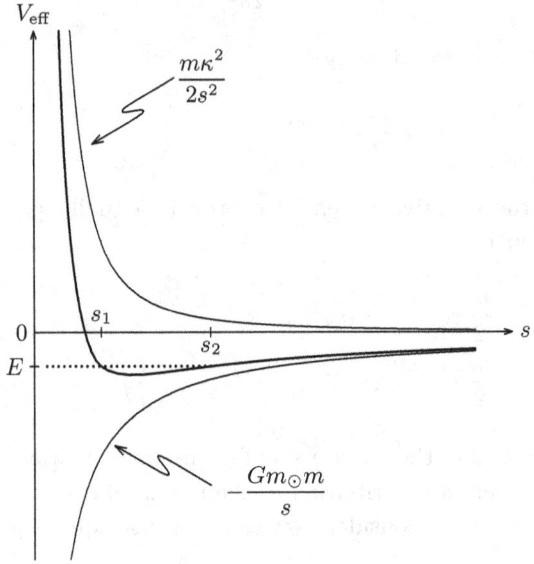

(5.3.14)

For negative energy $E = V_{\text{eff}}(s_1) = V_{\text{eff}}(s_2) < 0$, we have the analog of one-dimensional motion between two turning points, but now it is not motion along a line, it is motion in a plane and s_1, s_2 specify the minimal and maximal distances, the distance $s_1 = (1 - \epsilon)a$ to the perihelion of the ellipse and the distance $s_2 = (1 + \epsilon)a$ to the aphelion of the ellipse.

More generally, for another physical potential energy $V(s)$ we will typically get an effective potential $V_{\text{eff}}(s)$ with the same general features as depicted in (5.3.14), with an energy range for which we have *bounded motion* between two turning points in s. What is particular for the potential energy of the Kepler problem ($V \propto s^{-1}$) is that the orbits are closed. For almost all other potential energies, the point mass will not accumulate exactly 2π (or a multiple thereof) for the change in azimuth φ between two instants with $s(t_1) = s(t_2) = s_2$, say. As a rule, the situation is as depicted here:

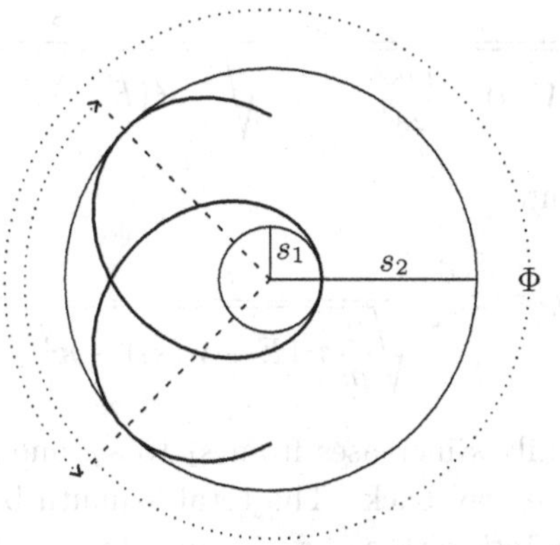

(5.3.15)

The orbit is confined by the circles with radii s_1 and s_2 and oscillates periodically between them, while progressing in azimuth φ. Typical orbits will not be closed, as in the case of Kepler's ellipse, which is really quite exceptional.

In the general situation, we can calculate the time it takes to get from $s = s_1$ to $s = s_2$ in full analogy with the calculation of the period of the motion between two turning points in the one-dimensional situation. First, we note that

$$\frac{\mathrm{d}s}{\mathrm{d}t} = \pm\sqrt{\frac{2}{m}\bigl(E - V_{\mathrm{eff}}(s)\bigr)} \tag{5.3.16}$$

with "+" for the stretches from the inner boundary to the outer boundary, and "−" on the way back. Then we integrate to find

$$T = 2\int_{s_1}^{s_2} \frac{\mathrm{d}s}{\sqrt{\frac{2}{m}\bigl(E - V_{\mathrm{eff}}(s)\bigr)}} \tag{5.3.17}$$

for the *radial period*, that is: the lapse of time between two successive instants with $s = s_1$, twice the time between an instant with $s = s_1$ and the next instant with $s = s_2$. We have, of course,

$$E = V_{\mathrm{eff}}(s_1) = V_{\mathrm{eff}}(s_2)\,, \tag{5.3.18}$$

here.

Similarly, we get the *angular period* Φ by combining (5.3.16) in the form

$$\mathrm{d}t = \pm\frac{\mathrm{d}s}{\sqrt{\frac{2}{m}\left(E - V(s) - \frac{m\kappa^2}{2s^2}\right)}} = \pm\frac{s\mathrm{d}s}{\sqrt{\frac{2}{m}s^2\big(E - V(s)\big) - \kappa^2}} \tag{5.3.19}$$

with $\mathrm{d}\varphi = (\kappa/s^2)\mathrm{d}t$ into

$$\mathrm{d}\varphi = \pm\frac{\mathrm{d}s}{s}\frac{\kappa}{\sqrt{\frac{2}{m}s^2\big(E - V(s)\big) - \kappa^2}}\,, \tag{5.3.20}$$

where "+" applies while s increases from s_1 to s_2, and "−" applies when we have $\mathrm{d}s < 0$ on the way back. The total azimuth between two points with $s = s_2$, or two points with $s = s_1$, is given by

$$\Phi = 2\int_{s_1}^{s_2}\frac{\mathrm{d}s}{s}\frac{\kappa}{\sqrt{\frac{2}{m}s^2\big(E - V(s)\big) - \kappa^2}}\,. \tag{5.3.21}$$

As in (5.3.17), the argument of the square root vanishes for the limiting values of $s = s_1$ and $s = s_2$ (since $\mathrm{d}s/\mathrm{d}t = 0$ at the circular boundaries) and is positive for $s_1 < s < s_2$.

As a check, let us verify that this gives the right answer for the situation studied by Newton,

$$V(s) = -\frac{C}{s} \quad \text{with} \quad C = Gm_{\odot}m > 0 \tag{5.3.22}$$

and $E < 0$, which we parameterize by

$$E = -\frac{m}{2}u_0^2 \quad \text{with} \quad u_0 > 0\,. \tag{5.3.23}$$

Note that the parameter u_0 has the metrical dimension of a velocity, which is for convenient parameterization, but it is not a velocity of the moving point mass at any time. The argument of the square root is then

$$\frac{2}{m}s^2\big(E - V(s)\big) - \kappa^2 = -u_0^2s^2 + \frac{2C}{m}s - \kappa^2 = u_0^2(s_2 - s)(s - s_1)\,, \tag{5.3.24}$$

with

$$u_0^2(s_1 + s_2) = \frac{2C}{m} \quad \text{and} \quad u_0^2s_1s_2 = \kappa^2\,. \tag{5.3.25}$$

This gives us

$$\Phi = 2\int\limits_{s_1}^{s_2} \frac{ds}{s}\frac{\kappa/u_0}{\sqrt{(s_2-s)(s-s_1)}} = 2\pi\frac{\kappa/u_0}{\sqrt{s_1 s_2}} = 2\pi \tag{5.3.26}$$

for the angular period, the correct result for the closed Kepler's ellipses.

In (5.3.26) we made use of

$$\int\limits_b^a \frac{dx}{x\sqrt{(a-x)(x-b)}} = \frac{\pi}{\sqrt{ab}} \qquad \text{for } a > b > 0\,, \tag{5.3.27}$$

a companion integral of the one in (3.1.40). The substitution used then works here as well, and the resulting integral over α can be expressed in terms of

$$\int\limits_{-\pi/2}^{\pi/2} \frac{d\alpha}{A + B\sin\alpha} = \int\limits_0^{\pi} \frac{d\alpha}{A + B\cos\alpha} = \frac{\pi}{\sqrt{A^2 - B^2}} \qquad \text{for } A > |B|\,, \tag{5.3.28}$$

which are well-known integrals themselves. One way of demonstrating (5.3.28) uses the method mentioned at (5.2.53); complex integration with the residue theorem is another standard method.

5.3.2 *Unbounded motion*

Continuing with the attractive inverse-squared-distance force, let us take a look at the situation of positive energy

$$E = \frac{m}{2}\left(\frac{ds}{dt}\right)^2 - \frac{C}{s} + \frac{m\kappa^2}{2s^2} > 0\,, \tag{5.3.29}$$

where $C = Gm_\odot m$ for the gravitational attraction between a planet with mass m and the sun with mass $m_\odot$. Figure (5.3.14) tells us that there is a minimal distance s_1 for $E \geq 0$, but no maximal distance. The orbit will then be that of a body approaching the sun from beyond the solar system, finally receding to outer space again, never to come back. The equation that expresses the energy in terms of s and $\dot{s}$ has the same structure for $E \geq 0$, as it has for $E < 0$. We can, therefore, look for solutions $s(t)$ of the form

$$s(t) = \frac{1+\epsilon}{1+\epsilon\cos\varphi}s_1 \quad \text{with} \quad \dot{\varphi} = \frac{\kappa}{s^2}\,, \tag{5.3.30}$$

where — as we know from above — ϵ is in the range $0 \le \epsilon < 1$ for Kepler's ellipses with $E < 0$. But now

$$E = -\frac{C}{s_1} + \frac{m\kappa^2}{2s_1^2} \ge 0\,, \tag{5.3.31}$$

so that

$$s_1 = \frac{C}{2E}\left(\sqrt{1 + 2m\kappa^2 E/C^2} - 1\right) \le \frac{m\kappa^2}{2C}\,, \tag{5.3.32}$$

and there is no second turning distance s_2.

With the ansatz (5.3.30) for $s(t)$, we have first

$$\begin{aligned}\frac{\mathrm{d}s}{\mathrm{d}t} &= -s^2 \frac{\mathrm{d}}{\mathrm{d}t}\frac{1}{s} = -s^2 \frac{\mathrm{d}}{\mathrm{d}t}\frac{1 + \epsilon\cos\varphi}{(1+\epsilon)s_1} \\ &= \frac{s^2\dot{\varphi}\,\epsilon\sin\varphi}{(1+\epsilon)s_1} = \frac{\kappa\epsilon\sin\varphi}{(1+\epsilon)s_1}\end{aligned} \tag{5.3.33}$$

and then

$$\begin{aligned}\left(\frac{\mathrm{d}s}{\mathrm{d}t}\right)^2 &= \left(\frac{\kappa\epsilon}{(1+\epsilon)s_1}\right)^2 - \left(\frac{\kappa\epsilon\cos\varphi}{(1+\epsilon)s_1}\right)^2 \\ &= \left(\frac{\kappa\epsilon}{(1+\epsilon)s_1}\right)^2 - \left(\frac{\kappa}{s} - \frac{\kappa}{(1+\epsilon)s_1}\right)^2\end{aligned} \tag{5.3.34}$$

after using (5.3.30) to express $\epsilon\cos\varphi$ differently. We compare this with

$$\left(\frac{\mathrm{d}s}{\mathrm{d}t}\right)^2 = \frac{2E}{m} + \frac{2C}{ms} - \frac{\kappa^2}{s^2} \tag{5.3.35}$$

and conclude that we need

$$\frac{C}{m} = \frac{\kappa^2}{(1+\epsilon)s_1}\,. \tag{5.3.36}$$

and

$$\frac{2E}{m} = \left(\frac{\kappa\epsilon}{(1+\epsilon)s_1}\right)^2 - \left(\frac{\kappa}{(1+\epsilon)s_1}\right)^2 = \frac{(\epsilon-1)\kappa^2}{(\epsilon+1)s_1^2}\,. \tag{5.3.37}$$

Together they repeat (5.3.31), so that only (5.3.36) tells us something new, namely that

$$\epsilon = \frac{m\kappa^2}{Cs_1} - 1 = \sqrt{1 + \frac{2m\kappa^2 E}{C^2}}\,. \tag{5.3.38}$$

This expression for the numerical eccentricity applies for $E \geq 0$ as well as $E < 0$, because the sign of E plays no role in the derivation that we went through. Since E cannot take on arbitrarily large negative values but is restricted by

$$E \geq \min_{s}\{V_{\text{eff}}(s)\} = -\frac{C^2}{2m\kappa^2}, \tag{5.3.39}$$

the argument of the square root in (5.3.38), and earlier in (5.3.32), is always nonnegative. We have

$$\begin{aligned} 0 \leq \epsilon < 1 \quad &\text{for} \quad 0 > E \geq -\frac{C^2}{2m\kappa^2}, \\ \epsilon = 1 \quad &\text{for} \quad E = 0, \\ \epsilon > 1 \quad &\text{for} \quad E > 0, \end{aligned} \tag{5.3.40}$$

which states that the orbit is an ellipse for $E < 0$, a parabola for $E = 0$, and a hyperbola for $E > 0$ — all of them are conic sections (see Appendix A).

5.3.3 *Scattering*

The hyperbolas for $E > 0$ have straight-line asymptotes:

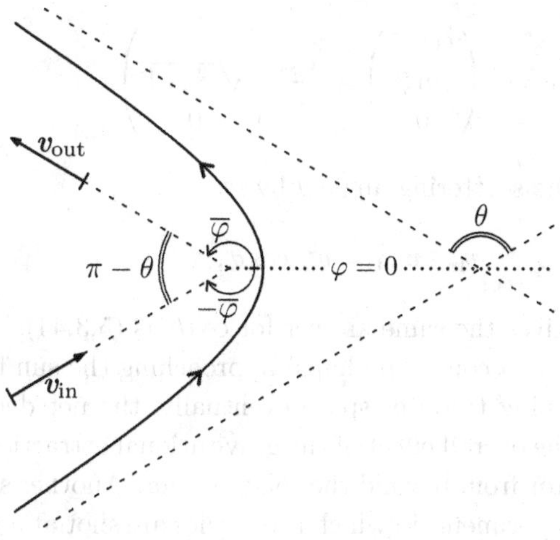

(5.3.41)

in the directions specified by the azimuthal angles for which $1 + \epsilon \cos\varphi = 0$ so that $s \to \infty$ as φ approaches these limiting values. These particular values of the azimuth are $\pm\overline{\varphi}$ with

$$\frac{1}{2}\pi < \overline{\varphi} = \cos^{-1}\left(-\frac{1}{\epsilon}\right) < \pi\,. \tag{5.3.42}$$

The resulting deflection is, therefore, by the *scattering angle* θ,

$$\theta = 2\overline{\varphi} - \pi\,, \tag{5.3.43}$$

or

$$\cos\theta = -\cos(2\overline{\varphi}) = 1 - 2\cos(\overline{\varphi})^2 = 1 - \frac{2}{\epsilon^2}\,, \tag{5.3.44}$$

as we can read off (5.3.41). Another, perhaps more physical way, of arriving at this statement, is to consider the velocity vectors long before and long after the point mass is in the vicinity of the center of the force at $\boldsymbol{r} = 0$. If we denote the terminal speed by v_∞, so that $E = \frac{1}{2}mv_\infty^2$, then the velocity in the distant past was

$$\boldsymbol{v}_{\text{in}} \widehat{=} v_\infty \begin{pmatrix} -\cos\overline{\varphi} \\ \sin\overline{\varphi} \\ 0 \end{pmatrix} = \frac{v_\infty}{\epsilon}\begin{pmatrix} 1 \\ \sqrt{\epsilon^2 - 1} \\ 0 \end{pmatrix}, \tag{5.3.45}$$

and the velocity in the distant future will be

$$\boldsymbol{v}_{\text{out}} \widehat{=} v_\infty \begin{pmatrix} \cos\overline{\varphi} \\ \sin\overline{\varphi} \\ 0 \end{pmatrix} = \frac{v_\infty}{\epsilon}\begin{pmatrix} -1 \\ \sqrt{\epsilon^2 - 1} \\ 0 \end{pmatrix}. \tag{5.3.46}$$

From these, we get the scattering angle θ by

$$\boldsymbol{v}_{\text{in}} \cdot \boldsymbol{v}_{\text{out}} = v_\infty^2 \cos\theta\,, \tag{5.3.47}$$

which, as it should, gives the same answer for $\cos\theta$ as (5.3.44).

For a massive body (a comet, perhaps) approaching the sun from outer space and then returning to outer space eventually, the net deflection by angle θ summarizes the overall effect of the gravitational attraction between the sun and that visitor from beyond the solar system. Another situation is that of a scattering experiment, in which projectiles are shot at a target and their deviation from the undisturbed straight-line motion is recorded. The projectiles arrive with a well-defined speed v_∞, which is the speed they have both long before and long after the period during which the target exerts a force on the projectile:

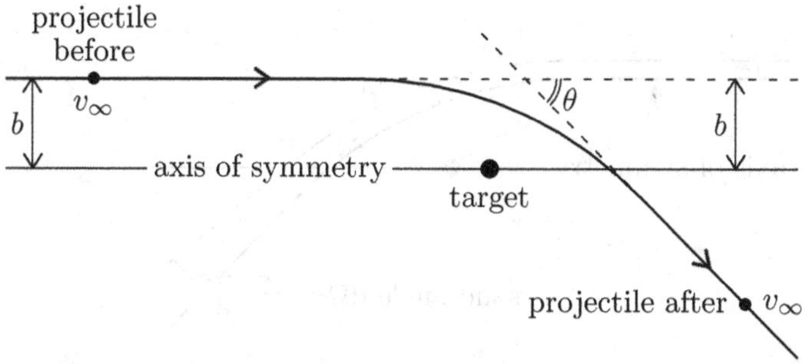

(5.3.48)

In the typical experimental situation, many projectiles are shot at the target (actually, there would also be many spatially separated targets), such as in the historic experiment by Rutherford,* in which charged projectiles (α particles) were scattered by charged targets (the nuclei of gold atoms in an amorphous foil) — the circumstance being such that each projectile got close enough to only one of the scattering centers and multiple scattering events were negligibly rare. For the individual projectile one has no knowledge about the so-called *impact parameter* b, the distance by which the projectile would miss the target if there were no forces between them.

Therefore, we consider a range of impact parameters, and also a range of azimuthal angles ϕ around the axis of symmetry, as depicted in this frontal view:

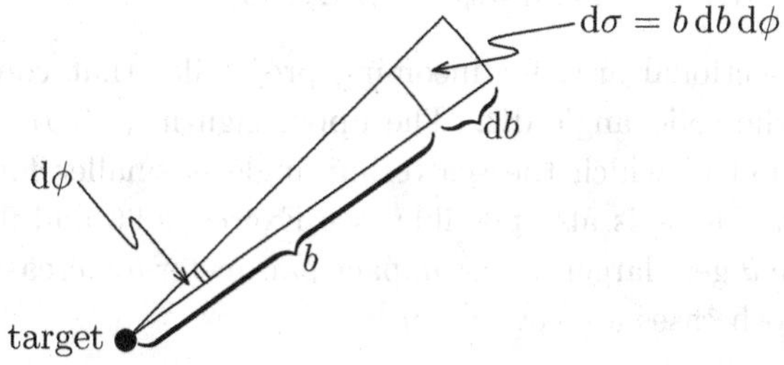

(5.3.49)

where we identify the small area $d\sigma$ that will contribute to the scattering into a corresponding small range of scattering angles $\theta \cdots \theta + d\theta$:

---

*Lord Ernest RUTHERFORD, Baron of Nelson (1871–1937)

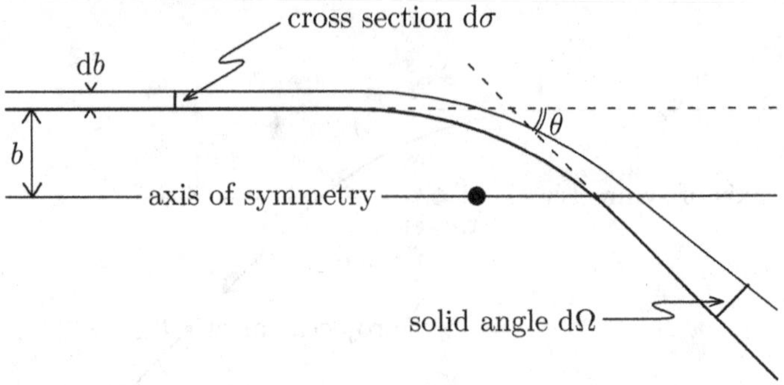

(5.3.50)

In (5.3.49) and (5.3.50), the angles θ and ϕ are the polar angle and the azimuth, respectively, of spherical coordinates that have the axis of symmetry as the z axis and the target at $z = 0$.

For the detector at some large distance from the scattering center, much farther away than indicated in (5.3.50), the part of the detector opening that supports the solid angle element $\mathrm{d}\Omega = \sin\theta \mathrm{d}\theta\, \mathrm{d}\phi$ for the projectiles arriving in the azimuthal range $\phi \cdots \phi + \mathrm{d}\phi$ and deflected by angles $\theta \cdots \theta + \mathrm{d}\theta$ has associated with it the corresponding cross-sectional area $\mathrm{d}\sigma = b\,\mathrm{d}b\,\mathrm{d}\phi$ with impact parameters in the range $b \cdots b + \mathrm{d}b$. Their ratio is the *differential cross section*

$$\frac{\mathrm{d}\sigma}{\mathrm{d}\Omega} = \mp \frac{b\,\mathrm{d}b\,\mathrm{d}\phi}{\sin\theta\,\mathrm{d}\theta\,\mathrm{d}\phi} = \pm\frac{1}{2}\frac{\mathrm{d}b^2}{\mathrm{d}\cos\theta}, \tag{5.3.51}$$

the effective cross-sectional area for incoming projectiles that contribute to scattering into the solid angle $\mathrm{d}\Omega$. The upper sign in (5.3.51) applies in the usual situation in which the scattering angle is smaller for larger impact parameters. But it is also possible (see Exercises 96 and 99) that the scattering angle θ gets larger as the impact parameter b increases, over some range of b. Both cases are covered in

$$\frac{\mathrm{d}\sigma}{\mathrm{d}\Omega} = \left|\frac{b\,\mathrm{d}b}{\sin\theta\,\mathrm{d}\theta}\right| = \frac{1}{2}\left|\frac{\mathrm{d}b^2}{\mathrm{d}\cos\theta}\right|, \tag{5.3.52}$$

where $\dfrac{\mathrm{d}\sigma}{\mathrm{d}\Omega} > 0$ is ensured.

The differential cross section is a function of the deflection angle θ. Usually, the solution of the equations of motion tells us how θ depends on the impact parameter b; we must invert this relation and express b as a function of θ when calculating $\frac{\mathrm{d}\sigma}{\mathrm{d}\Omega}$ in accordance with (5.3.52).

It can happen that there is more than one value of b for a given θ (see Exercises 96 and 99). Then the mapping $\theta \mapsto b(\theta)$ has several branches and we get one $\left|\frac{b\,\mathrm{d}b}{\mathrm{d}\cos\theta}\right|$ term for each branch, which we sum up to obtain $\frac{\mathrm{d}\sigma}{\mathrm{d}\Omega}$.

The differential cross section is the information that is acquired in a scattering experiment, such as the one sketched in (5.3.50): If ν projectiles are incoming per unit area and unit time, $\nu\mathrm{d}\sigma$ will pass through the cross sectional area $\mathrm{d}\sigma$ per unit time, and as many as

$$\int\limits_{\text{detector}} \mathrm{d}\Omega\, \nu \frac{\mathrm{d}\sigma}{\mathrm{d}\Omega} \tag{5.3.53}$$

projectiles will reach the detector per unit time, where the integration is over the finite solid angle of the detector opening. In

$$\int\limits_{\text{detector}} \mathrm{d}\Omega\, \frac{\mathrm{d}\sigma}{\mathrm{d}\Omega} = \int\limits_{\text{beam}} \mathrm{d}\sigma \tag{5.3.54}$$

the integration over the solid angle of the detector is expressed as the corresponding integral over the cross-sectional area of the beam of incoming projectiles from which we get deflection into the direction of the detector.

Since neither the incoming flux of projectiles, nor the geometry of the detector enters the differential cross section $\frac{\mathrm{d}\sigma}{\mathrm{d}\Omega}$, we recognize that it is a property of the scattering forces, or the potential energy associated with them. This is why differential cross sections are so important in many branches of physics, in particular of course for the scattering experiments of atomic, nuclear, and particle physics, but also for electrodynamics and optics where, for example, the scattering of sun light by the molecules in the atmosphere gives rise to the blue sky and other phenomena.

For the case of scattering by line-of-sight forces that are proportional to $1/(\text{distance})^2$, we have the situation of (5.3.41). The value of $\cos\theta$ is expressed in terms of the numerical eccentricity in (5.3.44), and ϵ itself is a known function of the energy E and the angular momentum $m\kappa$. The energy

$$E = \frac{m}{2} v_\infty^2 \tag{5.3.55}$$

is determined by the kinetic energy before (and after) the scattering period. Regarding κ, we recall from Section 5.3.1 that its value is given by $|\boldsymbol{r} \times \boldsymbol{v}|$, so that we have

$$\kappa = bv_\infty \tag{5.3.56}$$

in the scattering situation with impact parameter b. It follows that

$$m\kappa^2 = m(bv_\infty)^2 = 2Eb^2\,, \tag{5.3.57}$$

and the relation between $\cos\theta$ and b is

$$\frac{2}{1-\cos\theta} = \epsilon^2 = 1 + \left(\frac{2E}{C}\right)^2 b^2\,, \tag{5.3.58}$$

the appropriate combination of (5.3.44) and (5.3.38). Accordingly, we have

$$\left(\frac{2E}{C}\right)^2 \mathrm{d}b^2 = \frac{2}{(1-\cos\theta)^2}\mathrm{d}\cos\theta = \frac{\mathrm{d}\cos\theta}{2\sin\left(\frac{1}{2}\theta\right)^4} \tag{5.3.59}$$

and obtain

$$\frac{\mathrm{d}\sigma}{\mathrm{d}\Omega} = \left(\frac{C}{4E}\right)^2 \frac{1}{\sin\left(\frac{1}{2}\theta\right)^4} \tag{5.3.60}$$

for the differential cross section.

This is the famous *Rutherford differential cross section* for the scattering by an inverse-square force. In Rutherford's experiment of 1911, the force was electric rather than gravitational, but that does not change the essential physics beyond having us use another expression for the strength C of the potential energy. Also, as we learn from Exercise 92, the Rutherford cross section is the same for attractive forces ($C > 0$) and repulsive forces ($C < 0$).

The Rutherford cross section, as derived here and also historically by Rutherford himself, is a result of classical mechanics. But for the proper description of Rutherford's scattering experiment one should use quantum mechanics. It turns out that the differential cross section derived by quantum-mechanical methods is exactly the same. There are no quantum corrections at all to the classical-mechanical expression. More amazingly even, the simplest quantum-mechanical approximation — the so-called first-order Born* approximation — gives the correct result, all higher-order terms add up in such a way that the resulting differential cross section is identical with Rutherford's result. Yes, the inverse-squared-distance force is unique in many ways.

---

*Max BORN (1882–1970)

Chapter 6

Gravitating Mass Distributions

6.1 Gravitational potential

After this digression into the realm of scattering, we return to masses with gravitational attraction but lift the restriction to situations in which we treat all bodies as point masses. Imagine now that we have a probe mass m, small enough to be regarded as a point-size object, in proximity of an extended mass distribution of total mass M, which we describe by a *mass density* $\rho(\boldsymbol{r})$. The volume element $(\mathrm{d}\boldsymbol{r}')$ contains the mass element $(\mathrm{d}\boldsymbol{r}')\rho(\boldsymbol{r}')$:

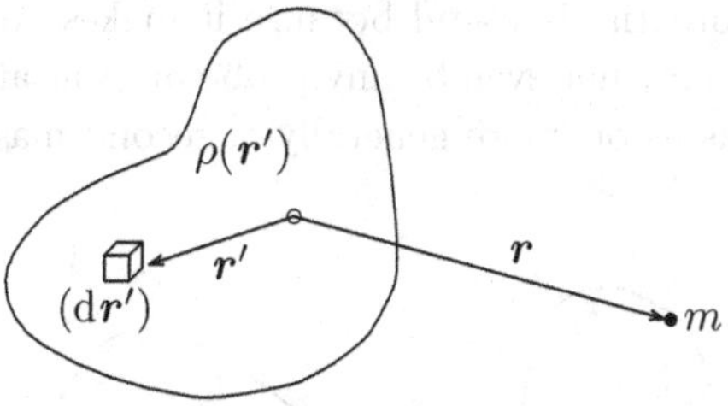

(6.1.1)

and the total mass is obtained by summing up all contributions,

$$M = \int (\mathrm{d}\boldsymbol{r}')\, \rho(\boldsymbol{r}') . \tag{6.1.2}$$

The force on the point-like probe that results from the gravitational attraction between the probe and the mass element at $\boldsymbol{r}'$ is

$$\mathrm{d}\boldsymbol{F} = -Gm \underbrace{(\mathrm{d}\boldsymbol{r}')\rho(\boldsymbol{r}')}_{\substack{\text{mass}\\ \text{element}\\ \text{at } \boldsymbol{r}'}} \underbrace{\frac{\boldsymbol{r}-\boldsymbol{r}'}{|\boldsymbol{r}-\boldsymbol{r}'|^3}}_{\text{unit vector}/(\text{distance})^2} , \tag{6.1.3}$$

and the total force is the vector sum of all such contributions,

$$\boldsymbol{F} = -mG\int (\mathrm{d}\boldsymbol{r}')\,\rho(\boldsymbol{r}')\frac{\boldsymbol{r}-\boldsymbol{r}'}{|\boldsymbol{r}-\boldsymbol{r}'|^3} = -\boldsymbol{\nabla}\left(m\Phi(\boldsymbol{r})\right). \tag{6.1.4}$$

It is the gradient of the potential energy

$$m\Phi(\boldsymbol{r}) = -mG\int (\mathrm{d}\boldsymbol{r}')\frac{\rho(\boldsymbol{r}')}{|\boldsymbol{r}-\boldsymbol{r}'|} \tag{6.1.5}$$

of the probe m in the *gravitational potential* $\Phi(\boldsymbol{r})$ of the mass distribution with density $\rho(\boldsymbol{r}')$. What is accomplished here, is the separation of the probe mass m from the quantity that characterizes the extended mass distribution: the potential

$$\Phi(\boldsymbol{r}) = -G\int (\mathrm{d}\boldsymbol{r}')\frac{\rho(\boldsymbol{r}')}{|\boldsymbol{r}-\boldsymbol{r}'|}. \tag{6.1.6}$$

Pay attention to the terminology: The extended massive body has the gravitational *potential* $\Phi(\boldsymbol{r})$, in which the probe mass m at position $\boldsymbol{r}$ has the *potential energy* $m\Phi(\boldsymbol{r})$.

The notion of potential is useful because it makes no reference to the probe mass — there may not even be any probe mass at all. Or there could be several probe masses or, more generally, a second mass distribution at some distance:

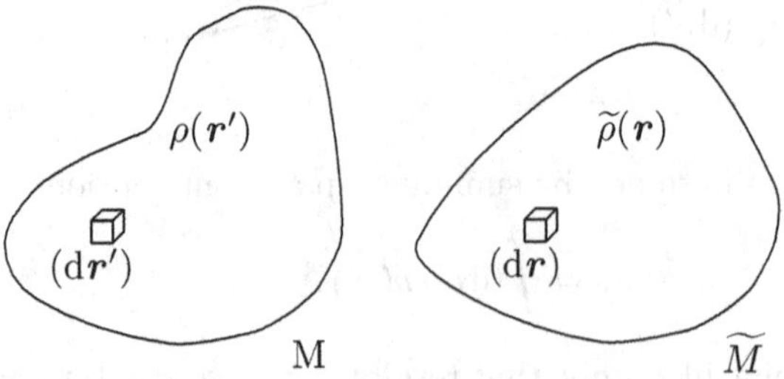

(6.1.7)

The total potential energy of the body with mass $\widetilde{M}$ in the potential of the body M is then

$$\int (\mathrm{d}\boldsymbol{r})\,\widetilde{\rho}(\boldsymbol{r})\Phi(\boldsymbol{r}), \tag{6.1.8}$$

where we sum the contributions of all mass elements $(\mathrm{d}\boldsymbol{r})\widetilde{\rho}(\boldsymbol{r})$ in the second body. Upon inserting the integral expression (6.1.6) for $\Phi(\boldsymbol{r})$, this is

$$E_{\text{int}} = -G\int(\mathrm{d}\boldsymbol{r})(\mathrm{d}\boldsymbol{r}')\frac{\widetilde{\rho}(\boldsymbol{r})\rho(\boldsymbol{r}')}{|\boldsymbol{r}-\boldsymbol{r}'|}, \tag{6.1.9}$$

the gravitational *interaction energy* of the two mass distributions, the sum of the contributions for all pairs of mass elements $(\mathrm{d}\boldsymbol{r})\widetilde{\rho}(\boldsymbol{r})$ and $(\mathrm{d}\boldsymbol{r}')\rho(\boldsymbol{r}')$, clearly the obvious and natural generalization of the gravitational interaction energy between two point masses.

While we are at it, let us ask how we can view two separate mass distributions as part of one single mass distribution:

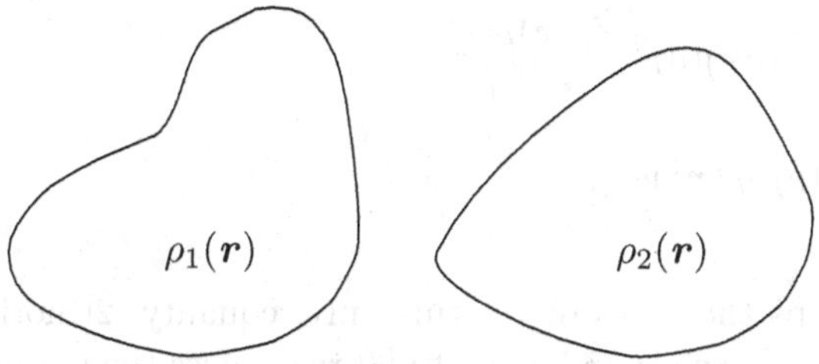

$$\rho(\boldsymbol{r}) = \rho_1(\boldsymbol{r}) + \rho_2(\boldsymbol{r}). \tag{6.1.10}$$

The interaction energy

$$-G\int(\mathrm{d}\boldsymbol{r})(\mathrm{d}\boldsymbol{r}')\frac{\rho_1(\boldsymbol{r})\rho_2(\boldsymbol{r}')}{|\boldsymbol{r}-\boldsymbol{r}'|} = -G\int(\mathrm{d}\boldsymbol{r})(\mathrm{d}\boldsymbol{r}')\frac{\rho_2(\boldsymbol{r})\rho_1(\boldsymbol{r}')}{|\boldsymbol{r}-\boldsymbol{r}'|} \tag{6.1.11}$$

is half of the 1,2-cross term in

$$-G\int(\mathrm{d}\boldsymbol{r})(\mathrm{d}\boldsymbol{r}')\frac{\left[\rho_1(\boldsymbol{r})+\rho_2(\boldsymbol{r})\right]\left[\rho_1(\boldsymbol{r}')+\rho_2(\boldsymbol{r}')\right]}{|\boldsymbol{r}-\boldsymbol{r}'|}. \tag{6.1.12}$$

Therefore, we have

$$E[\rho] = -\frac{1}{2}G\int(\mathrm{d}\boldsymbol{r})(\mathrm{d}\boldsymbol{r}')\frac{\rho(\boldsymbol{r})\rho(\boldsymbol{r}')}{|\boldsymbol{r}-\boldsymbol{r}'|} \tag{6.1.13}$$

for the total gravitational energy of the mass distribution with mass density $\rho(\boldsymbol{r})$. The factor $\frac{1}{2}$ in front avoids the double counting of the contributions from the many pairs of mass elements, $(\mathrm{d}\boldsymbol{r})\rho(\boldsymbol{r})$ at $\boldsymbol{r}$ and $(\mathrm{d}\boldsymbol{r}')\rho(\boldsymbol{r}')$ at $\boldsymbol{r}'$.

We write $E[\rho]$ for this *functional* of the mass density $\rho(\boldsymbol{r})$, rather than $E(\rho)$, just to remind ourselves by the notation that we do not have an ordinary function that assigns a number to a variable, or a set of variables, but rather a functional that assigns a number to a whole function, here: the function $\boldsymbol{r} \mapsto \rho(\boldsymbol{r})$.

In the context of the energy functional $E[\rho]$, we encounter another profound meaning of the potential $\Phi(\boldsymbol{r})$, when we consider the first-order response $\delta E[\rho]$ to infinitesimal changes of ρ, $\rho \to \rho + \delta\rho$,

$$\begin{aligned}
\delta E[\rho] &= E[\rho + \delta\rho] - E[\rho] \\
&= -\frac{1}{2}G \int (\mathrm{d}\boldsymbol{r})(\mathrm{d}\boldsymbol{r}') \frac{[\rho(\boldsymbol{r}) + \delta\rho(\boldsymbol{r})]\,[\rho(\boldsymbol{r}') + \delta\rho(\boldsymbol{r}')] - \rho(\boldsymbol{r})\rho(\boldsymbol{r}')}{|\boldsymbol{r} - \boldsymbol{r}'|} \\
&\overset{[1]}{=} -\frac{1}{2}G \int (\mathrm{d}\boldsymbol{r})(\mathrm{d}\boldsymbol{r}') \frac{\rho(\boldsymbol{r})\delta\rho(\boldsymbol{r}') + \delta\rho(\boldsymbol{r})\rho(\boldsymbol{r}')}{|\boldsymbol{r} - \boldsymbol{r}'|} \\
&\overset{[2]}{=} -G \int (\mathrm{d}\boldsymbol{r})(\mathrm{d}\boldsymbol{r}') \frac{\delta\rho(\boldsymbol{r})\rho(\boldsymbol{r}')}{|\boldsymbol{r} - \boldsymbol{r}'|} \\
&\overset{[3]}{=} \int (\mathrm{d}\boldsymbol{r})\, \delta\rho(\boldsymbol{r})\Phi(\boldsymbol{r})\,.
\end{aligned} \tag{6.1.14}$$

Equality [1] keeps the first-order terms only; equality [2] notices that we have the same term twice; and equality [3] recognizes the potential $\Phi(\boldsymbol{r})$ of (6.1.6). The overall statement

$$\delta E[\rho] = \int (\mathrm{d}\boldsymbol{r})\, \Phi(\boldsymbol{r})\delta\rho(\boldsymbol{r}) \tag{6.1.15}$$

identifies the potential as the first-order response of the energy to small changes in the mass density. If we like, we can introduce the concept of a functional derivative by writing

$$\frac{\delta E[\rho]}{\delta\rho(\boldsymbol{r})} = \Phi(\boldsymbol{r})\,, \tag{6.1.16}$$

but this has no more meaning than what is conveyed by (6.1.15). Do not miss the obvious analogy of (6.1.15) and (6.1.16) with

$$\delta f(x) = f(x + \delta x) - f(x) = f'(x)\,\delta x\,, \tag{6.1.17}$$

the familiar identification of the derivative $f'(x)$ as the first-order response of $f(x)$ to small changes of the argument.

6.2 Monopole moment and quadrupole moment dyadic

At large distances from the region where $\rho(\boldsymbol{r}')$ is nonzero,

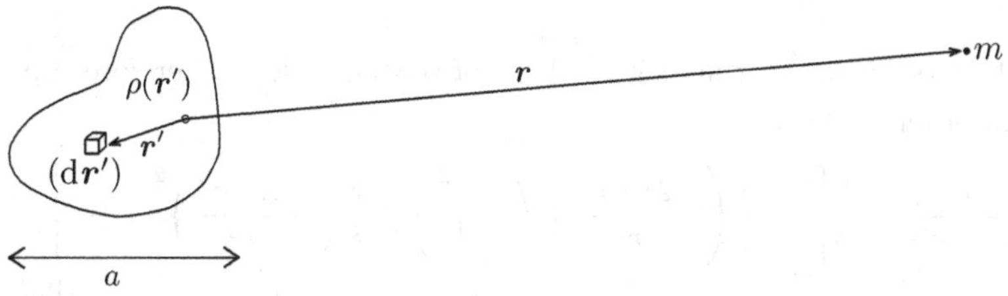

(6.2.1)

that is: $r \gg a$, where a is a characteristic length of $\rho(\boldsymbol{r}')$, we have $r \gg r'$ for all r' values in (6.1.6). In saying this, we tacitly assume what is indicated in (6.2.1), namely that the coordinate origin from which we measure the distances r and r' is in the vicinity of the "$\rho(\boldsymbol{r}') \neq 0$" region. But we can be more specific about this, by a deliberate choice of the point $\boldsymbol{r}' = 0$: We put it into the center-of-mass of the massive body with mass density $\rho(\boldsymbol{r}')$.

The center-of-mass position $\boldsymbol{R}$ is given by

$$M\boldsymbol{R} = \int (\mathrm{d}\boldsymbol{r}')\, \rho(\boldsymbol{r}')\boldsymbol{r}' , \tag{6.2.2}$$

the immediate generalization of (4.5.6), or equivalently

$$0 = \int (\mathrm{d}\boldsymbol{r}')\, \rho(\boldsymbol{r}')(\boldsymbol{r}' - \boldsymbol{R}) = \int (\mathrm{d}\boldsymbol{r}')\, \rho(\boldsymbol{R} + \boldsymbol{r}')\boldsymbol{r}' . \tag{6.2.3}$$

In the latter expression $(\mathrm{d}\boldsymbol{r}')\, \rho(\boldsymbol{R} + \boldsymbol{r}')$ is the mass element at position $\boldsymbol{r}'$ relative to $\boldsymbol{R}$. If we now agree, as we shall do in the present context, to use a coordinate system for which $\boldsymbol{R} = 0$, we have

$$\int (\mathrm{d}\boldsymbol{r}')\, \rho(\boldsymbol{r}')\boldsymbol{r}' = 0 . \tag{6.2.4}$$

Then all $\boldsymbol{r}'$ for which $\rho(\boldsymbol{r}') \neq 0$ are bounded in length by the characteristic length a, $r' < a$. The potential $\Phi(\boldsymbol{r})$ at a distant reference point $\boldsymbol{r}$, with $r \gg a \gtrsim r'$, can be approximated by the terms we get when expanding

$$\frac{1}{|\boldsymbol{r} - \boldsymbol{r}'|} = \left(r^2 - 2\boldsymbol{r} \cdot \boldsymbol{r}' + r'^2\right)^{-\frac{1}{2}} = \frac{1}{r}\left(1 - \frac{2\boldsymbol{r} \cdot \boldsymbol{r}'}{r^2} + \left(\frac{r'}{r}\right)^2\right)^{-\frac{1}{2}} \tag{6.2.5}$$

in powers of the small ratio $\frac{r'}{r}$. We note that

$$\frac{\boldsymbol{r}\cdot\boldsymbol{r}'}{r^2} = \frac{r'}{r}\cos\theta \qquad (\text{for } \boldsymbol{r}\cdot\boldsymbol{r}' = rr'\cos\theta) \tag{6.2.6}$$

is of first order in $\frac{r'}{r}$, whereas $\left(\frac{r'}{r}\right)^2$ is of second order. Therefore, up to second order we have

$$\frac{1}{|\boldsymbol{r}-\boldsymbol{r}'|} = \frac{1}{r}\left[1 - \frac{1}{2}\left(-\frac{2\boldsymbol{r}\cdot\boldsymbol{r}'}{r^2} + \left(\frac{r'}{r}\right)^2\right) + \frac{3}{8}\left(-\frac{2\boldsymbol{r}\cdot\boldsymbol{r}'}{r^2}\right)^2 + \cdots\right], \tag{6.2.7}$$

where the ellipsis stands for terms of higher order, that is: third, fourth, fifth, ... order in $\frac{r'}{r}$. This gives first

$$\begin{aligned}\frac{1}{|\boldsymbol{r}-\boldsymbol{r}'|} &= \frac{1}{r} + \frac{\boldsymbol{r}}{r^3}\cdot\boldsymbol{r}' + \frac{1}{2r^3}\left(3\frac{\boldsymbol{r}}{r}\cdot\boldsymbol{r}'\boldsymbol{r}'\cdot\frac{\boldsymbol{r}}{r} - r'^2\right) + \cdots \\ &= \frac{1}{r} + \frac{\boldsymbol{r}}{r^3}\cdot\boldsymbol{r}' + \frac{1}{2r^5}\boldsymbol{r}\cdot(3\boldsymbol{r}'\boldsymbol{r}' - r'^2\mathbf{1})\cdot\boldsymbol{r} + \cdots,\end{aligned} \tag{6.2.8}$$

where $\mathbf{1}$ is the unit dyadic of (3.2.63) and (3.2.64). We remember that in the dyadic product $\boldsymbol{r}'\boldsymbol{r}'$ we should think of the left vector as of column type and of the right vector as row type: $\boldsymbol{r}'\boldsymbol{r}'^{\mathrm{T}}$ if we are pedantic. Speaking of powers of r, the terms in (6.2.8) are proportional to $\frac{1}{r}$, $\frac{1}{r^2}$, and $\frac{1}{r^3}$, respectively, and we get

$$\Phi(\boldsymbol{r}) = -G\int(\mathrm{d}\boldsymbol{r}')\,\rho(\boldsymbol{r}')\left[\frac{1}{r} + \frac{\boldsymbol{r}}{r^3}\cdot\boldsymbol{r}' + \frac{1}{2r^5}\boldsymbol{r}\cdot(3\boldsymbol{r}'\boldsymbol{r}' - r'^2\mathbf{1})\cdot\boldsymbol{r} + \ldots\right] \tag{6.2.9}$$

for the gravitational potential. The $\boldsymbol{r}'$ integral of the $\frac{1}{r}$ term gives the total mass as in (6.1.2); the $\boldsymbol{r}'$ integral of the $\frac{1}{r^2}$ term is (6.2.4) and vanishes in view of our judicious choice of the coordinate origin; and the $\boldsymbol{r}'$ integral of the $\frac{1}{r^3}$ term identifies the so-called *quadrupole moment* dyadic $\mathbf{Q}$ of the mass distribution,

$$\mathbf{Q} = \int(\mathrm{d}\boldsymbol{r}')\,\rho(\boldsymbol{r}')\left(3\boldsymbol{r}'\boldsymbol{r}' - r'^2\mathbf{1}\right). \tag{6.2.10}$$

In the terminology of this hierarchy of *multipole moments*, the total mass M is the monopole moment (a good mouthful for something so simple), the dipole moment that equals the left-hand side of (6.2.4) vanishes, and the next in line after the quadrupole moment would be the octopole moment.

In summary, we have

$$\Phi(\boldsymbol{r}) = -G\frac{M}{r} - G\frac{\boldsymbol{r}\cdot\mathbf{Q}\cdot\boldsymbol{r}}{2r^5} + \cdots \tag{6.2.11}$$

for reference points $\boldsymbol{r}$ that are sufficiently far away from the mass distribution. The first term is the one that we have relied upon when discussing the Kepler orbits of planets attracted by the sun. This monopole term is *as if* all the mass were concentrated at $\boldsymbol{r} = 0$, which is the location of the center-of-mass. The leading correction is the quadrupole term which is smaller by a factor $(a/r)^2$, where a is the size of the mass distribution. As an example, let us consider the sun-earth system: The distance r is about 150×10^6 km, and the size of the sun is about 1.4×10^6 km, so that the ratio a/r is of the order of a percent, and its square is of the order of a percent of a percent, $(a/r)^2 \simeq 10^{-4}$. In addition, the components of the quadrupole moment dyadic $\mathbf{Q}$ are usually only a fraction of Ma^2. It follows that the quadrupole term represents a tiny correction to the $\frac{1}{r^2}$ Newtonian force, indeed.

6.3 Newton's shell theorem

A particular situation is that of an isotropic mass distribution, where $\rho(\boldsymbol{r}') = \rho(r')$ only depends on the distance $r' = |\boldsymbol{r}'|$ from the center-of-mass but not on the direction into which $\boldsymbol{r}'$ points:

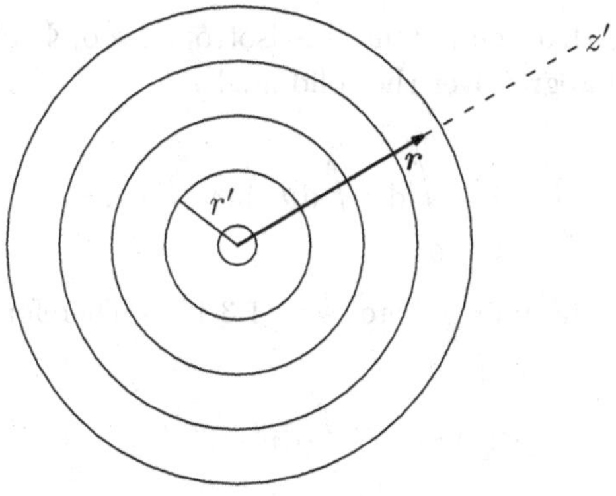

(6.3.1)

Spheres around the center-of-mass are constant-density surfaces. The gravitational potential for such a density is

$$\Phi(\boldsymbol{r}) = -G\int_0^\infty \mathrm{d}r'\, r'^2 \rho(r') \int_0^{2\pi} \mathrm{d}\varphi' \int_0^\pi \mathrm{d}\vartheta'\, \sin\vartheta' \frac{1}{\sqrt{r^2 - 2rr'\cos\vartheta' + r'^2}}\,, \tag{6.3.2}$$

if we use spherical coordinates — for which the volume element is $(\mathrm{d}\boldsymbol{r}') = \mathrm{d}r' r'^2 \mathrm{d}\varphi' \mathrm{d}\vartheta' \sin\vartheta'$ as in (1.3.15) — and orient the coordinate system such that the z' axis is along $\boldsymbol{r}$. The azimuthal integral gives a factor of 2π, and the polar angle integration is evaluated upon recognizing that the integrand is a total ϑ' derivative,

$$\begin{aligned}\int_0^\pi \mathrm{d}\vartheta' \frac{\sin\vartheta'}{\sqrt{r^2 - 2rr'\cos\vartheta' + r'^2}} &= \frac{1}{rr'}\sqrt{r^2 - 2rr'\cos\vartheta' + r'^2}\,\bigg|_{\vartheta'=0}^{\pi} \\ &= \frac{1}{rr'}\left(\sqrt{(r+r')^2} - \sqrt{(r-r')^2}\right) \\ &= \frac{r + r' - |r - r'|}{rr'} \\ &= \frac{(r_> + r_<) - (r_> - r_<)}{r_> r_<} = \frac{2}{r_>}\,,\end{aligned} \tag{6.3.3}$$

where $r_> = \mathrm{Max}\{r, r'\}$ and $r_< = \mathrm{Min}\{r, r'\}$ are the larger one and the smaller one of the two distances.

This takes us to

$$\Phi(r) = -G\int_0^\infty \mathrm{d}r'\, r'^2 4\pi\rho(r')\frac{1}{r_>}\,, \tag{6.3.4}$$

where we recognize that the potential is isotropic, too, $\Phi(\boldsymbol{r}) = \Phi(r)$. The factor of 4π is the integral over the solid angle,

$$\int \mathrm{d}\Omega = \int_0^{2\pi} \mathrm{d}\varphi \int_0^\pi \mathrm{d}\vartheta\, \sin\vartheta = 4\pi\,, \tag{6.3.5}$$

the surface area of the unit sphere; see (1.3.10). Therefore, we can also write

$$\Phi(r) = -G\int (\mathrm{d}\boldsymbol{r}')\, \frac{\rho(r')}{r_>} \tag{6.3.6}$$

for the isotropic gravitational potential of an isotropic mass density.

The remaining r' integration in (6.3.4) is split into the ranges $0 < r' < r$ and $r < r'$,

$$\Phi(r) = -G\int_0^r \mathrm{d}r'\, r'^2 4\pi\rho(r')\frac{1}{r} - G\int_r^\infty \mathrm{d}r'^2\, r' 4\pi\rho(r')\frac{1}{r'}$$

$$= -G\frac{M_r}{r} - G\int_r^\infty \mathrm{d}r'\, r' 4\pi\rho(r')\,, \tag{6.3.7}$$

with

$$M_r = \int_0^r \mathrm{d}r'\, r'^2 4\pi\rho(r') = \int_{(r'<r)} (\mathrm{d}\boldsymbol{r}')\, \rho(r')\,. \tag{6.3.8}$$

The first term in (6.3.7) is that of a point mass M_r at the center-of-mass, it is *as if* all the mass inside the sphere of radius r were at $r' = 0$. It follows that, for distances r so large that all of $\rho(r')$ is closer to $r' = 0$ than r,

$$\Phi(r) = -G\frac{M}{r} \quad \text{for} \quad M_r = M\,, \tag{6.3.9}$$

because then the integral over $r' > r$ vanishes as $\rho(r')$ is zero for all these r' values. This states that outside an isotropic mass distribution we only have the $-G\dfrac{M}{r}$ term, the monopole term of the mass distribution, while the quadrupole term and all higher multipole terms are exactly zero. In other words, the potential $\Phi(r)$ of an isotropic mass density has no further information about $\rho(r')$ for points r outside the massive body, other than the total mass of the monopole term.

The corresponding statement about the force on a point-like probe mass m is

$$\boldsymbol{F} = -m\boldsymbol{\nabla}\Phi(r) = -m\frac{\boldsymbol{r}}{r}\frac{\partial}{\partial r}\Phi(r) \tag{6.3.10}$$

with

$$\frac{\partial}{\partial r}\Phi(r) = G\frac{M_r}{r^2} - \frac{G}{r}\underbrace{\left[\frac{\partial}{\partial r}M_r + r\frac{\partial}{\partial r}\int_r^\infty \mathrm{d}r'\, r' 4\pi\rho(r')\right]}_{=0} = G\frac{M_r}{r^2}\,, \tag{6.3.11}$$

where

$$\frac{\partial}{\partial r} M_r = 4\pi r^2 \rho(r) = -r\frac{\partial}{\partial r} \int_r^\infty \mathrm{d}r'\, r' 4\pi\rho(r') \tag{6.3.12}$$

is used. The result

$$\boldsymbol{F} = -GmM_r \frac{\boldsymbol{r}}{r^3} \tag{6.3.13}$$

is *Newton's Shell Theorem*: For an isotropic mass density, the force at a distance r from the center is exactly equal to the force that one would have if all the mass inside the sphere with radius r were at the center.

Here is an application to terrestrial physics. If we assume that the mass distribution of the earth is isotropic — for many purposes, this is not a bad approximation — then the gravitational pull at the surface is given by mg with

$$g = \frac{GM}{R^2}\,, \tag{6.3.14}$$

where $G \cong \frac{2}{3} \times 10^{-10}\,\mathrm{m}^3/(\mathrm{kg\,s}^2)$ is the gravitational constant of (5.2.46), M is the mass of the earth, and

$$R = \frac{40\,000\,\mathrm{km}}{2\pi} = \frac{2}{\pi} \times 10^7\,\mathrm{m} \tag{6.3.15}$$

is the (equatorial) radius of the earth. With $g = 9.8\,\mathrm{m/s}^2 \cong \pi^2\,\mathrm{m/s}^2$, this gives

$$M = \frac{gR^2}{G} \cong \pi^2 \left(\frac{2}{\pi} \times 10^7\right)^2 \frac{3}{2} \times 10^{10}\,\mathrm{kg} = 6 \times 10^{24}\,\mathrm{kg} \tag{6.3.16}$$

for the mass of the earth (this value is actually quite accurate), and

$$\frac{M}{\frac{4\pi}{3}R^3} \cong 5.5\,\frac{\mathrm{g}}{\mathrm{cm}^3} \tag{6.3.17}$$

for the average mass density.

There is another consequence of $\Phi(r) = -\dfrac{GM}{r}$ for $r > R$. A massive body that is moving about the earth, possibly thrown up from the ground,

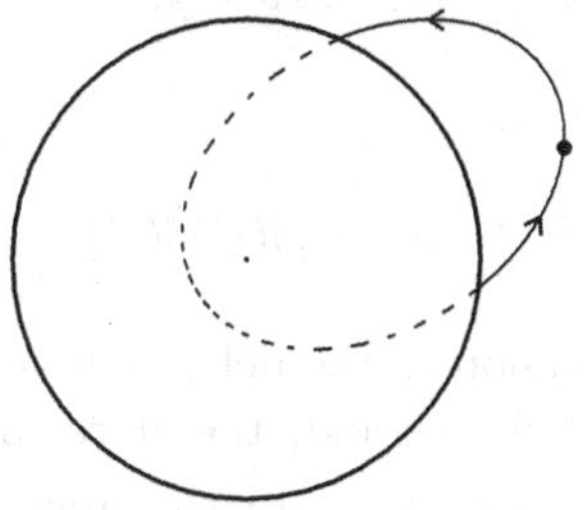

(6.3.18)

will follow a Kepler's ellipse with the center of the earth at one focus — well, at least until it hits the surface again, and only under circumstances that allow us to ignore the effects of air drag in the atmosphere. Satellites in low orbits (a few hundred km about the surface) do show the effects of the quadrupole moment of the earth, but this is finer detail. Also, there are the effects that stem from the rotation of the earth about the south-north axis; more about this in Chapter 12.

6.4 Green's function of the Laplacian differential operator

Returning to the general integral expression (6.1.6) for the gravitational potential of a given mass density, we note that it has the structure of an integration with a Green's function,

$$\Phi(\boldsymbol{r}) = \int (\mathrm{d}\boldsymbol{r}')\, G(\boldsymbol{r} - \boldsymbol{r}')\rho(\boldsymbol{r}'), \tag{6.4.1}$$

where the double use of the letter G for the Green's function and the gravitational constant is a bit unfortunate, but occasionally we are facing the fact that the alphabet has a finite number of letters with conventions about quite a few of them. Having this Green's function structure, we expect that there is an inverse relation that gives $\rho(\boldsymbol{r})$ by differentiating $\Phi(\boldsymbol{r})$ in the right way.

As a preparation for that discussion, let us first ask which mass density would give us the point-mass potential

$$\Phi(\boldsymbol{r}) = -\frac{GM}{r} \tag{6.4.2}$$

for *all* $r > 0$, not only for those distances beyond the radius of an isotropic mass density. Clearly, we need

$$\rho(\boldsymbol{r}') = M\delta(\boldsymbol{r}'), \tag{6.4.3}$$

where

$$\delta(\boldsymbol{r}') = \delta(x')\delta(y')\delta(z') \tag{6.4.4}$$

is the three-dimensional version of the delta function. The differential operator that we are looking for should, therefore, be such that it gives a multiple of $\delta(\boldsymbol{r})$ when applied to $\frac{1}{r}$. In the first place, this means that we need to get zero for $r > 0$. Since $\boldsymbol{\nabla}$ is a vector, the simplest scalar differential operator that we can try is its square, $\boldsymbol{\nabla}^2 = \boldsymbol{\nabla} \cdot \boldsymbol{\nabla}$, the Laplacian differential operator of (3.2.85). Recalling that $\boldsymbol{\nabla} f(r) = \frac{\boldsymbol{r}}{r} f'(r)$ and $\boldsymbol{\nabla} \cdot \boldsymbol{r} = 3$, we have

$$\begin{aligned} \boldsymbol{\nabla} \cdot \boldsymbol{\nabla} f(r) &= \frac{3}{r} f'(r) + \boldsymbol{r} \cdot \frac{\boldsymbol{r}}{r} \frac{\partial}{\partial r} \left[\frac{1}{r} f'(r) \right] = \frac{2}{r} f'(r) + f''(r) \\ &= \frac{1}{r^2} \frac{\partial}{\partial r} \left(r^2 \frac{\partial}{\partial r} f(r) \right) = \frac{1}{r} \frac{\partial^2}{\partial r^2} \left(r f(r) \right) \end{aligned} \tag{6.4.5}$$

for the Laplacian differentiating an isotropic function. In particular, for $f(r) = \frac{1}{r}$ this gives

$$\boldsymbol{\nabla}^2 \frac{1}{r} = 0 \quad \text{for} \quad r > 0 \tag{6.4.6}$$

as we wished, but we cannot say anything about $\boldsymbol{r} = 0$ from this argument other than that there should be a singularity.

We regard $\frac{1}{r}$ as the limit of a nonsingular isotropic function, for which

$$f(r) = \frac{1}{\sqrt{r^2 + r_0^2}} \tag{6.4.7}$$

is a convenient choice, where $r_0 > 0$ with $r_0 \to 0$ eventually. Here,

$$\boldsymbol{\nabla}^2 f(r) = \frac{1}{r} \frac{\partial^2}{\partial r^2} \frac{r}{\sqrt{r^2 + r_0^2}} = -\frac{3r_0^2}{\sqrt{r^2 + r_0^2}^{\,5}}, \tag{6.4.8}$$

which has the correct null limit for $r > 0$ and $r_0 \to 0$, while it becomes largely negative for $r = 0$ and $r_0 \to 0$. These are the hallmarks of the three-dimensional delta function, provided that the integral of $\nabla^2 f(r)$ is finite for $r_0 \to 0$. Well, let us check this. First, we note that

$$\int (\mathrm{d}\boldsymbol{r})\,\nabla^2 f(r) = 4\pi \int_0^\infty \mathrm{d}r\,r^2\,\frac{1}{r^2}\frac{\partial}{\partial r}\left(r^2\frac{\partial}{\partial r}f(r)\right) = 4\pi r^2\frac{\partial f(r)}{\partial r}\bigg|_{r\,=\,0}^{\infty}, \tag{6.4.9}$$

where we use (6.4.5) for $\nabla^2 f(r)$, and the factor of 4π is the integral over the solid angle of (6.3.5). Now we apply this to the $f(r)$ in (6.4.7),

$$\int (\mathrm{d}\boldsymbol{r})\,\nabla^2 f(r) = -4\pi\frac{r^3}{\sqrt{r^2+r_0^2}^{\,3}}\bigg|_{r\,=\,0}^{\infty} = -4\pi\,. \tag{6.4.10}$$

Not only is this finite in the limit $r_0 \to 0$, it does not even depend on r_0. We conclude that

$$\nabla^2\frac{1}{\sqrt{r^2+r_0^2}}\bigg|_{r_0\,\to\,0} = -4\pi\delta(\boldsymbol{r})\,, \tag{6.4.11}$$

which is to say that $-\dfrac{1}{4\pi}\nabla^2\dfrac{1}{\sqrt{r^2+r_0^2}} = \dfrac{3}{4\pi}\dfrac{r_0^2}{\sqrt{r^2+r_0^2}^{\,5}}$ is a model for the three-dimensional delta function, much like we had models for the one-dimensional delta function in Section 2.2.6(b).

It follows that

$$\nabla^2\frac{1}{r} = -4\pi\delta(\boldsymbol{r})\,, \tag{6.4.12}$$

which is an important identity that is worth memorizing. Its immediate consequence

$$\nabla^2\frac{1}{|\boldsymbol{r}-\boldsymbol{r}'|} = -4\pi\delta(\boldsymbol{r}-\boldsymbol{r}') \tag{6.4.13}$$

tells us that

$$\begin{aligned}\nabla^2\Phi(\boldsymbol{r}) &= -G\int(\mathrm{d}\boldsymbol{r}')\,\rho(\boldsymbol{r}')\nabla^2\frac{1}{|\boldsymbol{r}-\boldsymbol{r}'|}\\ &= -G\int(\mathrm{d}\boldsymbol{r}')\,\rho(\boldsymbol{r}')\big[-4\pi\delta(\boldsymbol{r}-\boldsymbol{r}')\big]\\ &= 4\pi G\rho(\boldsymbol{r})\,.\end{aligned} \tag{6.4.14}$$

In summary, we have the pair of equations

$$\Phi(\boldsymbol{r}) = -G \int (\mathrm{d}\boldsymbol{r}') \, \frac{\rho(\boldsymbol{r}')}{|\boldsymbol{r}-\boldsymbol{r}'|} \,,$$

$$\rho(\boldsymbol{r}) = \frac{1}{4\pi G} \boldsymbol{\nabla}^2 \Phi(\boldsymbol{r}) \,, \tag{6.4.15}$$

which state how to find the gravitational potential $\Phi(\boldsymbol{r})$ if we know the mass density $\rho(\boldsymbol{r})$, and how to find the mass density if we know the potential. In terms of Green's functions, the mathematical statement is that $|\boldsymbol{r}-\boldsymbol{r}'|^{-1}$ is the Green's function for the Laplacian differential operator, where we need to remember the factor of -4π in (6.4.13).

Chapter 7

Variational Problems

7.1 Johann Bernoulli's challenge: The brachistochrone

Starting from rest at $(x, y) = (0, 0)$ a point mass m slides without friction on a curve $y(x)$ that takes it to point $(x, y) = (a, b)$ below,with $a > 0$and $b > 0$:

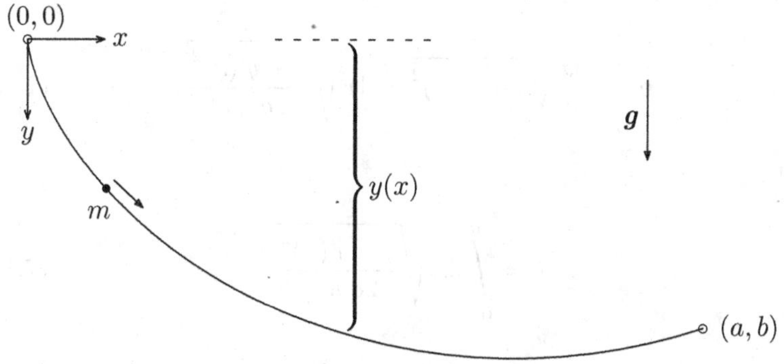

(7.1.1)

Along which curve will the mass get there in the shortest time? This question about the *brachistochrone* was first asked by Johann Bernoulli* in 1696 who challenged his contemporaries by it. It is said that only five solutions were received, among them one by an anonymous author. Upon seeing this solution, however, Bernoulli recognized immediately that its author was Newton himself and exclaimed *tanquam ex ungue leonem* (just like the lion is known by his paw). The story is that Newton received Bernoulli's personal letter with the challenge in the afternoon and solved the problem before going to bed that night.

*Johann BERNOULLI (1667–1748)

In terms of the unknown path function $y(x)$, we have the speed at $(x, y(x))$ given by

$$\frac{m}{2}v^2 = mgy \quad \text{or} \quad v = \sqrt{2gy(x)}\,, \tag{7.1.2}$$

which is an immediate consequence of energy conservation. We combine this with

$$v^2 = \left(\frac{\mathrm{d}x}{\mathrm{d}t}\right)^2 + \left(\frac{\mathrm{d}y}{\mathrm{d}t}\right)^2 = \left(\frac{\mathrm{d}x}{\mathrm{d}t}\right)^2\left[1 + \left(\frac{\mathrm{d}y}{\mathrm{d}x}\right)^2\right] \tag{7.1.3}$$

or

$$v = \frac{\mathrm{d}x}{\mathrm{d}t}\sqrt{1 + y'(x)^2} \tag{7.1.4}$$

to express the time differential in terms of the x differential and the x derivative of $y(x)$,

$$\mathrm{d}t = \frac{\mathrm{d}x}{v}\sqrt{1 + y'(x)^2} = \mathrm{d}x\,\sqrt{\frac{1 + y'(x)^2}{2gy(x)}}\,, \tag{7.1.5}$$

so that

$$T = \int_0^a \mathrm{d}x\,\sqrt{\frac{1 + y'(x)^2}{2gy(x)}} \tag{7.1.6}$$

is the time it takes to get from $(0,0)$ to (a,b) along the path specified by $y(x)$. The challenge is now to determine for which $y(x)$ this travel time is shortest, whereby all $y(x)$s that enter the competition must respect the constraints

$$y(0) = 0 \quad \text{and} \quad y(a) = b\,. \tag{7.1.7}$$

This is an example of a *variational problem* for which one has the so-called calculus of variations as a tool, the generalization of the ordinary calculus of functions of one or several variables to functionals of functions. We got a first taste of that in Section 6.1 when we noted that the gravitational energy of a mass distribution is a functional of the mass density.

7.2 Euler–Lagrange equations

Let us regard the brachistochrone problem as a special case of problems of the structure,

$$\int_{x_0}^{x_1} \mathrm{d}x\, f\big(x, y(x), y'(x)\big) \overset{!}{=} \text{extremum}, \tag{7.2.1}$$

where the extremum can be a maximum or a minimum (as is the case for the brachistochrone) or a saddle point of sorts. The integrand $f(x, y, y')$ that specifies the problem assigns a number to the triplet x, y, y' of numbers, and the permissible functions $y(x)$ are restricted by preassigned values for $x = x_0$ and $x = x_1$,

$$y(x_0) = y_0\,, \qquad y(x_1) = y_1\,. \tag{7.2.2}$$

Let us denote by $\overline{y}(x)$ the solution of the problem, then all other permissible $y(x)$ can be written as

$$y(x) = \overline{y}(x) + \delta y(x)\,, \tag{7.2.3}$$

where the deviation $\delta y(x)$ is restricted by

$$\delta y(x_0) = 0 = \delta y(x_1)\,. \tag{7.2.4}$$

The situation is depicted here:

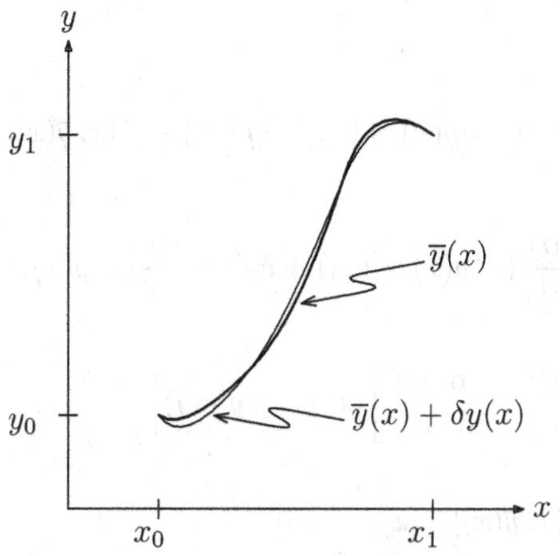

(7.2.5)

where the thick curve symbolizes the solution $\overline{y}(x)$ and the thin curve the trial function $\overline{y}(x) + \delta y(x)$.

That we have an extremum for $y = \overline{y}$ is to say that

$$\int_{x_0}^{x_1} \mathrm{d}x\, f\big(x, \overline{y}(x) + \delta y(x), \overline{y}'(x) + \delta y'(x)\big) \tag{7.2.6}$$

differs from the extremal value

$$\int_{x_0}^{x_1} \mathrm{d}x\, f\big(x, \overline{y}(x), \overline{y}'(x)\big) \tag{7.2.7}$$

by terms that are second-order in δy. Put differently: There is no first-order difference.

We expand $f(x, \overline{y} + \delta y, \overline{y}' + \delta y')$ to first order in δy,

$$\begin{aligned} f\big(x, \overline{y}(x) + \delta y(x), \overline{y}'(x) + \delta y'(x)\big) &= f\big(x, \overline{y}(x), \overline{y}'(x)\big) \\ &\quad + \delta y(x) \frac{\partial f}{\partial y}\big(x, \overline{y}(x), \overline{y}'(x)\big) \\ &\quad + \delta y'(x) \frac{\partial f}{\partial y'}\big(x, \overline{y}(x), \overline{y}'(x)\big)\,, \end{aligned} \tag{7.2.8}$$

where $\dfrac{\partial f}{\partial y}$ and $\dfrac{\partial f}{\partial y'}$ denote the usual partial derivatives of $f(x, y, y')$ with respect to the second and the third argument. The first-order change of the functional is then

$$\begin{aligned} &\delta \int_{x_0}^{x_1} \mathrm{d}x\, f\big(x, y(x), y'(x)\big)\bigg|_{y=\overline{y}} \\ &= \int_{x_0}^{x_1} \mathrm{d}x \Big[f\big(x, \overline{y}(x) + \delta y(x), \overline{y}'(x) + \delta y'(x)\big) - f\big(x, \overline{y}(x), \overline{y}'(x)\big)\Big] \\ &= \int_{x_0}^{x_1} \mathrm{d}x \left[\delta y(x) \frac{\partial f}{\partial y}\big(x, y(x), y'(x)\big) + \delta y'(x) \frac{\partial f}{\partial y'}\big(x, y(x), y'(x)\big)\right]\Bigg|_{y=\overline{y}} \\ &= \int_{x_0}^{x_1} \mathrm{d}x\, \delta y(x) \left[\frac{\partial f}{\partial y} - \frac{\mathrm{d}}{\mathrm{d}x}\frac{\partial f}{\partial y'}\right]\big(x, y(x), y'(x)\big)\bigg|_{y=\overline{y}} \\ &\quad + \delta y(x) \frac{\partial f}{\partial y'}\big(x, \overline{y}(x), \overline{y}'(x)\big)\bigg|_{x_0}^{x_1}\,, \end{aligned} \tag{7.2.9}$$

where the last step is an integration by parts. The boundary term that it produces vanishes as a consequence of (7.2.4), and we arrive at

$$\delta \int_{x_0}^{x_1} \mathrm{d}x\, f(x, y(x), y'(x))\Big|_{y=\overline{y}}$$

$$= \int_{x_0}^{x_1} \mathrm{d}x\, \delta y(x) \left[\frac{\partial f}{\partial y} - \frac{\mathrm{d}}{\mathrm{d}x}\frac{\partial f}{\partial y'}\right](x, \overline{y}(x), \overline{y}'(x)) . \qquad (7.2.10)$$

At an extremum of the functional, the difference $\frac{\partial f}{\partial y} - \frac{\mathrm{d}}{\mathrm{d}x}\frac{\partial f}{\partial y'}$ must vanish for all x. For, if it were nonzero, we could choose $\delta y(x)$ such that it has the same sign as this difference everywhere and get a definite first-order increase, or with the opposite sign everywhere to get a definite decrease, both contradicting the extremum property that there are no first-order changes. It follows that the extremum $\overline{y}(x)$ must obey the *Euler*–Lagrange† equation*

$$\left(\frac{\partial f}{\partial y} - \frac{\mathrm{d}}{\mathrm{d}x}\frac{\partial f}{\partial y'}\right)(x, y(x), y'(x)) = 0 , \qquad (7.2.11)$$

a second-order differential equation for the unknown solution $y(x)$ (we drop the bar on $\overline{y}(x)$ now), to be solved subject to the constraints (7.2.2) that state the values of $y(x)$ at the boundaries.

Note that $\frac{\mathrm{d}}{\mathrm{d}x}$ in $\frac{\mathrm{d}}{\mathrm{d}x}\frac{\partial f}{\partial y'}$ is the total derivative of $\frac{\partial f}{\partial y'}(x, y(x), y'(x))$ with respect to x; explicitly it is

$$\frac{\mathrm{d}}{\mathrm{d}x}\frac{\partial f}{\partial y'} = \frac{\partial^2 f}{\partial x \partial y'} + y'\frac{\partial^2 f}{\partial y \partial y'} + y''\frac{\partial^2 f}{\partial y'^2} \qquad (7.2.12)$$

after two applications of the chain rule. Careful distinction of $\frac{\partial}{\partial x}$ and $\frac{\mathrm{d}}{\mathrm{d}x}$ is called for.

We are almost ready for applying the formalism to the brachistochrone problem, but it helps to establish an alternative form of the Euler–Lagrange equation first. Consider this expression:

$$\frac{\mathrm{d}}{\mathrm{d}x}\left(f - y'\frac{\partial f}{\partial y'}\right) = \frac{\partial f}{\partial x} + y'\frac{\partial f}{\partial y} + \underbrace{y''\frac{\partial f}{\partial y'} - y''\frac{\partial f}{\partial y'}}_{=0} - y'\frac{\mathrm{d}}{\mathrm{d}x}\frac{\partial f}{\partial y'}$$

$$= \frac{\partial f}{\partial x} + y'\underbrace{\left(\frac{\partial f}{\partial y} - \frac{\mathrm{d}}{\mathrm{d}x}\frac{\partial f}{\partial y'}\right)}_{=0}, \qquad (7.2.13)$$

*Leonhard EULER (1707–1783) †Joseph Louis de LAGRANGE (1736–1813) where the Euler–Lagrange equation (7.2.11) is used. We thus have

$$\frac{\mathrm{d}}{\mathrm{d}x}\left(f - y'\frac{\partial f}{\partial y'}\right) = \frac{\partial f}{\partial x} \tag{7.2.14}$$

as an equivalent statement, which is particularly useful if there is no parametric x dependence in f but only the implicit dependence that results from the x dependence of $y(x)$ and $y'(x)$.

7.3 Solution of the brachistochrone problem

This is indeed the situation in the problem of the brachistochrone, for which

$$f(x, y, y') = \sqrt{\frac{1 + y'^2}{2gy}} \tag{7.3.1}$$

has no x dependence, $\dfrac{\partial f}{\partial x} = 0$. The second version (7.2.14) of the Euler–Lagrange equation then tells us that

$$\begin{aligned} f - y'\frac{\partial f}{\partial y'} &= \frac{1}{\sqrt{2gy}}\left(\sqrt{1+y'^2} - \frac{y'^2}{\sqrt{1+y'^2}}\right) \\ &= \frac{1}{\sqrt{2gy}}\frac{1}{\sqrt{1+y'^2}} \end{aligned} \tag{7.3.2}$$

is constant in x for the solution $y(x)$. Upon writing $1/\sqrt{4gR}$ for this constant, we have

$$\left(1 + y'^2\right) y = 2R\,, \tag{7.3.3}$$

a first-order differential equation for $y(x)$. We separate the variables,

$$\mathrm{d}x = \pm \mathrm{d}y\sqrt{\frac{y}{2R - y}}\,, \tag{7.3.4}$$

where both signs can occur if the trajectory $y(x)$ is as depicted in (7.1.1) with a largest y value between $x = 0$ and $x = a$. The expression on the right motivates the ansatz

$$y = 2R\sin\left(\frac{\phi}{2}\right)^2\,, \tag{7.3.5}$$

for which

$$dy = R\,d\phi\,\sin\phi \quad \text{and} \quad \frac{y}{2R-y} = \tan\left(\frac{\phi}{2}\right)^2, \tag{7.3.6}$$

and we take care of the sign assignment by choosing

$$\sqrt{\frac{y}{2R-y}} = \pm\tan\left(\frac{\phi}{2}\right) = \pm\frac{1-\cos\phi}{\sin\phi} \tag{7.3.7}$$

to arrive at

$$dx = R(1-\cos\phi)d\phi\,. \tag{7.3.8}$$

Together with (7.3.5) and $(x, y) = (0, 0)$ for $\phi = 0$, this gives the parameterization

$$\begin{aligned} x &= R(\phi - \sin\phi)\,, \\ y &= R(1-\cos\phi)\,, \end{aligned} \tag{7.3.9}$$

for the brachistochrone. One verifies easily that (7.3.3) holds, confirming that the ansatz (7.3.5) works.

It is not necessary, nor easy or helpful to solve the x equation in (7.3.9) for ϕ and then express y as a function of x. The parameterization specifies the extremal path all right. In fact, it tells us that the path is a *cycloid*, the trajectory of a point on a circle of radius R that is rolling along a line:

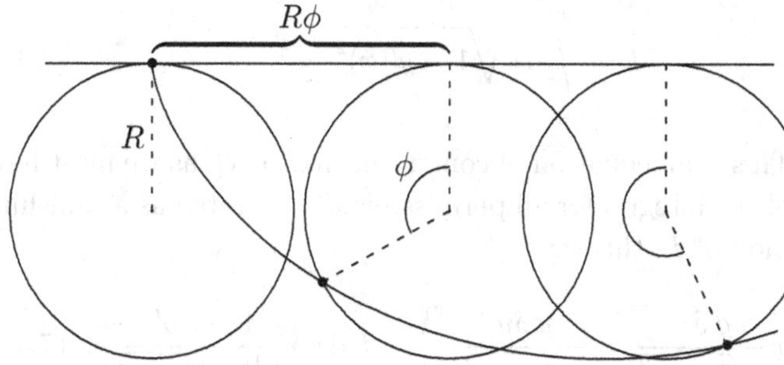

(7.3.10)

For the given endpoint $(x, y) = (a, b)$, we have

$$a = R(\phi_0 - \sin\phi_0)\,, \quad b = R(1-\cos\phi_0) \tag{7.3.11}$$

to determine the final turning angle ϕ_0 and the radius R of the circle, with $0 < \phi_0 < 2\pi$ and $R > 0$. For obvious physical reasons, these equations have no solutions when $b < 0$, that is: when the end point is above the start point in (7.1.1).

7.4 Jakob Bernoulli's problem: The catenary

Here is another problem, this one invented by Johann Bernoulli's older brother Jakob,* the problem of the *catenary*: What is the shape of the curve of a hanging homogeneous chain (rope, cable, ...)

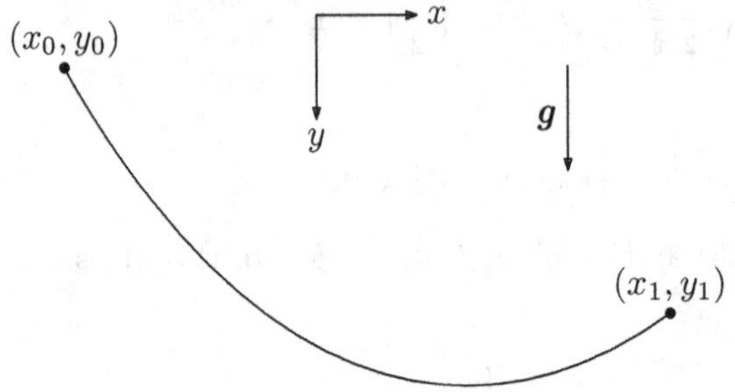

(7.4.1)

In this problem the length L and the mass M of the chain are given as well as the positions of the suspension points, (x_0, y_0) and (x_1, y_1). Since the length of the piece of chain between neighboring points (x, y) and $(x + \mathrm{d}x, y + \mathrm{d}y)$ is

$$\sqrt{(\mathrm{d}x)^2 + (\mathrm{d}y)^2} = \mathrm{d}x\sqrt{1 + y'(x)^2}\,, \tag{7.4.2}$$

we have

$$L = \int_{x_0}^{x_1} \mathrm{d}x\,\sqrt{1 + y'(x)^2} \tag{7.4.3}$$

for the length. This is an equation of constraint, inasmuch as we must have the same value of the integral for all permissible $y(x)$. Stated as a vanishing first-order variation of L, this says

$$0 = \delta L = \int_{x_0}^{x_1} \mathrm{d}x\,\frac{y'\delta y'}{\sqrt{1 + y'^2}} = \underbrace{\frac{y'\delta y}{\sqrt{1 + y'^2}}\bigg|_{x_0}^{x_1}}_{=0} - \int_{x_0}^{x_1} \mathrm{d}x\,\delta y \frac{\mathrm{d}}{\mathrm{d}x}\frac{y'}{\sqrt{1 + y'^2}}\,, \tag{7.4.4}$$

where the usual integration by parts converts $\delta y'$ into δy, and the boundary term does not contribute since $y(x_0) = y_0$ and $y(x_1) = y_1$ are fixed values.

---

*Jakob BERNOULLI (1655–1705)

Accordingly, the restriction

$$\int_{x_0}^{x_1} \mathrm{d}x\,\delta y \frac{\mathrm{d}}{\mathrm{d}x}\frac{y'}{\sqrt{1+y'^2}} = 0 \tag{7.4.5}$$

selects the permissible deviations $\delta y(x)$ in addition to $\delta y(x_0) = 0$ and $\delta y(x_1) = 0$, for all $y(x)$ that are allowed to enter the competition.

The segment of length $\mathrm{d}s = \mathrm{d}x\sqrt{1+y'^2}$ has mass $\frac{M}{L}\mathrm{d}s$ and contributes $-\frac{Mg}{L}y\mathrm{d}s$ to the potential energy of the chain in the homogeneous gravitational field near the surface of the earth. Therefore, the total potential energy of the chain is

$$E_{\mathrm{pot}} = -\frac{Mg}{L}\int_{x_0}^{x_1} \mathrm{d}x\, y\sqrt{1+y'^2}\,. \tag{7.4.6}$$

This potential energy is smallest for the actual $y(x)$, for which $\delta E_{\mathrm{pot}} = 0$ must hold. After yet another integration by parts that gives us a vanishing contribution from the boundaries, the response of E_{pot} to variations of $y(x)$ is

$$\delta E_{\mathrm{pot}} = -\frac{Mg}{L}\int_{x_0}^{x_1} \mathrm{d}x\,\delta y\left[\sqrt{1+y'^2} - \frac{\mathrm{d}}{\mathrm{d}x}\frac{yy'}{\sqrt{1+y'^2}}\right], \tag{7.4.7}$$

where we recognize, of course, the two terms of the Euler–Lagrange equation (7.2.11),

$$\begin{aligned} \frac{\partial}{\partial y}\left(y\sqrt{1+y'^2}\right) &= \sqrt{1+y'^2}\,, \\ \frac{\mathrm{d}}{\mathrm{d}x}\frac{\partial}{\partial y'}\left(y\sqrt{1+y'^2}\right) &= \frac{\mathrm{d}}{\mathrm{d}x}\frac{yy'}{\sqrt{1+y'^2}}\,. \end{aligned} \tag{7.4.8}$$

This time, however, we cannot conclude that the factor $[\cdots]$ in (7.4.7) that multiplies $\delta y(x)$ vanishes; rather it must be a multiple of the corresponding factor in the equation of constraint (7.4.5),

$$\sqrt{1+y'^2} - \frac{\mathrm{d}}{\mathrm{d}x}\frac{yy'}{\sqrt{1+y'^2}} = -\lambda\frac{\mathrm{d}}{\mathrm{d}x}\frac{y'}{\sqrt{1+y'^2}}\,, \tag{7.4.9}$$

where λ does not depend on x; this λ is an example of a *Lagrange multiplier.*

Upon combining the two $\frac{\mathrm{d}}{\mathrm{d}x}$ terms,

$$\sqrt{1+y'^2} - \frac{\mathrm{d}}{\mathrm{d}x}\frac{(y-\lambda)y'}{\sqrt{1+y'^2}} = 0\,, \tag{7.4.10}$$

we have the Euler–Lagrange equation for

$$f(x,y,y') = y\sqrt{1+y'^2} - \lambda\sqrt{1+y'^2} = (y-\lambda)\sqrt{1+y'^2}\,, \tag{7.4.11}$$

and since there is no explicit x dependence, the second form (7.2.14) of the Euler–Lagrange equation states that $f - y'\frac{\partial f}{\partial y'}$ is constant,

$$(y-\lambda)\sqrt{1+y'^2} - y'\frac{(y-\lambda)y'}{\sqrt{1+y'^2}} = \frac{y-\lambda}{\sqrt{1+y'^2}} = \text{constant}\,. \tag{7.4.12}$$

We could now separate variables and so forth, but the structure of this constant combination of $y(x)$ and $y'(x)$ is quite suggestive and invites us to try the ansatz

$$y(x) = \lambda - a\cosh\left(\frac{x-\overline{x}}{a}\right) \tag{7.4.13}$$

with constant parameters a and $\overline{x}$ whose values are to be determined. This gives

$$1 + y'^2 = 1 + \sinh\left(\frac{x-\overline{x}}{a}\right)^2 = \cosh\left(\frac{x-\overline{x}}{a}\right)^2 = \left(\frac{y-\lambda}{a}\right)^2, \tag{7.4.14}$$

so that

$$\frac{y-\lambda}{\sqrt{1+y'^2}} = -a \tag{7.4.15}$$

is indeed constant, which confirms that the ansatz (7.4.13) works.

In summary, then, the curve of the hanging chain, the catenary, is the hyperbolic function (7.4.13). The three parameters of this solution — $\overline{x}$, a, and λ — are to be determined from the boundary conditions

$$\begin{aligned} y(x_0) &= y_0 = \lambda - a\cosh\left(\frac{\overline{x}-x_0}{a}\right), \\ y(x_1) &= y_1 = \lambda - a\cosh\left(\frac{x_1-\overline{x}}{a}\right), \end{aligned} \tag{7.4.16}$$

and the constraint (7.4.3),

$$L = \int_{x_0}^{x_1} \mathrm{d}x \sqrt{1 + y'(x)^2} = \int_{x_0}^{x_1} \mathrm{d}x \cosh\left(\frac{x - \overline{x}}{a}\right), \tag{7.4.17}$$

that is

$$L = a \sinh\left(\frac{x_1 - \overline{x}}{a}\right) + a \sinh\left(\frac{\overline{x} - x_0}{a}\right). \tag{7.4.18}$$

The meaning of $\overline{x}$ is clear: It is the x coordinate of the lowest point of the chain. In a typical configuration, we expect $x_0 < \overline{x} < x_1$ but both $\overline{x} < x_0 < x_1$ and $x_0 < x_1 < \overline{x}$ are possible as well; steel cables that support cable cars on steep slopes are real-life examples. The chain must be long enough,

$$L > \sqrt{(x_1 - x_0)^2 + (y_1 - y_0)^2}\,, \tag{7.4.19}$$

otherwise the are no values for $\overline{x}$, a, and λ that obey (7.4.16) and (7.4.18); see Exercise 114.

A particular situation is the symmetric one with $x_0 = -b$, $x_1 = b$, and $y_0 = y_1 = 0$, for which $\lambda = a\cosh(b/a)$, $\overline{x} = 0$, and

$$y(x) = a\left[\cosh\left(\frac{b}{a}\right) - \cosh\left(\frac{x}{a}\right)\right] \tag{7.4.20}$$

with $2a\sinh(b/a) = L > 2b$. Clearly, there is no solution when the chain is too short, that is $L < 2b$.

7.5 Handling constraints: Lagrange multipliers

Having seen how a variational problem with an integral constraint is handled with the aid of a Lagrange multiplier, let us now consider the general scenario. It is characterized by the request to find a function $y(x)$ for which the functional of $f(x, y, y')$ is extremal,

$$\int_{x_0}^{x_1} \mathrm{d}x\, f\big(x, y(x), y'(x)\big) \overset{!}{=} \text{extremum}\,, \tag{7.5.1}$$

while $y(x)$ is subject to J constraints of the form

$$\int_{x_0}^{x_1} \mathrm{d}x\, g_j(x, y(x), y'(x)) = G_j \tag{7.5.2}$$

with $j = 1, 2, \ldots, J$. In addition, there are the fixed boundary values

$$y(x_0) = y_0\,, \quad y(x_1) = y_1\,, \tag{7.5.3}$$

so that the variation $\delta y(x)$ must respect the boundary constraints

$$\delta y(x_0) = 0\,, \quad \delta y(x_1) = 0 \tag{7.5.4}$$

as well as the integral constraints

$$\int_{x_0}^{x_1} \mathrm{d}x\, \delta y(x) \left[\frac{\partial}{\partial y} g_j(x, y, y') - \frac{\mathrm{d}}{\mathrm{d}x} \frac{\partial g_j}{\partial y'}(x, y, y') \right] = 0 \tag{7.5.5}$$

for $j = 1, 2, \ldots, J$. These constraints jointly state which variations are permissible.

The extremum condition for the f-functional (7.5.1) then says that

$$\int_{x_0}^{x_1} \mathrm{d}x\, \delta y(x) \left[\frac{\partial f}{\partial y}(x, y, y') - \frac{\mathrm{d}}{\mathrm{d}x} \frac{\partial f}{\partial y'}(x, y, y') \right] = 0 \tag{7.5.6}$$

for all *permissible* variations. It follows that

$$\frac{\partial f}{\partial y} - \frac{\mathrm{d}}{\mathrm{d}x} \frac{\partial f}{\partial y'} = \sum_{j=1}^{J} \lambda_j \left[\frac{\partial g_j}{\partial y} - \frac{\mathrm{d}}{\mathrm{d}x} \frac{\partial g_j}{\partial y'} \right] \tag{7.5.7}$$

is obeyed by the function $y(x)$ for which the f-functional acquires an extremal value. Here, the λ_js are constants, the *Lagrange multipliers*, one for each constraint. The solution of the second-order differential equation for $y(x)$ will contain two more constants, so that we have exactly the right number of constants to satisfy the values of the boundary constraints and the integral constraints with values $G_1, G_2, \ldots, G_J$. In the example of the catenary in Section 7.4, there was one constraint (the fixed length of the chain) and we had a total of three parameters in the general solution $y(x)$ of (7.4.13).

The equation for $y(x)$ can be rearranged to read

$$\frac{\partial}{\partial y}\left(f - \sum_{j=1}^{J}\lambda_j g_j\right) - \frac{\mathrm{d}}{\mathrm{d}x}\frac{\partial}{\partial y'}\left(f - \sum_{j=1}^{J}\lambda_j g_j\right) = 0\,, \qquad (7.5.8)$$

which is the Euler–Lagrange equation for the *unconstrained* variation of the integrand

$$f - \sum_{j=1}^{J}\lambda_j g_j\,. \qquad (7.5.9)$$

We can derive this systematically by replacing the original problem

$$\delta\int_{x_0}^{x_1}\mathrm{d}x\, f\bigl(x, y(x), y'(x)\bigr) = 0\,, \qquad (7.5.10)$$

subject to the integral constraints, by the new problem

$$\delta\left[\int_{x_0}^{x_1}\mathrm{d}x\, f\bigl(x, y(x), y'(x)\bigr) + \sum_{j=1}^{J}\lambda_j\left(G_j - \int_{x_0}^{x_1}\mathrm{d}x\, g_j\bigl(x, y(x), y'(x)\bigr)\right)\right] = 0\,, \qquad (7.5.11)$$

where δ means variations of the looked-for function $y(x)$ *and* of all the Lagrange multipliers $\lambda_1, \lambda_2, \ldots, \lambda_J$. For the extremal $y(x)$, the λ_js multiply vanishing expressions, so that the value of the functional is not changed by adding these terms. But now we get (7.5.8) upon varying $y(x)$, and the variations of the λ_js give the equations of constraint (7.5.2),

$$\delta\lambda_j:\quad G_j - \int_{x_0}^{x_1}\mathrm{d}x\, g_j\bigl(x, y(x), y'(x)\bigr) = 0\,. \qquad (7.5.12)$$

In summary, the inclusion of the constraints into the functional with the aid of Lagrange multipliers, one for each constraint, leads to the correct differential equation for the extremal $y(x)$ when requiring that the extended functional — that is: the contents of $[\cdots]$ in (7.5.11) — is extremal for variations of $y(x)$, and we get the integral equations of constraint when requiring that this extended functional is stationary under variations of the λ_js.

Chapter 8

Principle of Stationary Action

8.1 Lagrange function

8.1.1 *One coordinate*

For the variational problem

$$\delta \int_{x_0}^{x_1} \mathrm{d}x\, f(x, y, y') = 0 \tag{8.1.1}$$

we have the Euler–Lagrange equation (7.2.11),

$$\frac{\mathrm{d}}{\mathrm{d}x}\frac{\partial f}{\partial y'} = \frac{\partial f}{\partial y}\,, \tag{8.1.2}$$

a second-order differential equation that determines the extremal $y(x)$ for which the integral in (8.1.1) is stationary, that is: there are no first-order changes if $y(x)$ is replaced by a neighboring $y(x) + \delta y(x)$. Now, in the context of classical mechanics, we have Newton's equation of motion

$$\frac{\mathrm{d}}{\mathrm{d}t} m\dot{x} = F(x, t) = -\frac{\partial}{\partial x} V(x, t)\,, \tag{8.1.3}$$

obeyed by the actual trajectory $x(t)$, the position of the point mass as a function of time (one-dimensional motion for now). The comparison

$$\begin{array}{r|c|c} \text{variable} & x & t \\ \text{function} & y(x) & x(t) \\ \text{derivative} & y'(x) & \dot{x}(t) \end{array} \tag{8.1.4}$$

suggests strongly that there is a variational problem that has the trajectory $x(t)$ as the extremal solution, with Newton's equation of motion stated as the Euler–Lagrange equation of a suitable *Lagrange function* $L(t, x, \dot{x})$,

$$\frac{\mathrm{d}}{\mathrm{d}t}\frac{\partial L}{\partial \dot{x}} = \frac{\partial L}{\partial x} . \tag{8.1.5}$$

It follows that $\dfrac{\partial L}{\partial \dot{x}} = m\dot{x}$ and $\dfrac{\partial L}{\partial x} = F = -\dfrac{\partial V}{\partial x}$ would do, so that

$$L(t, x, \dot{x}) = \frac{m}{2}\dot{x}^2 - V(x, t) \tag{8.1.6}$$

is one option. We could multiply this L by a numerical constant, because $3L$ or $-7L$, say, would eventually give the same differential equation for $x(t)$, but there are good reasons to stick to the standard form of (8.1.6):

$$\text{Lagrange function} = (\text{kinetic energy}) - (\text{potential energy}) . \tag{8.1.7}$$

Its ingredients have immediate, and familiar, physical significance.

We then have the Lagrange form of the *action* W_{01},

$$W_{01} = \int_{t_0}^{t_1} \mathrm{d}t\, L\big(t, x(t), \dot{x}(t)\big) = \int_{t_0}^{t_1} \mathrm{d}t \left[\frac{m}{2}\dot{x}(t)^2 - V\big(x(t), t\big)\right] \tag{8.1.8}$$

and the *Principle of Stationary Action* (PSA),

$$\delta W_{01} = 0 \quad \text{for} \quad \delta x(t_0) = 0 \quad \text{and} \quad \delta x(t_1) = 0 , \tag{8.1.9}$$

where all $x(t)$ that are allowed to enter the competition must have the same initial and final position,

$$x(t_0) = x_0 , \quad x(t_1) = x_1 , \tag{8.1.10}$$

as depicted here:

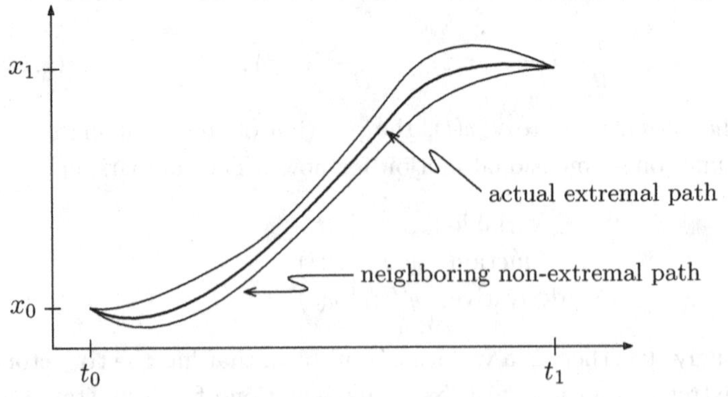

(8.1.11)

8.1.2 *More coordinates*

Physics happens in three-dimensional space, so we need to supplement the equation of motion for the x coordinate with those for the y and z coordinates, which correctly suggests the Lagrange function

$$L = \frac{m}{2}\dot{\boldsymbol{r}}^2 - V(\boldsymbol{r}, t)\,. \tag{8.1.12}$$

We have one Euler–Lagrange equation for each coordinate,

$$\frac{\mathrm{d}}{\mathrm{d}t}\frac{\partial L}{\partial \dot{x}} = \frac{\partial L}{\partial x}\,, \quad \frac{\mathrm{d}}{\mathrm{d}t}\frac{\partial L}{\partial \dot{y}} = \frac{\partial L}{\partial y}\,, \quad \frac{\mathrm{d}}{\mathrm{d}t}\frac{\partial L}{\partial \dot{z}} = \frac{\partial L}{\partial z}\,, \tag{8.1.13}$$

giving

$$m\ddot{x} = -\frac{\partial V}{\partial x}\,, \quad m\ddot{y} = -\frac{\partial V}{\partial y}\,, \quad m\ddot{z} = -\frac{\partial V}{\partial z} \tag{8.1.14}$$

as three coupled second-order differential equations for the cartesian coordinates or, if we recombine the components into a vector,

$$m\ddot{\boldsymbol{r}} = -\boldsymbol{\nabla} V(\boldsymbol{r}, t)\,, \tag{8.1.15}$$

one second-order differential equation for $\boldsymbol{r}(t)$. This extension from one coordinate to three is, indeed, immediate as soon as we realize that, when considering paths in the vicinity of the actual one,

$$\boldsymbol{r}(t) \to \boldsymbol{r}(t) + \delta\boldsymbol{r}(t)\,, \tag{8.1.16}$$

we can separately consider variations of $x(t)$ only, $\delta\boldsymbol{r}(t) = \boldsymbol{e}_x\delta x(t)$, or of $y(t)$ only, $\delta\boldsymbol{r}(t) = \boldsymbol{e}_y\delta y(t)$, or of $z(t)$ only, $\delta\boldsymbol{r}(t) = \boldsymbol{e}_z\delta z(t)$.

Rather than arguing cartesian coordinate by cartesian coordinate, let us look at the variation of the action as a whole,

$$\begin{aligned}
\delta W_{01} &= \delta \int_{t_0}^{t_1} \mathrm{d}t \left[\frac{m}{2}\boldsymbol{v}^2 - V(\boldsymbol{r}, t)\right] \\
&= \int_{t_0}^{t_1} \mathrm{d}t \left[m \underbrace{\boldsymbol{v}\cdot\delta\boldsymbol{v}}_{= \frac{\mathrm{d}}{\mathrm{d}t}(\boldsymbol{v}\cdot\delta\boldsymbol{r}) - \delta\boldsymbol{r}\cdot\frac{\mathrm{d}}{\mathrm{d}t}\boldsymbol{v}} - \delta\boldsymbol{r}\cdot\boldsymbol{\nabla} V(\boldsymbol{r}, t)\right] \\
&= m\boldsymbol{v}\cdot\delta\boldsymbol{r}\Big|_{t\,=\,t_0}^{t_1} - \int_{t_0}^{t_1} \mathrm{d}t\,\delta\boldsymbol{r}\cdot\left[m\frac{\mathrm{d}}{\mathrm{d}t}\boldsymbol{v} + \boldsymbol{\nabla} V(\boldsymbol{r}, t)\right].
\end{aligned} \tag{8.1.17}$$

Once again the boundary term vanishes,

$$m\boldsymbol{v}\cdot\delta\boldsymbol{r}\Big|_{t\,=\,t_0}^{t_1} = m\boldsymbol{v}(t_1)\cdot\delta\boldsymbol{r}(t_1) - m\boldsymbol{v}(t_0)\cdot\delta\boldsymbol{r}(t_0) = 0\,, \tag{8.1.18}$$

because the initial and final positions are fixed, $\delta\boldsymbol{r}(t_0) = 0$ and $\delta\boldsymbol{r}(t_1) = 0$. The PSA then implies (8.1.15), since $\delta W_{01} = 0$ must be true for all permissible $\delta\boldsymbol{r}(t)$. The equation of motion (8.1.15) and the path variation (8.1.16) make no reference to a particular coordinate system, nor does the PSA: As (8.1.17) and (8.1.18) show, we can carry out the full calculation without specifying a coordinate system.

The extension to more point masses with more coordinates is equally immediate. We have the general scenario in which there is an *initial configuration* at $t = t_0$, specified by stating the initial values of all coordinates of all point masses, and a final configuration at $t = t_1$, analogously specified by stating the final values of all coordinates. If we denote by $X(t)$ the collection of all coordinates at time t, and by $\dot{X}(t)$ the collection of their time derivatives, then the PSA reads

$$\delta W_{01} = 0 \quad \text{for} \quad W_{01} = \int_{t_0}^{t_1} \mathrm{d}t\, L\big(t, X(t), \dot{X}(t)\big) \tag{8.1.19}$$

with $X(t_0) = X_0$ and $X(t_1) = X_1$ fixed:

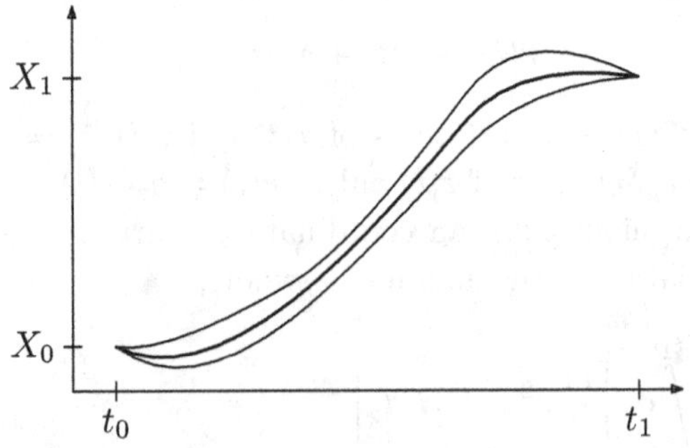

(8.1.20)

The graph now symbolizes a multidimensional path. The Euler–Lagrange equation for the kth coordinate is then

$$\frac{\mathrm{d}}{\mathrm{d}t}\frac{\partial L}{\partial \dot{x}_k} = \frac{\partial L}{\partial x_k}\,, \tag{8.1.21}$$

where x_k is any one of the coordinates in the collection $X = (x_1, x_2, x_3, \dots)$ and $\dot{x}_k$ is its velocity.

8.1.3 *Change of description, cyclic coordinates*

In this context, we make one very important observation: What matters is the action W_{01} and its first-order response to variations of the path, with the value of W_{01} determined by the path, whereby it does not matter how the path is parameterized. Any set of coordinates can be used, there is no requirement at all to stick to cartesian coordinates. The coordinate-free reasoning of (8.1.17) and (8.1.18) illustrates the matter.

As an explicit example, let us consider a point mass that moves in the xy plane whereby its potential energy is a function of $s = \sqrt{x^2 + y^2}$ only. The Lagrange function is

$$L = \frac{m}{2}(\dot{x}^2 + \dot{y}^2) - V\Bigl(\sqrt{x^2 + y^2}\Bigr) \tag{8.1.22}$$

if we use cartesian coordinates or, equivalently,

$$L = \frac{m}{2}\Bigl[\dot{s}^2 + (s\dot{\varphi})^2\Bigr] - V(s) \tag{8.1.23}$$

if we use polar coordinates, $x = s\cos\varphi$ and $y = s\sin\varphi$. The equations of motion that we get for cartesian coordinates are

$$\begin{aligned} \frac{\mathrm{d}}{\mathrm{d}t}\frac{\partial L}{\partial \dot{x}} = \frac{\partial L}{\partial x} &: \; m\ddot{x} = -\frac{x}{\sqrt{x^2 + y^2}} V'\Bigl(\sqrt{x^2 + y^2}\Bigr), \\ \frac{\mathrm{d}}{\mathrm{d}t}\frac{\partial L}{\partial \dot{y}} = \frac{\partial L}{\partial y} &: \; m\ddot{y} = -\frac{y}{\sqrt{x^2 + y^2}} V'\Bigl(\sqrt{x^2 + y^2}\Bigr), \end{aligned} \tag{8.1.24}$$

and the equations of motion that we get for polar coordinates are

$$\begin{aligned} \frac{\mathrm{d}}{\mathrm{d}t}\frac{\partial L}{\partial \dot{s}} = \frac{\partial L}{\partial s} &: \; m\ddot{s} = ms\dot{\varphi}^2 - V'(s)\,, \\ \frac{\mathrm{d}}{\mathrm{d}t}\frac{\partial L}{\partial \dot{\varphi}} = \frac{\partial L}{\partial \varphi} &: \; ms(2\dot{s}\dot{\varphi} + s\ddot{\varphi}) = 0\,. \end{aligned} \tag{8.1.25}$$

We verify that (8.1.24) and (8.1.25) are equivalent by first expressing, for example, $m\ddot{x}$ in terms of the polar coordinates, then use (8.1.25), followed by converting back to cartesian coordinates, finally arriving at the first equation in (8.1.24):

$$\begin{aligned} m\ddot{x} &= m(\ddot{s} - s\dot{\varphi}^2)\cos\varphi - m(2\dot{s}\dot{\varphi} + s\ddot{\varphi})\sin\varphi \\ &= -V'(s)\cos\varphi = -\frac{x}{\sqrt{x^2 + y^2}} V'\Bigl(\sqrt{x^2 + y^2}\Bigr); \end{aligned} \tag{8.1.26}$$

and analogously for $m\ddot{y}$.

Clearly, in this case the polar coordinates are a better fit to the problem. In particular, we note that L does not depend on φ and, therefore the equation of motion for φ gives us a constant of motion, $\frac{\mathrm{d}}{\mathrm{d}t}\frac{\partial L}{\partial \dot{\varphi}} = 0$, or

$$\frac{\partial L}{\partial \dot{\varphi}} = ms^2\dot{\varphi} = m\kappa = \text{constant}, \tag{8.1.27}$$

which we can then use in the equation for $s(t)$,

$$m\ddot{s} = \frac{m}{s^3}\left(s^2\dot{\varphi}\right)^2 - V'(s) = \frac{m\kappa^2}{s^3} - V'(s) = -\frac{\partial}{\partial s}\left(\frac{m\kappa^2}{2s^2} + V(s)\right), \tag{8.1.28}$$

or

$$m\ddot{s} = -\frac{\partial}{\partial s}V_{\text{eff}}(s) \quad \text{with} \quad V_{\text{eff}}(s) = \frac{m\kappa^2}{2s^2} + V(s)\,, \tag{8.1.29}$$

a structure that we have already met in (5.3.8).

More generally, whenever there is a so-called *cyclic coordinate*, which is the technical term for a coordinate on which the Lagrange function does not depend, the derivative of the Lagrange function with respect to the velocity of the cyclic coordinate is constant in time. It follows that we have a constant of motion for each cyclic coordinate.

The recognition of a cyclic coordinate requires the choice of appropriate coordinates, which is usually suggested by a symmetry property of the physical situation. In the example above, we have a cyclic coordinate in the polar-coordinate version of the Lagrange function in (8.1.23), but not in the cartesian-coordinate version of (8.1.22). Yes, the polar coordinates fit to the rotationally invariant potential energy $V(s)$.

8.2 Time and energy

The time argument of $L(t, X, \dot{X})$ is not on the same footing as the coordinates X, at least it appears so, but when there is no parametric t dependence in L, we also have a conserved quantity — the energy. To see this we briefly return to the variational problem

$$\delta \int_{x_0}^{x_1} \mathrm{d}x\, f\bigl(x, y(x), y'(x)\bigr) = 0 \tag{8.2.1}$$

and recall the second form (7.2.14) of the Euler–Lagrange equation,

$$\frac{\mathrm{d}}{\mathrm{d}x}\left(f - y'\frac{\partial f}{\partial y'}\right) = \frac{\partial f}{\partial x}\,. \tag{8.2.2}$$

The analog for the Lagrange function $L(t, X, \dot{X})$ is

$$\frac{\mathrm{d}}{\mathrm{d}t}\left(\sum_k \dot{x}_k\frac{\partial L}{\partial \dot{x}_k} - L\right) = -\frac{\partial L}{\partial t}\,, \tag{8.2.3}$$

where, in addition to a switch of the overall sign, the main difference is the replacement of $y'\frac{\partial f}{\partial y'}$ by the sum $\sum_k \dot{x}_k\frac{\partial L}{\partial \dot{x}_k}$ that accounts for one such term for each coordinate. Let us verify this statement:

$$\begin{aligned}\frac{\mathrm{d}}{\mathrm{d}t}\left(\sum_k \dot{x}_k\frac{\partial L}{\partial \dot{x}_k} - L\right) &= \sum_k\left(\ddot{x}_k\frac{\partial L}{\partial \dot{x}_k} + \dot{x}_k\frac{\mathrm{d}}{\mathrm{d}t}\frac{\partial L}{\partial \dot{x}_k}\right) - \frac{\mathrm{d}L}{\mathrm{d}t}\\ &= \sum_k\left(\ddot{x}_k\frac{\partial L}{\partial \dot{x}_k} + \dot{x}_k\frac{\partial L}{\partial x_k}\right) - \frac{\mathrm{d}L}{\mathrm{d}t}\,,\end{aligned} \tag{8.2.4}$$

where the Euler–Lagrange equation (8.1.21) for the kth coordinate has been used. Then, by the chain rule

$$\frac{\mathrm{d}}{\mathrm{d}t}L(t, X, \dot{X}) = \frac{\partial L}{\partial t} + \sum_k\left(\dot{x}_k\frac{\partial L}{\partial x_k} + \ddot{x}_k\frac{\partial L}{\partial \dot{x}_k}\right), \tag{8.2.5}$$

and upon combining these two statements we arrive at (8.2.3).

As an immediate consequence of (8.2.3), we have

$$\sum_k \dot{x}_k\frac{\partial L}{\partial \dot{x}_k} - L = \text{constant} \quad \text{if} \quad \frac{\partial L}{\partial t} = 0\,, \tag{8.2.6}$$

that is: if there is no parametric time dependence in the Lagrange function. Employing terminology used earlier, this says that the mechanical system is *conservative* if $\frac{\partial L}{\partial t} = 0$. That the conserved quantity is the energy indeed, is quite obvious if we use cartesian coordinates,

$$L = \sum_k \frac{1}{2}m_k\dot{x}_k^2 - V(X, t)\,, \tag{8.2.7}$$

for which

$$\frac{\partial L}{\partial \dot{x}_k} = m\dot{x}_k \tag{8.2.8}$$

and

$$\sum_k \dot{x}_k \frac{\partial L}{\partial \dot{x}_k} - L = \sum_k \frac{1}{2} m_k \dot{x}_k^2 + V(X,t) \tag{8.2.9}$$

is the *sum* of the kinetic energy and the potential energy. "No parametric time dependence" requires that the potential energy depends solely on the configuration, on the values of the coordinates, $V = V(X)$, and there should also be no time dependence in the masses m_k, which is usually the case.

8.3 Examples

8.3.1 *Two masses strung up*

Here is an example that illustrates some of these points:

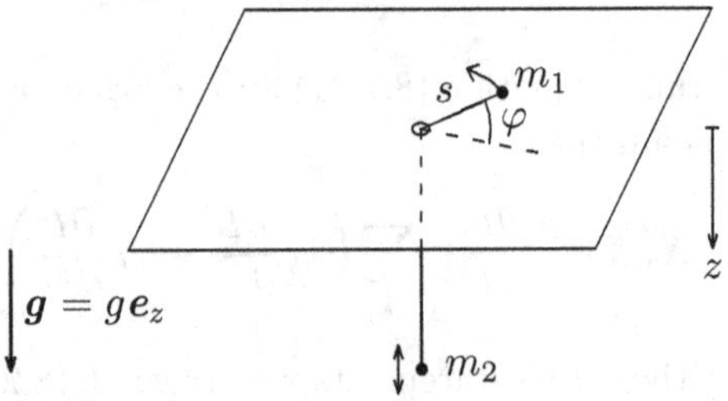

(8.3.1)

Mass m_1 moves without friction on a horizontal table; the table has a small hole through which a string of length a that is attached to mass m_1 holds mass m_2 vertically below the hole. For simplicity, we consider only the vertical up-down motion of mass m_2 although it could, of course, also swing like a pendulum.

The coordinates that suggest themselves are polar coordinates s, φ for mass m_1 and the distance z below the table for mass m_2. We then have

$$\frac{m_1}{2}\left[\dot{s}^2 + (s\dot{\varphi})^2\right] + \frac{m_2}{2}\dot{z}^2 \tag{8.3.2}$$

for the kinetic energy, and

$$-m_2 g z \tag{8.3.3}$$

for the potential energy. In addition, there is the constraint that the string has a fixed length L, which means

$$s + z = a\,, \tag{8.3.4}$$

so that we can get rid of z right away by the replacements

$$z = a - s\,, \quad \dot{z} = -\dot{s}\,. \tag{8.3.5}$$

Then, assuming that the string has so little mass that its contributions to the kinetic energy and the potential energy are negligible, the Lagrange function is

$$L = \frac{m_1}{2}\left[\dot{s}^2 + (s\dot{\varphi})^2\right] + \frac{m_2}{2}\dot{s}^2 + m_2 g(a - s)\,, \tag{8.3.6}$$

where the constant $m_2 g a$ is of no consequence and could be dropped because the equations of motion involve only the derivatives of the Lagrange function.

These equations of motion are here

$$\begin{aligned} \frac{\mathrm{d}}{\mathrm{d}t}\frac{\partial L}{\partial \dot{s}} = \frac{\partial L}{\partial s} &: \ (m_1 + m_2)\ddot{s} = m_1 s\dot{\varphi}^2 - m_2 g\,, \\ \frac{\mathrm{d}}{\mathrm{d}t}\frac{\partial L}{\partial \dot{\varphi}} = \frac{\partial L}{\partial \varphi} &: \ \frac{\mathrm{d}}{\mathrm{d}t}\left(m_1 s^2 \dot{\varphi}\right) = 0\,. \end{aligned} \tag{8.3.7}$$

There is the constant of motion $s^2\dot{\varphi} = \kappa$ associated with the cyclic coordinate φ, which we use, as usual, to eliminate $\dot{\varphi}$ from the equation for s,

$$(m_1 + m_2)\ddot{s} = \frac{m_1 \kappa^2}{s^3} - m_2 g = -\frac{\partial}{\partial s} V_{\text{eff}}(s) \tag{8.3.8}$$

with the effective potential energy

$$V_{\text{eff}}(s) = \frac{m_1 \kappa^2}{2s^2} + m_2 g s\,. \tag{8.3.9}$$

The conserved energy is

$$\begin{aligned} E &= \dot{s}\frac{\partial L}{\partial \dot{s}} + \dot{\varphi}\frac{\partial L}{\partial \dot{\varphi}} - L \\ &= \frac{m_1 + m_2}{2}\dot{s}^2 + \frac{m_1}{2}(s\dot{\varphi})^2 + m_2 g(s - a)\,, \end{aligned} \tag{8.3.10}$$

and the verification of $\frac{d}{dt}E = 0$ is a simple exercise. Here, too, we can eliminate $\dot{\varphi}$ with the help of $s^2\dot{\varphi} = \kappa$,

$$E = \frac{m_1 + m_2}{2}\dot{s}^2 + \frac{m_1\kappa^2}{2s^2} + m_2 g(s - a)$$
$$= \frac{m_1 + m_2}{2}\dot{s}^2 + V_{\text{eff}}(s) - m_2 ga\,, \tag{8.3.11}$$

so that the remaining s motion is one-dimensional motion in the effective potential energy $V_{\text{eff}}(s)$:

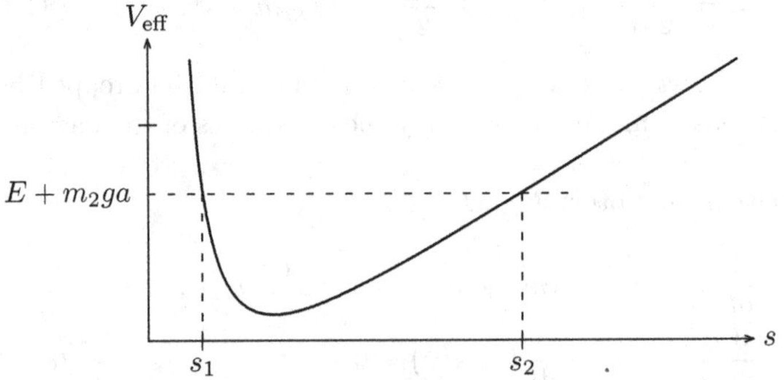

(8.3.12)

It is periodic motion between two turning points — here: the smallest and the largest distance from the hole — accompanied by azimuthal motion around the hole in accordance with $\dot{\varphi} = \kappa/s^2$. Mass m_2 moves vertically up and down following the periodic $s(t)$ dependence as enforced by the fixed length a of the connecting string, $z(t) = a - s(t)$. All in all, the Lagrange function made it quite simple to establish the correct equations of motion and find two constants of motion, each associated with a variable on which the Lagrange function does not depend (φ and t).

The great efficiency of the formalism does come at a price, however: Some of the physical details remain hidden or implicit. In the present example, we do not get information about the tension in the string. This detail could be important if there is danger that the string tears in two because it is too weak to supply the required force. This information can be extracted if we incorporate the constraint $s + z = a$ into the action rather than using it for an immediate elimination of $z(t)$ from the Lagrange function. If we keep $z(t)$, then we are asked to find the extremum of the action

$$W_{01} = \int_{t_0}^{t_1} \mathrm{d}t \left(\frac{m_1}{2} \left[\dot{s}^2 + (s\dot{\varphi})^2 \right] + \frac{m_2}{2} \dot{z}^2 + m_2 g z \right) \tag{8.3.13}$$

under the constraint

$$s(t) + z(t) - a = 0 \,. \tag{8.3.14}$$

This is not an integral constraint like the one we met in the catenary problem in Section 7.4, but a condition that has to hold *at each instant* t. We can think of this as a continuum of constraints, and then use a continuum of Lagrange multipliers to incorporate it into the functional,

$$\int_{t_0}^{t_1} \mathrm{d}t \, L^{(\lambda)} \overset{!}{=} \text{extremum}, \tag{8.3.15}$$

with

$$\begin{aligned} L^{(\lambda)} = {} & \frac{m_1}{2} \left(\dot{s}(t)^2 + \left[s(t)\dot{\varphi}(t) \right]^2 \right) + \frac{m_2}{2} \dot{z}(t)^2 + m_2 g z(t) \\ & - \lambda(t) \left[s(t) + z(t) - a \right], \end{aligned} \tag{8.3.16}$$

where we now consider variations of $s(t)$, $\varphi(t)$, $z(t)$, and also of the Lagrange-multiplier function $\lambda(t)$. The resulting Euler–Lagrange equations are

$$\begin{aligned} \frac{\mathrm{d}}{\mathrm{d}t} \frac{\partial L^{(\lambda)}}{\partial \dot{s}} &= \frac{\partial L^{(\lambda)}}{\partial s} : \; m_1 \ddot{s} = m_1 s \dot{\varphi}^2 - \lambda \,, \\ \frac{\mathrm{d}}{\mathrm{d}t} \frac{\partial L^{(\lambda)}}{\partial \dot{\varphi}} &= \frac{\partial L^{(\lambda)}}{\partial \varphi} : \; m_1 s^2 \dot{\varphi} = \text{constant}, \\ \frac{\mathrm{d}}{\mathrm{d}t} \frac{\partial L^{(\lambda)}}{\partial \dot{z}} &= \frac{\partial L^{(\lambda)}}{\partial z} : \; m_2 \ddot{z} = m_2 g - \lambda \,, \\ \frac{\mathrm{d}}{\mathrm{d}t} \frac{\partial L^{(\lambda)}}{\partial \dot{\lambda}} &= \frac{\partial L^{(\lambda)}}{\partial \lambda} : \; 0 = s + z - a \,. \end{aligned} \tag{8.3.17}$$

There is no equation of motion for $\lambda(t)$, since $\dot{\lambda}$ does not show up in the extended Lagrange function $L^{(\lambda)}$, the Euler–Lagrange equation for $\lambda(t)$ is the constraint (8.3.14), for which the Lagrange-multiplier function $\lambda(t)$ is introduced. The equations of motion for $s(t)$ and $z(t)$ identify the physical significance of $\lambda(t)$: It is the force by which the string pulls at the masses, toward the hole for mass m_1 (if $\lambda > 0$), upwards for mass m_2.

We can now go ahead and use the Euler–Lagrange equations for $z(t)$ and $\lambda(t)$ to eliminate both of them from the equations for $s(t)$ and $\varphi(t)$, which takes us back to the equations in (8.3.7). After solving these and thus finding $s(t)$, we get $\lambda(t)$ from

$$\lambda(t) = m_2\big(g - \ddot{z}(t)\big) = m_2\big(g + \ddot{s}(t)\big)\,. \tag{8.3.18}$$

This tells us that the force in the string compensates for the weight $m_2 g$ and provides for the acceleration force $m_2\ddot{z}(t)$, where the minus sign accounts for the fact that a positive string tension ($\lambda > 0$) gives an acceleration in the $-z$ direction. All of this is, of course, as could have been anticipated.

A word of warning is in order. We made repeated use of $s^2\dot{\varphi} = \kappa$ to eliminate $\dot{\varphi}$ from the equation of motion for $s(t)$, and also from the expression for the energy E. This is all right because the equations of motion refer to the actual $s(t)$ and the energy is to be evaluated for the actual path of motion. In the Lagrange function, however, $s(t)$, $\varphi(t)$, ... stand for coordinates along *any* path that is permissible, including paths along which the equations of motion do not hold. It would, therefore, be *incorrect* to use $s^2\dot{\varphi} = \kappa$ to eliminate $\dot{\varphi}$ from the Lagrange function. It is a different matter for the constraint $s + z = a$ because it must be obeyed by *all permissible* $s(t)$ and $z(t)$, whether they are those of the actual path or not.

8.3.2 *Two coupled harmonic oscillators*

The next example is a system of two coupled harmonic oscillators, as depicted here:

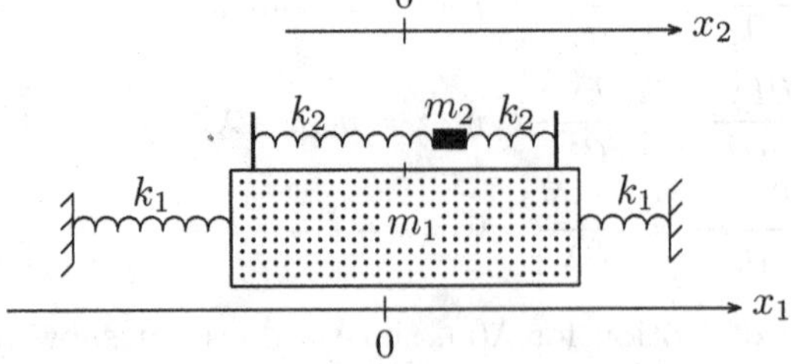

(8.3.19)

Mass m_1 is held by two equal springs with spring constant k_1 and can move left-right in the x direction. Attached to it are two springs, each with spring constant k_2, that hold mass m_2 relative to m_1. We measure the position x_1 of mass m_1 from the middle point between the two k_1 springs and the position x_2 of mass m_2 from the middle of mass m_1. The kinetic energy is then

$$\frac{m_1}{2}\dot{x}_1^2 + \frac{m_2}{2}(\dot{x}_1 + \dot{x}_2)^2\,, \tag{8.3.20}$$

and for the potential energy we have

$$k_1 x_1^2 + k_2 x_2^2 \tag{8.3.21}$$

(no factors $\frac{1}{2}$ because it is two springs for each mass). This gives us the Lagrange function

$$L = \frac{m_1}{2}\dot{x}_1^2 + \frac{m_2}{2}(\dot{x}_1 + \dot{x}_2)^2 - k_1 x_1^2 - k_2 x_2^2\,, \tag{8.3.22}$$

and the resulting equations of motion are

$$\begin{aligned}
\frac{\mathrm{d}}{\mathrm{d}t}\frac{\partial L}{\partial \dot{x}_1} = \frac{\partial L}{\partial x_1} &: \; m_1\ddot{x}_1 + m_2(\ddot{x}_1 + \ddot{x}_2) = -2k_1 x_1\,, \\
\frac{\mathrm{d}}{\mathrm{d}t}\frac{\partial L}{\partial \dot{x}_2} = \frac{\partial L}{\partial x_2} &: \; m_2(\ddot{x}_1 + \ddot{x}_2) = -2k_2 x_2\,,
\end{aligned} \tag{8.3.23}$$

which are two coupled linear second-order differential equations with constant coefficients for the two looked-for functions $x_1(t)$ and $x_2(t)$. We could solve this pair of equations by applying the lessons of Exercise 27 to an equivalent fourth-order equation for one of the functions. Although the details of this procedure are not really necessary for the method that we'll eventually use for solving (8.3.23), it is instructive to see the conversion to that fourth-order equation once, and the example of (8.3.23) is fine for this purpose.

Here is how it goes. We subtract the bottom equation from the top equation in (8.3.23) and solve for x_2,

$$x_2 = \frac{1}{2k_2}\left[m_1\left(\frac{\mathrm{d}}{\mathrm{d}t}\right)^2 + 2k_1\right]x_1\,, \tag{8.3.24}$$

and then use this in the bottom equation

$$\left[m_2\left(\frac{\mathrm{d}}{\mathrm{d}t}\right)^2 + 2k_2\right]x_2 + m_2\left(\frac{\mathrm{d}}{\mathrm{d}t}\right)^2 x_1 = 0\,, \tag{8.3.25}$$

which takes us to

$$\left[m_2\left(\frac{\mathrm{d}}{\mathrm{d}t}\right)^2 + 2k_2\right]\left[m_1\left(\frac{\mathrm{d}}{\mathrm{d}t}\right)^2 + 2k_1\right]x_1 + 2m_2k_2\left(\frac{\mathrm{d}}{\mathrm{d}t}\right)^2 x_1 = 0\,. \tag{8.3.26}$$

The observation of Exercise 27 for $n = 4$ applies to this equation: The ansatz $x_1(t) = c\,\mathrm{e}^{\lambda t}$ is appropriate, and we get $x_2(t) = \dfrac{c}{2k_2}(m_1\lambda^2 + 2k_1)\,\mathrm{e}^{\lambda t}$ from (8.3.24).

It follows that the exponential ansatz

$$\begin{pmatrix} x_1(t) \\ x_2(t) \end{pmatrix} = \begin{pmatrix} c_1 \\ c_2 \end{pmatrix} \mathrm{e}^{\lambda t} \tag{8.3.27}$$

works for (8.3.23), and we can apply it directly without the detour that took us to (8.3.26). For this purpose, we rewrite (8.3.23) compactly as

$$M\ddot{X} = -KX \tag{8.3.28}$$

with the two-component column $X(t) = \begin{pmatrix} x_1(t) \\ x_2(t) \end{pmatrix}$ and two symmetric 2×2 matrices — the mass matrix

$$M = \begin{pmatrix} m_1 + m_2 & m_2 \\ m_2 & m_2 \end{pmatrix} \tag{8.3.29}$$

and the matrix of spring constants

$$K = \begin{pmatrix} 2k_1 & 0 \\ 0 & 2k_2 \end{pmatrix}. \tag{8.3.30}$$

The ansatz (8.3.27) then leads us to

$$(\lambda^2 M + K)\begin{pmatrix} c_1 \\ c_2 \end{pmatrix} = 0\,. \tag{8.3.31}$$

This always has the solution $\begin{pmatrix} c_1 \\ c_2 \end{pmatrix} = \begin{pmatrix} 0 \\ 0 \end{pmatrix}$, but that is of no use other than reminding us that $x_1(t) = x_2(t) = 0$ for all times is one solution of (8.3.23). Of course, it is; the two masses in (8.3.19) could just be at rest at their equilibrium positions.

We really need, however, solutions of (8.3.31) with $\begin{pmatrix} c_1 \\ c_2 \end{pmatrix} \neq \begin{pmatrix} 0 \\ 0 \end{pmatrix}$, which is then an eigencolumn of the matrix $\lambda^2 M + K$ with eigenvalue zero. Accordingly, the possible values of λ^2 are those for which the determinant of $\lambda^2 M + K$ vanishes,

$$\det\{\lambda^2 M + K\} = 0\,, \tag{8.3.32}$$

so that we have the quadratic equation

$$\big((m_1 + m_2)\lambda^2 + 2k_1\big)(m_2\lambda^2 + 2k_2) - (m_2\lambda^2)^2 = 0 \tag{8.3.33}$$

or

$$m_1 m_2 \lambda^4 + 2\big((m_1 + m_2)k_2 + m_2 k_1\big)\lambda^2 + 4k_1 k_2 = 0 \tag{8.3.34}$$

for λ^2. After expressing the left-hand side as

$$m_1 m_2 (\lambda^2 - \lambda_1^2)(\lambda^2 - \lambda_2^2) = 0\,, \tag{8.3.35}$$

we find that the solutions λ_1^2 and λ_2^2 are such that

$$\begin{aligned} \lambda_1^2 + \lambda_2^2 &= -\frac{2(k_1 + k_2)}{m_1} - \frac{2k_2}{m_2} < 0\,, \\ \lambda_1^2 \lambda_2^2 &= \frac{4k_1 k_2}{m_1 m_2} > 0\,, \end{aligned} \tag{8.3.36}$$

which imply that λ_1^2 and λ_2^2 are negative. The four λ values for which we have $\begin{pmatrix} c_1 \\ c_2 \end{pmatrix} \neq \begin{pmatrix} 0 \\ 0 \end{pmatrix}$ in (8.3.31) are, therefore, composed of two pairs of complex conjugate numbers, $\pm\mathrm{i}\sqrt{-\lambda_1^2}$ and $\pm\mathrm{i}\sqrt{-\lambda_2^2}$, and the solutions of (8.3.23) are oscillatory functions with the circular frequencies of $\sqrt{-\lambda_1^2}$ and $\sqrt{-\lambda_2^2}$. Further details are the subject matter of Exercise 129.

Chapter 9

Small-Amplitude Oscillations

9.1 Near an equilibrium: Lagrange function and equations of motion

Having seen this example of the two coupled oscillations, let us now consider the general situation of small-amplitude oscillations about an equilibrium configuration. At equilibrium the forces are vanishing, so that we can choose the coordinates such that

$$\left.\frac{\partial}{\partial x_k} V(X)\right|_{X=0} = 0 \quad \text{for} \quad k = 1, 2, \ldots, n, \tag{9.1.1}$$

if we are dealing with n independent coordinates, corresponding to n degrees of freedom. Near $X = 0$, the potential energy is then given by a quadratic form

$$V(X) = V_0 + \frac{1}{2} \sum_{j,k=1}^{n} x_j K_{jk} x_k + \cdots, \tag{9.1.2}$$

where

$$K_{jk} = \left.\frac{\partial}{\partial x_j} \frac{\partial}{\partial x_k} V(X)\right|_{X=0} = K_{kj} \tag{9.1.3}$$

and the ellipsis stands for terms of cubic and higher order which are negligible as long as all x_ks remain sufficiently small. Under these circumstances, we can approximate $V(X)$ by the quadratic term, and since the constant $V_0 = V(X)\big|_{X=0}$ is irrelevant for the equations of motion, we drop it as well, so that

$$V(X) = \frac{1}{2} \sum_{j,k=1}^{n} x_j K_{jk} x_k \tag{9.1.4}$$

is the form of the potential energy that we shall use when studying small-amplitude oscillations about $X = 0$.

It is expedient to think of X as a n-component column

$$X = \begin{pmatrix} x_1 \\ x_2 \\ \vdots \\ x_n \end{pmatrix} \tag{9.1.5}$$

rather than just a symbol for the collection of all coordinates, and then $X^{\mathrm{T}} = (x_1, x_2, ..., x_n)$ is its transpose, a n-component row. When assembling the potential constants K_{jk}, the analogs of the spring constants in (8.3.30), in a $n \times n$ matrix

$$K = \begin{pmatrix} K_{11} & K_{12} & \cdots & K_{1n} \\ K_{21} & K_{22} & \cdots & K_{2n} \\ \vdots & \vdots & \ddots & \vdots \\ K_{n1} & K_{n2} & \cdots & K_{nn} \end{pmatrix} = K^{\mathrm{T}}, \tag{9.1.6}$$

we have the compact expression

$$V = \frac{1}{2} X^{\mathrm{T}} K X \tag{9.1.7}$$

for the potential energy. What we have here is the immediate generalization of $V(\boldsymbol{r}) = V(\boldsymbol{r}_0) + \frac{1}{2}\boldsymbol{\epsilon} \cdot \boldsymbol{\nabla\nabla} V(\boldsymbol{r}_0) \cdot \boldsymbol{\epsilon}$ of (3.2.67) to the situation of n independent coordinates, which may or may not be cartesian coordinates.

Likewise, for the kinetic energy we have

$$\frac{1}{2} \sum_{j,k=1}^{n} \dot{x}_j m_{jk} \dot{x}_k = \frac{1}{2} \dot{X}^{\mathrm{T}} M \dot{X} \tag{9.1.8}$$

for small $\dot{x}_j$s and small x_ks. The $n \times n$ mass matrix M can be chosen to be symmetric — that is: $M = M^{\mathrm{T}}$, $M_{jk} = M_{kj}$ — because if we should first have an asymmetric matric $\widetilde{M}$, then the observation that

$$\begin{aligned} \frac{1}{2} \dot{X}^{\mathrm{T}} \widetilde{M} \dot{X} &= \left(\frac{1}{2} \dot{X}^{\mathrm{T}} \widetilde{M} \dot{X}\right)^{\mathrm{T}} = \frac{1}{2} \dot{X}^{\mathrm{T}} \widetilde{M}^{\mathrm{T}} \dot{X} \\ &= \frac{1}{2} \dot{X}^{\mathrm{T}} \left(\frac{1}{2} \widetilde{M} + \frac{1}{2} \widetilde{M}^{\mathrm{T}}\right) \dot{X} \\ &= \frac{1}{2} \dot{X}^{\mathrm{T}} M \dot{X} \end{aligned} \tag{9.1.9}$$

enables us to replace $\widetilde{M}$ by $M = \frac{1}{2}(\widetilde{M} + \widetilde{M}^{\mathrm{T}}) = M^{\mathrm{T}}$ without changing the kinetic energy in its dependence on the velocities $\dot{x}_j$.

The Lagrange function is now

$$L = \frac{1}{2}\dot{X}^{\mathrm{T}}M\dot{X} - \frac{1}{2}X^{\mathrm{T}}KX\,. \tag{9.1.10}$$

The Euler–Lagrange equation for the jth coordinate,

$$\frac{\mathrm{d}}{\mathrm{d}t}\frac{\partial L}{\partial \dot{x}_j} = \frac{\partial L}{\partial x_j}\,, \tag{9.1.11}$$

needs

$$\frac{\partial L}{\partial x_j} = -\frac{\partial}{\partial x_j}\frac{1}{2}X^{\mathrm{T}}KX = -\frac{1}{2}\frac{\partial}{\partial x_j}\sum_{k,l=1}^{n} x_k K_{kl} x_l\,, \tag{9.1.12}$$

where we switch from summation indices jk to summation indices kl to avoid a double use of index j. Since

$$\frac{\partial}{\partial x_j}x_k = \left\{\begin{matrix} 1 & \text{if} & j = k \\ 0 & \text{if} & j \neq k \end{matrix}\right\} = \delta_{jk}\,, \tag{9.1.13}$$

where δ_{jk} is the Kronecker delta symbol of (1.1.9), we have

$$\begin{aligned}\frac{\partial L}{\partial x_j} &= -\frac{1}{2}\sum_{k,l=1}^{n}(\delta_{jk}K_{kl}x_l + x_k K_{kl}\delta_{lj}) \\ &= -\frac{1}{2}\sum_{l=1}^{n}K_{jl}x_l - \frac{1}{2}\sum_{k=1}^{n}\underbrace{x_k K_{kj}}_{=K_{jk}x_k} \\ &= -\sum_{l=1}^{n}K_{jl}x_l = -(KX)_j\,,\end{aligned} \tag{9.1.14}$$

where $K = K^{\mathrm{T}}$ has entered. Likewise,

$$\frac{\partial L}{\partial \dot{x}_j} = (M\dot{X})_j\,, \tag{9.1.15}$$

so that

$$\frac{\mathrm{d}}{\mathrm{d}t}(M\dot{X})_j = (M\ddot{X})_j = -(KX)_j \tag{9.1.16}$$

is the equation of motion for the jth coordinate. It follows that

$$M\ddot{X} = -KX \tag{9.1.17}$$

is the equation of motion for the column of coordinates $X(t)$.

We could have obtained this more directly, without the detour of considering a particular coordinate first, by recalling the PSA,

$$\delta\int_{t_0}^{t_1}\mathrm{d}t\left(\frac{1}{2}\dot{X}^{\mathrm{T}}M\dot{X}-\frac{1}{2}X^{\mathrm{T}}KX\right)=0\,. \tag{9.1.18}$$

Here we need

$$\begin{aligned}\delta\left(\frac{1}{2}X^{\mathrm{T}}KX\right)&=\frac{1}{2}\delta X^{\mathrm{T}}KX+\frac{1}{2}X^{\mathrm{T}}K\delta X\\&=\frac{1}{2}\delta X^{\mathrm{T}}(K+K^{\mathrm{T}})X=\delta X^{\mathrm{T}}KX\end{aligned} \tag{9.1.19}$$

and

$$\begin{aligned}\delta\left(\frac{1}{2}\dot{X}^{\mathrm{T}}M\dot{X}\right)&=\delta\dot{X}^{\mathrm{T}}M\dot{X}\\&=\frac{\mathrm{d}}{\mathrm{d}t}\left(\delta X^{\mathrm{T}}M\dot{X}\right)-\delta X^{\mathrm{T}}M\ddot{X}\,,\end{aligned} \tag{9.1.20}$$

which give

$$0=\delta X^{\mathrm{T}}M\dot{X}\Big|_{t=t_0}^{t_1}-\int_{t_0}^{t_1}\mathrm{d}t\,\delta X^{\mathrm{T}}(M\ddot{X}+KX)\,. \tag{9.1.21}$$

Once more, the boundary term vanishes as a consequence of

$$\delta X(t_0)=0 \quad\text{and}\quad \delta X(t_1)=0\,, \tag{9.1.22}$$

and the PSA then implies (9.1.17).

9.2 Characteristic frequencies and normal modes

In the example of Section 8.3.2, we had $n=2$ coordinates and the 2×2 matrices of (8.3.29) and (8.3.30), and the ansatz (8.3.27) was useful. Analogously, we use the exponential ansatz

$$X(t)=X^{(l)}\,\mathrm{e}^{\mathrm{i}\omega_l t} \tag{9.2.1}$$

for (9.1.17), where $X^{(l)}$ is a constant n-component column and $\mathrm{i}\omega_l$ plays the role of λ in (8.3.27). Since we found $\lambda_1^2<0$ and $\lambda_2^2<0$ in Section 8.3.2, we expect $(\mathrm{i}\omega_l)^2=-\omega_l^2<0$ here as well, and this is why it is convenient to switch from λ to $\mathrm{i}\omega_l$. Usually, there will be n different values for ω_l^2 and one n-component column $X^{(l)}$ for each of them, and we label them by $l = 1, 2, \ldots, n$. We note that the possible degeneracy discussed in Exercise 27, which requires solutions with $t\,e^{i\omega t}$, $t^2\,e^{i\omega t}$, ... as their time dependences, does not occur in the current context because the mass matrix must be positive, $M > 0$, to ensure a positive kinetic energy when $\dot{X} \neq 0$.

The ansatz (9.2.1) turns (9.1.17) into

$$\big(\omega_l^2 M - K\big) X^{(l)} = 0 \tag{9.2.2}$$

and — repeating the argument that took us from (8.3.31) to (8.3.32) — we conclude that $X^{(l)} \neq 0$ is an eigencolumn of $\omega_l^2 M - K$ with eigenvalue zero. Therefore, the ω_l^2s are the solutions of

$$\det\big\{\omega^2 M - K\big\} = 0\,, \tag{9.2.3}$$

where the left-hand side is a polynomial of degree n in ω^2. Once the ω_l^2s are known, we get the $X^{(l)}$s from (9.2.2).

Eventually, the small-amplitude oscillations are given by

$$X(t) = \sum_{l=1}^{n} \big[a_l \cos(\omega_l t) + b_l \sin(\omega_l t)\big] X^{(l)} = \sum_{l=1}^{n} x^{(l)}(t) X^{(l)}\,, \tag{9.2.4}$$

where the coefficients a_l and b_l are determined by the initial values of X and $\dot{X}$ at $t = 0$, say; see Exercise 127. In many applications, however, one is not interested in this finer detail, but one cares about the *characteristic frequencies* $\omega_l/2\pi$ and their eigencolumns $X^{(l)}$. Together, they constitute the *normal modes* of the mechanical system. The time-dependent coefficients

$$x^{(l)}(t) = a_l \cos(\omega_l t) + b_l \sin(\omega_l t) = \epsilon_l \cos(\omega_l t - \phi_l) \tag{9.2.5}$$

are often called the *normal coordinates*, alternatively specified by their amplitudes $\epsilon_l = \sqrt{a_l^2 + b_l^2}$ and their phases ϕ_l.

The characteristic frequencies of the normal modes tell us the vibrational response of the mechanical system when it is *slightly* perturbed from equilibrium. For example, when you ring a bell or hit a string, you will hear a characteristic sound made up by the characteristic frequencies in proportions that depend on where you hit. Knowing the characteristic frequencies of a mechanical system is crucial in many engineering applications. For instance, if you design a support structure for a generator that runs at 1000 rpm (rounds per minute, that is), the characteristic frequencies of the casing should be well above 20 Hz to avoid any resonance of potentially disastrous consequences.

Finally, we note the following: If the configuration with $X = 0$ is a potential-energy maximum, rather than a minimum, all ω_l^2s will be negative. And in the case of a saddle point, some ω_l^2s will be positive, some others negative. Then the small-X approximation is valid for rather short times only. But the analysis above is still useful because it identifies the values of the negative ω_l^2 and so tells us what times can be regarded as short.

9.3 Examples

9.3.1 *Pendulum*

An elementary example is the pendulum, a point mass, attached to the ceiling with a string of length a:

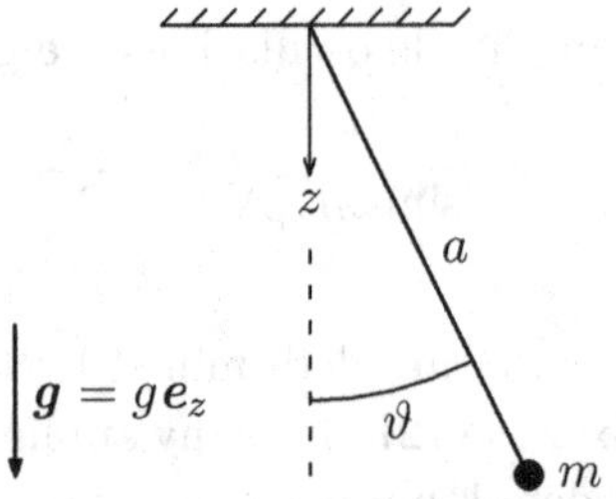

(9.3.1)

We use coordinates with the z-direction downwards and $(x, y, z) = 0$ where the string is attached. The kinetic energy is

$$\frac{m}{2}(\dot{x}^2 + \dot{y}^2 + \dot{z}^2) = \frac{m}{2}\dot{\boldsymbol{r}}^2 \tag{9.3.2}$$

and the potential energy is

$$mg(a - z) = mg(a - \boldsymbol{e}_z \cdot \boldsymbol{r}) \tag{9.3.3}$$

if we agree that it vanishes when the mass is vertically below $\boldsymbol{r} = 0$. There is also the constraint

$$r = \sqrt{x^2 + y^2 + z^2} = a\,, \tag{9.3.4}$$

which we incorporate by a Lagrange-multiplier function $f(t)$.

The extended Lagrange function is then

$$L = \frac{m}{2}\dot{\boldsymbol{r}}^2 - mg(a - \boldsymbol{e}_z \cdot \boldsymbol{r}) - f(t)(r - a)\,, \tag{9.3.5}$$

which gives the equation of motion

$$m\ddot{\boldsymbol{r}} = mg\boldsymbol{e}_z - f(t)\frac{\boldsymbol{r}}{r}\,, \tag{9.3.6}$$

where we have the gravitational pull $mg\boldsymbol{e}_z$ downwards and the force $-f(t)\frac{\boldsymbol{r}}{r}$ exerted by the string toward the point of attachment at $\boldsymbol{r} = 0$, as well as the constraint $r - a = 0$ as the Euler–Lagrange equation for $f(t)$.

Since the motion is confined to the sphere $r = a$, spherical coordinates should fit the problem and, of course, they do. So, we write

$$\begin{aligned} x &= a\sin\vartheta\cos\varphi\,,\\ y &= a\sin\vartheta\sin\varphi\,,\\ z &= a\cos\vartheta\,, \end{aligned} \tag{9.3.7}$$

where $r = a$ is already incorporated, leaving the polar angle ϑ and the azimuth φ as the relevant coordinates. Then,

$$\dot{\boldsymbol{r}}^2 = (a\dot{\vartheta})^2 + (a\dot{\varphi}\sin\vartheta)^2 \tag{9.3.8}$$

and $a - \boldsymbol{e}_z\cdot\boldsymbol{r} = a(1-\cos\vartheta)$, giving

$$L = \frac{m}{2}a^2\left[\dot{\vartheta}^2 + \dot{\varphi}^2(\sin\vartheta)^2\right] - mga(1-\cos\vartheta) \tag{9.3.9}$$

for the Lagrange function in spherical coordinates. We recognize that φ is a cyclic coordinate,

$$\frac{\partial L}{\partial\dot{\varphi}} = ma^2\dot{\varphi}(\sin\vartheta)^2 = ma^2\kappa = \text{constant}, \tag{9.3.10}$$

and the equation of motion for the polar angle,

$$\frac{\mathrm{d}}{\mathrm{d}t}\frac{\partial L}{\partial\dot{\vartheta}} = \frac{\partial L}{\partial\vartheta}: \quad ma^2\ddot{\vartheta} = ma^2\dot{\varphi}^2\sin\vartheta\cos\vartheta - mga\sin\vartheta\,, \tag{9.3.11}$$

can then be cast into an equation that only contains ϑ,

$$\begin{aligned} \ddot{\vartheta} &= \left(\frac{\kappa}{(\sin\vartheta)^2}\right)^2\sin\vartheta\cos\vartheta - \frac{g}{a}\sin\vartheta\\ &= \kappa^2\frac{\cos\vartheta}{(\sin\vartheta)^3} - \frac{g}{a}\sin\vartheta\\ &= -\frac{\partial}{\partial\vartheta}\left[\frac{\kappa^2}{2(\sin\vartheta)^2} + \frac{g}{a}(1-\cos\vartheta)\right]. \end{aligned} \tag{9.3.12}$$

The implied statement

$$\frac{1}{2}\dot{\vartheta}^2 + \frac{\kappa^2}{2(\sin\vartheta)^2} + \frac{g}{a}(1-\cos\vartheta) = \text{constant} \tag{9.3.13}$$

expresses the conservation of energy, of course.

If we are interested in the small oscillations about the equilibrium position $\vartheta = 0$, we have to face the singularity that the spherical coordinates have on the z axis, where φ is not determined. In the present context, this has the consequence that, when keeping φ and ϑ only up to joint second order, the Lagrange function retains no trace of φ at all,

$$L \cong \frac{m}{2}a^2\dot{\vartheta}^2 - \frac{1}{2}mga\vartheta^2\,, \tag{9.3.14}$$

and even the appearance of what remains is misleading: This is not the Lagrange function of a harmonic oscillator, because ϑ is restricted to $0 \leq \vartheta \leq \pi$ and does not take on negative values. It follows that, as self-suggesting as they are, the spherical coordinates are ill-suited for an investigation of the small-amplitude oscillations about $(x, y, z) = (0, 0, a)$.

Instead, we can return to the cartesian coordinates and consider small deviations from $(x, y, z) = (0, 0, a)$. The coordinates x, y parameterize horizontal displacements, and for $a - z \ll a$ we write

$$\begin{aligned} a - z &= \frac{a^2 - z^2}{a+z} = \frac{x^2+y^2}{a+z} \\ &= \frac{x^2+y^2}{2a}\frac{1}{1-\dfrac{a-z}{2a}} = \frac{x^2+y^2}{2a}\left[1 + \frac{a-z}{2a} + \cdots\right] \\ &= \frac{x^2+y^2}{2a} + \text{terms of cubic and higher order.} \end{aligned} \tag{9.3.15}$$

Then $\dot{z}$ is of second order and $\dot{z}^2$ is of fourth order and thus negligible. Accordingly, we have

$$L = \frac{m}{2}(\dot{x}^2 + \dot{y}^2) - \frac{mg}{2a}(x^2 + y^2) \tag{9.3.16}$$

for the small-amplitude oscillations. This is the Lagrange function of an isotropic two-dimensional harmonic oscillator with angular frequency $\omega_0 = \sqrt{g/a}$. The period

$$T = \frac{2\pi}{\omega_0} = 2\pi\sqrt{\frac{a}{g}} \tag{9.3.17}$$

is about two seconds for a pendulum of length $a = 1\,\text{m}$ since $g \cong \pi^2\,\text{m/s}^2$ is quite a good approximate value for the gravitational acceleration near the surface of the earth.

9.3.2 *Double pendulum*

More complicated and more interesting is the plane double pendulum:

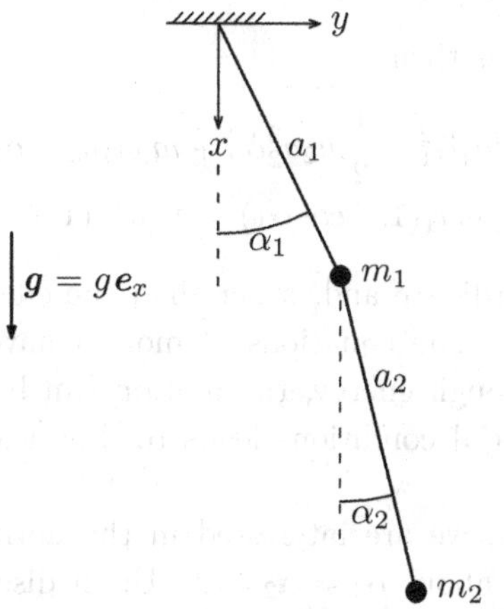

(9.3.18)

Mass m_2 hangs on a string of length a_2 from mass m_1, which itself hangs on a string of length a_1 from the ceiling, with the motion restricted to one vertical plane. We use the azimuth-type angles α_1 and α_2, which state the angles between the strings and the vertical direction of $\boldsymbol{g}$, as the coordinates. Then

$$\begin{aligned}(x_1, y_1) &= a_1(\cos\alpha_1, \sin\alpha_1)\,,\\ (x_2, y_2) &= (x_1, y_1) + a_2(\cos\alpha_2, \sin\alpha_2)\end{aligned} \tag{9.3.19}$$

are the positions of the two point masses. For the potential energy we take

$$V = m_1 g(a_1 - x_1) + m_2 g(a_1 + a_2 - x_2)\,, \tag{9.3.20}$$

which attains its minimal value when $x_1 = a$, and $x_2 = a_1 + a_2$, that is: $\alpha_1 = \alpha_2 = 0$. Expressed in terms of the angle coordinates, we have

$$\begin{aligned}V &= m_1 g a_1(1 - \cos\alpha_1) + m_2 g\big[a_1(1 - \cos\alpha_1) + a_2(1 - \cos\alpha_2)\big]\\ &= (m_1 + m_2) g a_1(1 - \cos\alpha_1) + m_2 g a_2(1 - \cos\alpha_2)\,.\end{aligned} \tag{9.3.21}$$

With the kinetic energy

$$\begin{aligned}
&\frac{m_1}{2}(\dot{x}_1^2+\dot{y}_1^2)+\frac{m_2}{2}(\dot{x}_2^2+\dot{y}_2^2)\\
&=\frac{m_1}{2}a_1^2\dot{\alpha}_1^2+\frac{m_2}{2}(-a_1\dot{\alpha}_1\sin\alpha_1-a_2\dot{\alpha}_2\sin\alpha_2)^2\\
&\qquad+\frac{m_2}{2}(a_1\dot{\alpha}_1\cos\alpha_1+a_2\dot{\alpha}_2\cos\alpha_2)^2\\
&=\frac{m_1+m_2}{2}a_1^2\dot{\alpha}_1^2+\frac{m_2}{2}a_2^2\dot{\alpha}_2^2+m_2a_1a_2\dot{\alpha}_1\dot{\alpha}_2\cos(\alpha_1-\alpha_2)\,, \qquad (9.3.22)
\end{aligned}$$

the Lagrange function is then

$$\begin{aligned}
L=&\frac{1}{2}(m_1+m_2)a_1^2\dot{\alpha}_1^2+\frac{1}{2}m_2a_2^2\dot{\alpha}_2^2+m_2a_1a_2\dot{\alpha}_1\dot{\alpha}_2\cos(\alpha_1-\alpha_2)\\
&-(m_1+m_2)ga_1(1-\cos\alpha_1)-m_2ga_2(1-\cos\alpha_2)\,. \qquad (9.3.23)
\end{aligned}$$

There is no cyclic coordinate and, other than the energy, we do not have a constant of motion. The equations of motion have quite complicated solutions; for large enough energy, the motion can be chaotic, that is: a tiny change in the initial conditions leads to drastically different motion after a while.

Currently, however, we are interested in the normal modes of small-amplitude oscillations about $\alpha_1=\alpha_2=0$. Upon discarding all contributions of cubic and higher joint order, we have

$$\begin{aligned}
L=&\frac{1}{2}(m_1+m_2)a_1^2\dot{\alpha}_1^2+\frac{1}{2}m_2a_2^2\dot{\alpha}_2^2+m_2a_1a_2\dot{\alpha}_1\dot{\alpha}_2\\
&-\frac{1}{2}(m_1+m_2)ga_1\alpha_1^2-\frac{1}{2}m_2ga_2\alpha_2^2 \qquad (9.3.24)
\end{aligned}$$

or

$$L=\frac{1}{2}(\,\dot{\alpha}_1\;\dot{\alpha}_2\,)M\begin{pmatrix}\dot{\alpha}_1\\ \dot{\alpha}_2\end{pmatrix}-\frac{1}{2}(\,\alpha_1\;\alpha_2\,)K\begin{pmatrix}\alpha_1\\ \alpha_2\end{pmatrix} \qquad (9.3.25)$$

with the 2×2 matrices

$$M=\begin{pmatrix}(m_1+m_2)a_1^2 & m_2a_1a_2\\ m_2a_1a_2 & m_2a_2^2\end{pmatrix},\quad K=\begin{pmatrix}(m_1+m_2)ga_1 & 0\\ 0 & m_2ga_2\end{pmatrix}. \qquad (9.3.26)$$

For simplicity, let us now just deal with the particular case of $m_1=m_2=m$ and $a_1=a_2=a$. With $g=\omega_0^2a$, then, these matrices are

$$M=ma^2\begin{pmatrix}2\,1\\ 1\,1\end{pmatrix},\quad K=m\omega_0^2a^2\begin{pmatrix}2\,0\\ 0\,1\end{pmatrix}, \qquad (9.3.27)$$

so that the characteristic polynomial for ω^2 is given by the determinant of

$$\omega^2 M - K = ma^2 \begin{pmatrix} 2(\omega^2 - \omega_0^2) & \omega^2 \\ \omega^2 & (\omega^2 - \omega_0^2) \end{pmatrix}, \tag{9.3.28}$$

whereby the constant prefactor of ma^2 is irrelevant. This states that the characteristic frequencies are the roots of

$$2(\omega^2 - \omega_0^2)^2 - \omega^4 = 0, \tag{9.3.29}$$

so that

$$\left.\begin{matrix} \omega_1^2 \\ \omega_2^2 \end{matrix}\right\} = 2\omega_0^2 \pm \sqrt{2}\,\omega_0^2 \quad \text{or} \quad \left.\begin{matrix} \omega_1 \\ \omega_2 \end{matrix}\right\} = \sqrt{2 \pm \sqrt{2}}\,\omega_0 \,. \tag{9.3.30}$$

The corresponding $X^{(1,2)}$ columns are now determined by

$$(\omega_{1,2}^2 M - K) X^{(1,2)} = 0, \tag{9.3.31}$$

which is here, after removing the common factor $ma^2\omega_0^2$,

$$\left[(2 \pm \sqrt{2})\begin{pmatrix} 2 & 1 \\ 1 & 1 \end{pmatrix} - \begin{pmatrix} 2 & 0 \\ 0 & 1 \end{pmatrix}\right] X^{(1,2)} = 0 \tag{9.3.32}$$

or

$$\begin{pmatrix} 2(1+\sqrt{2}) & (2+\sqrt{2}) \\ (2+\sqrt{2}) & (1+\sqrt{2}) \end{pmatrix} X^{(1)} = (\sqrt{2}+1)\begin{pmatrix} 2 & \sqrt{2} \\ \sqrt{2} & 1 \end{pmatrix} X^{(1)} = 0,$$

$$\begin{pmatrix} 2(1-\sqrt{2}) & (2-\sqrt{2}) \\ (2-\sqrt{2}) & (1-\sqrt{2}) \end{pmatrix} X^{(2)} = (\sqrt{2}-1)\begin{pmatrix} -2 & \sqrt{2} \\ \sqrt{2} & -1 \end{pmatrix} X^{(2)} = 0, \tag{9.3.33}$$

so that

$$X^{(1)} = \begin{pmatrix} 1 \\ -\sqrt{2} \end{pmatrix}, \quad X^{(2)} = \begin{pmatrix} 1 \\ \sqrt{2} \end{pmatrix}, \tag{9.3.34}$$

or multiples thereof.

The small-amplitude oscillations associated with the two normal modes are, therefore, of the form

$$\begin{pmatrix} \alpha_1 \\ \alpha_2 \end{pmatrix} = \epsilon_1 \cos(\omega_1 t - \phi_1)\begin{pmatrix} 1 \\ -\sqrt{2} \end{pmatrix} + \epsilon_2 \cos(\omega_2 t - \phi_2)\begin{pmatrix} 1 \\ \sqrt{2} \end{pmatrix} \tag{9.3.35}$$

where ϵ_1, ϵ_2 are the small amplitudes and ϕ_1, ϕ_2 are phases, with all four constants determined by the initial conditions. For $\epsilon_2 = 0$, we only have the eigenmode with angular frequency ω_1, and the minus sign in $X^{(1)} = \begin{pmatrix} 1 \\ -\sqrt{2} \end{pmatrix}$ tells us that the two pendulums are out of phase:

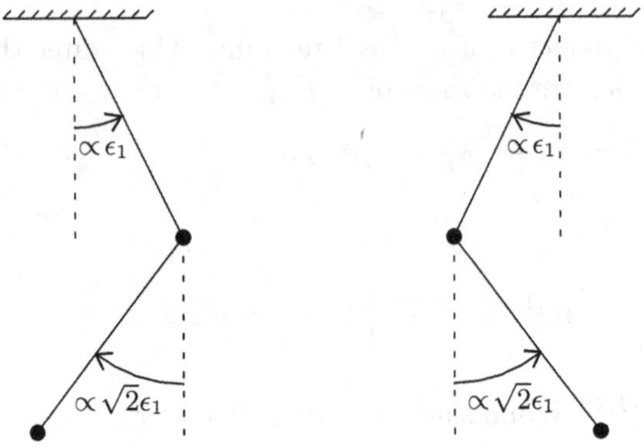

$$\text{when } \epsilon_1 \cos(\omega_1 t - \phi_1) > 0\,, \qquad \text{when } \epsilon_1 \cos(\omega_1 t - \phi_1) < 0\,, \tag{9.3.36}$$

whereas they are in phase for $\epsilon_1 = 0$ and $X^{(2)} = \begin{pmatrix} 1 \\ \sqrt{2} \end{pmatrix}$:

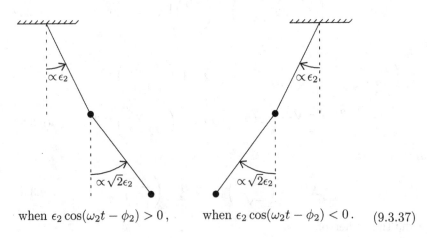

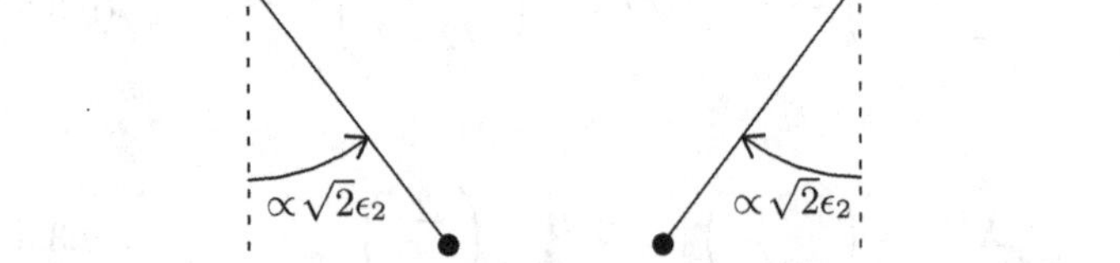

$$\text{when } \epsilon_2 \cos(\omega_2 t - \phi_2) > 0\,, \qquad \text{when } \epsilon_2 \cos(\omega_2 t - \phi_2) < 0\,. \tag{9.3.37}$$

9.3.3 *Linear triatomic molecule*

The next example is a model for the vibrational modes of linear molecules such as carbon dioxide, where the carbon atom is in the middle between the two oxygen atoms, and the simple model has them connected with springs as depicted here:

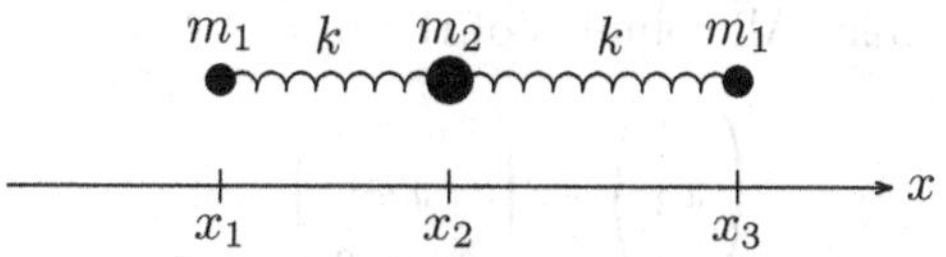

(9.3.38)

We shall only consider the motion along the x axis. Then, with the outer masses equal and both springs having the same spring constant k and natural length a, the Lagrange function is

$$\begin{aligned}L &= \frac{m_1}{2}\dot{x}_1^2 + \frac{m_2}{2}\dot{x}_2^2 + \frac{m_1}{2}\dot{x}_3^2 \\ &\quad - \frac{1}{2}k(x_2 - x_1 - a)^2 - \frac{1}{2}k(x_3 - x_2 - a)^2 \\ &= \frac{1}{2}(\dot{x}_1\ \dot{x}_2\ \dot{x}_3)M\begin{pmatrix}\dot{x}_1\\ \dot{x}_2\\ \dot{x}_3\end{pmatrix} - \frac{1}{2}(x_1\ x_2\ x_3)K\begin{pmatrix}x_1\\ x_2\\ x_3\end{pmatrix} \\ &\quad + ka(x_3 - x_1) - ka^2\end{aligned} \tag{9.3.39}$$

with the 3×3 mass matrix

$$M = \begin{pmatrix} m_1 & 0 & 0 \\ 0 & m_2 & 0 \\ 0 & 0 & m_1 \end{pmatrix} \tag{9.3.40}$$

and the matrix of the spring constants

$$K = \begin{pmatrix} k & -k & 0 \\ -k & 2k & -k \\ 0 & -k & k \end{pmatrix}. \tag{9.3.41}$$

All force components $\dfrac{\partial}{\partial x_j}L$ vanish at configurations for which

$$K\begin{pmatrix}x_1\\ x_2\\ x_3\end{pmatrix} + \begin{pmatrix}ka\\ 0\\ -ka\end{pmatrix} = 0, \tag{9.3.42}$$

that is

$$\begin{pmatrix} 1 & -1 & 0 \\ -1 & 2 & -1 \\ 0 & -1 & 1 \end{pmatrix}\begin{pmatrix}x_1\\ x_2\\ x_3\end{pmatrix} = \begin{pmatrix}-a\\ 0\\ a\end{pmatrix}. \tag{9.3.43}$$

But we cannot solve this equation for a unique $\begin{pmatrix} x_1 \\ x_2 \\ x_3 \end{pmatrix}$ because the 3×3 matrix is not invertible. All columns of the form

$$\begin{pmatrix} x_1 \\ x_2 \\ x_3 \end{pmatrix} = \begin{pmatrix} x_0 - a \\ x_0 \\ x_0 + a \end{pmatrix} \tag{9.3.44}$$

solve (9.3.43), irrespective of the value of x_0. This, of course, tells us the obvious: There is full translational invariance along the x axis, it does not matter at which distance x_0 from the origin of the coordinate system is the central mass m_2 located, as long as the outer masses are at distance a to the left and to the right, respectively. We could even have the three masses moving with constant velocity, $x_0 = v_0 t$, and would still have no forces acting.

So, let us now pick a certain x_0 as the equilibrium position of the central mass m_2 and consider small-amplitude oscillations about the configuration with $x_1 = x_0 - a$, $x_2 = x_0$, $x_3 = x_0 + a$ by writing

$$\begin{pmatrix} x_1 \\ x_2 \\ x_3 \end{pmatrix} = \begin{pmatrix} x_0 - a \\ x_0 \\ x_0 + a \end{pmatrix} + X\,. \tag{9.3.45}$$

Then

$$\begin{aligned} L &= \frac{1}{2}\dot{X}^{\mathrm{T}} M \dot{X} - \frac{1}{2} X^{\mathrm{T}} K X - \big(\, x_0 - a \;\; x_0 \;\; x_0 + a \,\big) K X \\ &\quad - \frac{1}{2}\big(\, x_0 - a \;\; x_0 \;\; x_0 + a \,\big) K \begin{pmatrix} x_0 - a \\ x_0 \\ x_0 + a \end{pmatrix} \\ &\quad - \big(\, ka \;\; 0 \;\; -ka \,\big) \left[\begin{pmatrix} x_0 - a \\ x_0 \\ x_0 + a \end{pmatrix} + X \right] - ka^2 \\ &= \frac{1}{2}\dot{X}^{\mathrm{T}} M \dot{X} - \frac{1}{2} X^{\mathrm{T}} K X\,, \end{aligned} \tag{9.3.46}$$

where the cancellation of the other terms is as expected and can be verified with the help of

$$\big(\, x_0 - a \;\; x_0 \;\; x_0 + a \,\big) K = \big(\, -ka \;\; 0 \;\; ka \,\big)\,. \tag{9.3.47}$$

The characteristic frequencies are now determined as the solutions of

$$\det\{\omega^2 M - K\} = 0 \tag{9.3.48}$$

with the 3×3 matrix

$$\omega^2 M - K = \begin{pmatrix} m_1\omega^2 - k & k & 0 \\ k & m_2\omega^2 - 2k & k \\ 0 & k & m_1\omega^2 - k \end{pmatrix}, \tag{9.3.49}$$

so that the characteristic polynomial is

$$(m_1\omega^2 - k)^2(m_2\omega^2 - 2k) - 2(m_1\omega^2 - k)k^2 = 0 \tag{9.3.50}$$

or

$$\omega^2(m_1\omega^2 - k)\big[m_1 m_2\omega^2 - (2m_1 + m_2)k\big] = 0\,. \tag{9.3.51}$$

The roots are

$$\omega_1^2 = 0\,, \quad \omega_2^2 = \frac{k}{m_1}\,, \quad \omega_3^2 = \frac{2m_1 + m_2}{m_1 m_2}k = \frac{2m_1 + m_2}{m_2}\omega_2^2 \tag{9.3.52}$$

or, with $k = m_1\omega_0^2$,

$$\omega_1 = 0\,, \quad \omega_2 = \omega_0\,, \quad \omega_3 = \sqrt{\frac{2m_1 + m_2}{m_2}}\,\omega_0\,. \tag{9.3.53}$$

The vanishing characteristic frequency $\omega_1 = 0$ has the eigencolumn $X^{(1)}$ such that

$$KX^{(1)} = 0\,, \tag{9.3.54}$$

that is

$$X^{(1)} = \begin{pmatrix} 1 \\ 1 \\ 1 \end{pmatrix} \tag{9.3.55}$$

or a multiple thereof. This is not really describing a small-amplitude oscillation, rather it reminds us of the translational invariance, inasmuch as it gives

$$\begin{pmatrix} x_1 \\ x_2 \\ x_3 \end{pmatrix} = \begin{pmatrix} x_0 - a \\ x_0 \\ x_0 + a \end{pmatrix} + (\epsilon_1 + v_1 t)\begin{pmatrix} 1 \\ 1 \\ 1 \end{pmatrix} \tag{9.3.56}$$

for this "zero-frequency oscillation," which is just a translation with constant velocity v_1. We recall in this context that the $\omega \to 0$ limit of the harmonic-oscillator position $x_0 \cos(\omega t) + (\dot{x}_0/\omega)\sin(\omega t)$ is $x_0 + \dot{x}_0 t$ with the linear-in-t dependence associated with a constant velocity.

Now turning to $\omega_2 = \sqrt{k/m_1} = \omega_0$, the eigencolumn of this normal mode is determined by

$$(\omega_2^2 M - K)X^{(2)} = \begin{pmatrix} 0 & k & 0 \\ k & \left(\frac{m_2}{m_1} - 2\right)k & k \\ 0 & k & 0 \end{pmatrix} X^{(2)} = 0\,, \tag{9.3.57}$$

which gives

$$X^{(2)} = \begin{pmatrix} 1 \\ 0 \\ -1 \end{pmatrix} \tag{9.3.58}$$

or a multiple thereof. Small-amplitude oscillations in this mode are then of the form

$$\begin{pmatrix} x_1 \\ x_2 \\ x_3 \end{pmatrix} = \begin{pmatrix} x_0 - a \\ x_0 \\ x_0 + a \end{pmatrix} + \epsilon_2 \cos(\omega_2 t - \phi_2) \begin{pmatrix} 1 \\ 0 \\ -1 \end{pmatrix}, \tag{9.3.59}$$

that is: x_2 is staying put at x_0, and x_1, x_3 oscillate around $x_0 - a$ and $x_0 + a$, respectively, with equal amplitude but *out of phase*:

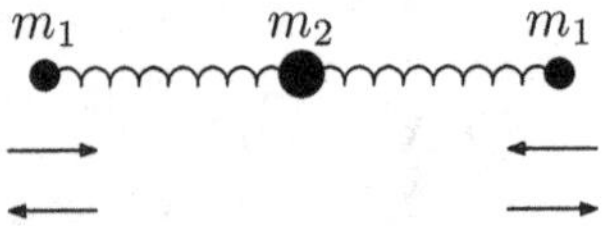

(9.3.60)

When the left-hand mass m_1 is displaced to the right, the right-hand mass m_2 is displaced to the left, and vice versa. A vibration of this kind is often referred to as a *breathing mode.*

Finally, we look at the normal mode with angular frequency $\omega_3 = \sqrt{1 + 2m_1/m_2}\,\omega_0$, for which

$$(\omega_3^2 M - K)X^{(3)} = k \begin{pmatrix} 2\frac{m_1}{m_2} & 1 & 0 \\ 1 & \frac{m_2}{m_1} & 1 \\ 0 & 1 & 2\frac{m_1}{m_2} \end{pmatrix} X^{(3)} = 0 \tag{9.3.61}$$

gives

$$X^{(3)} = \begin{pmatrix} 1 \\ -2\frac{m_1}{m_2} \\ 1 \end{pmatrix} \tag{9.3.62}$$

or a multiple thereof. The vibration in this mode has

$$\begin{pmatrix} x_1 \\ x_2 \\ x_3 \end{pmatrix} = \begin{pmatrix} x_0 - a \\ x_0 \\ x_0 + a \end{pmatrix} + \epsilon_3 \cos(\omega_3 t - \phi_3) \begin{pmatrix} 1 \\ -2\dfrac{m_1}{m_2} \\ 1 \end{pmatrix}, \tag{9.3.63}$$

that is: the outer masses m_1 oscillate with equal amplitude *in phase* while the central mass m_2 oscillates out of phase relative to the outer masses and with an amplitude that is $2m_1/m_2$ times the amplitude of the outer-mass oscillation:

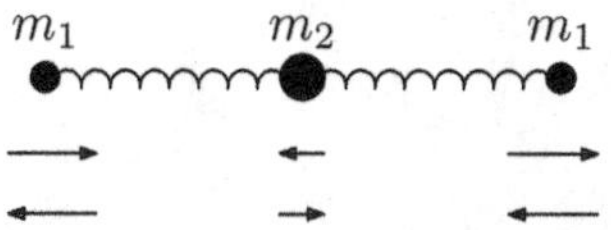

(9.3.64)

This vibrational mode is asymmetric and has a higher frequency than the symmetric breathing mode.

Chapter 10

From Lagrange to Hamilton

10.1 Time as a coordinate

After these examples that illustrate the use and the usefulness of the Lagrange-function method, let us return to the energy statement (8.2.3),

$$\frac{\mathrm{d}}{\mathrm{d}t}\left(\sum_k \dot{x}_k \frac{\partial L}{\partial \dot{x}_k} - L\right) = -\frac{\partial L}{\partial t}\,, \tag{10.1.1}$$

and put it on equal footing with the Euler–Lagrange equation (8.1.21) for one of the coordinates,

$$\frac{\mathrm{d}}{\mathrm{d}t}\frac{\partial L}{\partial \dot{x}_k} = \frac{\partial L}{\partial x_k}\,. \tag{10.1.2}$$

For this discussion, we shall consider only one coordinate, $x(t)$, for notational simplicity; the generalization to many coordinates will be immediate.

Thus, we reconsider the PSA for the action

$$W_{01} = \int_{t_0}^{t_1} \mathrm{d}t\, L\bigl(t, x(t), \dot{x}(t)\bigr)\,, \tag{10.1.3}$$

where

$$\delta W_{01} = 0 \tag{10.1.4}$$

for variations about the actual trajectory with fixed endpoints,

$$\delta x(t_0) = 0\,, \quad \delta x(t_1) = 0\,. \tag{10.1.5}$$

Now, rather than parameterizing the trajectory by stating x as a function of t, let us parameterize both t and x by a parameter α,

$$\bigl(t, x(t)\bigr) \;\to\; \bigl(t(\alpha), x(\alpha)\bigr)\,, \tag{10.1.6}$$

with $\alpha = 0$ and $\alpha = 1$ at the end points,

$$\begin{aligned}\bigl(t(\alpha), x(\alpha)\bigr)\Big|_{\alpha=0} &= (t_0, x_0)\,,\\ \bigl(t(\alpha), x(\alpha)\bigr)\Big|_{\alpha=1} &= (t_1, x_1)\,.\end{aligned} \tag{10.1.7}$$

Then

$$\mathrm{d}t = \mathrm{d}\alpha \frac{\mathrm{d}t(\alpha)}{\mathrm{d}\alpha} = \mathrm{d}\alpha\, t'(\alpha) \quad \text{and} \quad \frac{\mathrm{d}}{\mathrm{d}t} = \frac{1}{t'(\alpha)}\frac{\mathrm{d}}{\mathrm{d}\alpha}\,, \tag{10.1.8}$$

where we denote differentiation with respect to α by a prime to distinguish it from the differentiation with respect to t, denoted by the overdot. Accordingly,

$$\frac{\mathrm{d}}{\mathrm{d}t}x = \dot{x} = \frac{\mathrm{d}x}{\mathrm{d}\alpha}\bigg/\frac{\mathrm{d}t}{\mathrm{d}\alpha} = \frac{x'}{t'}\,, \tag{10.1.9}$$

and the action is now given by the integral

$$W_{01} = \int_0^1 \mathrm{d}\alpha\, t' L\bigl(t, x, x'/t'\bigr) = \int_0^1 \mathrm{d}\alpha\, f(t, t', x, x') \tag{10.1.10}$$

with the new integrand function

$$f(t, t', x, x') = t' L\bigl(t, x, x'/t'\bigr) \tag{10.1.11}$$

and the constraints

$$\begin{aligned}\delta t &= 0 \ \text{ for } \alpha = 0 \text{ and } \alpha = 1\,,\\ \delta x &= 0 \ \text{ for } \alpha = 0 \text{ and } \alpha = 1\,.\end{aligned} \tag{10.1.12}$$

In effect, the transition from (10.1.3) to (10.1.10) turned time t into one of the coordinates that vary along the path from (t_1, x_1) to (t_2, x_2).

For the action in (10.1.10), the Euler–Lagrange equation for x is

$$\frac{\mathrm{d}}{\mathrm{d}\alpha}\frac{\partial f}{\partial x'} = \frac{\partial f}{\partial x}\,, \tag{10.1.13}$$

which amounts to

$$\frac{\mathrm{d}}{\mathrm{d}\alpha}\Biggl(t'\,\frac{\partial L}{\partial \dot{x}}\,\underbrace{\frac{\partial}{\partial x'}\frac{x'}{t'}}_{=1/t'}\Biggr) = t'\frac{\partial L}{\partial x} \tag{10.1.14}$$

or

$$\frac{1}{t'}\frac{\mathrm{d}}{\mathrm{d}\alpha}\frac{\partial L}{\partial \dot{x}} = \frac{\partial L}{\partial x}, \tag{10.1.15}$$

that is

$$\frac{\mathrm{d}}{\mathrm{d}t}\frac{\partial L}{\partial \dot{x}} = \frac{\partial L}{\partial x} \tag{10.1.16}$$

in view of (10.1.8). In short, we get back the equation of motion in the form of the Euler–Lagrange equation for the Lagrange function $L(t, x, \dot{x})$.

The Euler–Lagrange equation for t is

$$\frac{\mathrm{d}}{\mathrm{d}\alpha}\frac{\partial f}{\partial t'} = \frac{\partial f}{\partial t} \tag{10.1.17}$$

with the ingredients

$$\frac{\partial f}{\partial t'} = L + t'\frac{\partial L}{\partial \dot{x}}\underbrace{\frac{\partial}{\partial t'}\frac{x'}{t'}}_{=-x'/t'^2} = L - \frac{x'}{t'}\frac{\partial L}{\partial \dot{x}} = L - \dot{x}\frac{\partial L}{\partial \dot{x}} \tag{10.1.18}$$

and

$$\frac{\partial f}{\partial t} = t'\frac{\partial L}{\partial t}, \tag{10.1.19}$$

so that

$$\frac{\mathrm{d}}{\mathrm{d}\alpha}\frac{\partial f}{\partial t'} = t'\frac{\mathrm{d}}{\mathrm{d}t}\left(L - \dot{x}\frac{\partial L}{\partial \dot{x}}\right) = t'\frac{\partial L}{\partial t} \tag{10.1.20}$$

or

$$\frac{\mathrm{d}}{\mathrm{d}t}\left(\dot{x}\frac{\partial L}{\partial \dot{x}} - L\right) = -\frac{\partial L}{\partial t}, \tag{10.1.21}$$

upon invoking (10.1.8) once more. This is the one-coordinate version of (10.1.1), of course. Indeed, we here get the second form of the Euler–Lagrange equation for x as the Euler–Lagrange equation for t, which tells us that this equation has really more to do with variations of t than with variations of x.

As announced above, the generalization to several coordinates is immediate, we just have to repeat the argument for $L(t, x_1, \dot{x}_1, x_2, \dot{x}_2, \ldots)$ and

$$f(t, t', x_1, x_1', x_2, x_2', \ldots) = t'L\bigl(t, x_1, x_1'/t', x_2, x_2'/t', \ldots\bigr) \tag{10.1.22}$$

and so find

$$\frac{\mathrm{d}}{\mathrm{d}t}\left(\sum_k \dot{x}_k \frac{\partial L}{\partial \dot{x}_k} - L\right) = -\frac{\partial L}{\partial t} \tag{10.1.23}$$

from the Euler–Lagrange equation for t as it results from the action in the form

$$W_{01} = \int_0^1 \mathrm{d}\alpha\, f(t, t', x_1, x_1', x_2, x_2', \ldots)\,. \tag{10.1.24}$$

As a consequence, energy conservation

$$E = \sum_k \dot{x}_k \frac{\partial L}{\partial \dot{x}_k} - L = \text{constant} \tag{10.1.25}$$

is the case if t is a cyclic coordinate of $f(t, t', x_1, x_1', \ldots)$, in full analogy with

$$\frac{\partial L}{\partial \dot{x}_k} = \text{constant} \tag{10.1.26}$$

if x_k is a cyclic coordinate of $L(t, x_1, \dot{x}_1, \ldots)$ and, as a consequence, also of $f(t, t', x_1, x_1', \ldots)$. We are now prepared for the next step, for which the f version of the action integral is truly useful.

10.2 Endpoint variations: Momentum and Hamilton function

We reconsider the action in the form (10.1.10) but now allow end-point variations:

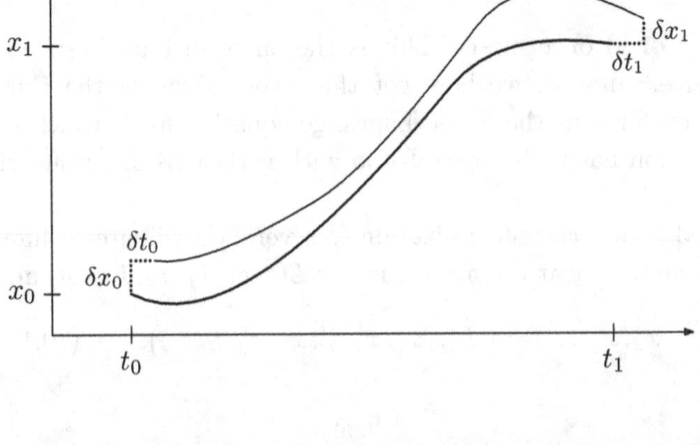

(10.2.1)

and ask: What is the resulting first-order change of the action? We calculate δW_{01} as the difference between the actions for the slightly altered path and the unaltered path,

$$\begin{aligned}\delta W_{01} &= \int_0^1 \mathrm{d}\alpha \left[f(t+\delta t, t'+\delta t', x+\delta x, x'+\delta x') - f(t,t',x,x')\right] \\ &= \int_0^1 \mathrm{d}\alpha \left(\delta t\frac{\partial f}{\partial t} + \delta t'\frac{\partial f}{\partial t'} + \delta x\frac{\partial f}{\partial x} + \delta x'\frac{\partial f}{\partial x'}\right) \\ &= \int_0^1 \mathrm{d}\alpha \left[\frac{\mathrm{d}}{\mathrm{d}\alpha}\left(\delta t\frac{\partial f}{\partial t'} + \delta x\frac{\partial f}{\partial x'}\right)\right. \\ &\qquad \left. + \delta t\left(\frac{\partial f}{\partial t} - \frac{\mathrm{d}}{\mathrm{d}\alpha}\frac{\partial f}{\partial t'}\right) + \delta x\left(\frac{\partial f}{\partial x} - \frac{\mathrm{d}}{\mathrm{d}\alpha}\frac{\partial f}{\partial x'}\right)\right], \end{aligned} \tag{10.2.2}$$

where the last step is essentially the integration by parts that we first encountered at (7.2.9). After completing this integration by parts by evaluating the total α derivative in terms of its endpoint values, we have

$$\begin{aligned}\delta W_{01} &= \left(\delta x\frac{\partial f}{\partial x'} + \delta t\frac{\partial f}{\partial t'}\right)\Bigg|_{\alpha=0}^{1} \\ &\quad + \int_0^1 \mathrm{d}\alpha \left[\delta x\left(\frac{\partial f}{\partial x} - \frac{\mathrm{d}}{\mathrm{d}\alpha}\frac{\partial f}{\partial x'}\right) + \delta t\left(\frac{\partial f}{\partial t} - \frac{\mathrm{d}}{\mathrm{d}\alpha}\frac{\partial f}{\partial t'}\right)\right]. \end{aligned} \tag{10.2.3}$$

For the actual path, $t(\alpha)$ and $x(\alpha)$ obey the Euler–Lagrange equations, so that δW_{01} involves only the endpoint variations,

$$\begin{aligned}\delta W_{01} &= \delta x_1\left(\frac{\partial f}{\partial x'}\right)_{\alpha=1} + \delta t_1\left(\frac{\partial f}{\partial t'}\right)_{\alpha=1} \\ &\quad - \delta x_0\left(\frac{\partial f}{\partial x'}\right)_{\alpha=0} - \delta t_0\left(\frac{\partial f}{\partial t'}\right)_{\alpha=0}, \end{aligned} \tag{10.2.4}$$

but has no contribution from intermediate α values. We can, therefore, rephrase the PSA:

Principle of Stationary Action
The response of the action to infinitesimal changes of the actual path results solely from the variations of the endpoints of the path. (10.2.5)

The proper technical formulation is the variant of (10.2.4) that is given in (10.2.15) below.

Standard notation is

$$\frac{\partial f}{\partial x'} = p \quad (\text{"momentum"}) \tag{10.2.6}$$

and

$$-\frac{\partial f}{\partial t'} = H \quad (\text{"Hamilton* function"}) \tag{10.2.7}$$

for the expressions that multiply δx and δt at the endpoints. We have

$$p = \frac{\partial f}{\partial x'} = \frac{\partial}{\partial x'}\Big[t' L(t, x, x'/t')\Big] = \frac{\partial L}{\partial \dot{x}}(t, x, x'/t') \tag{10.2.8}$$

or

$$p = \frac{\partial L}{\partial \dot{x}}(t, x, \dot{x}), \tag{10.2.9}$$

and

$$\begin{aligned} H = -\frac{\partial f}{\partial t'} &= -\frac{\partial}{\partial t'}\Big[t' L(t, x, x'/t')\Big] \\ &= \frac{x'}{t'}\frac{\partial L}{\partial \dot{x}}(t, x, x'/t') - L(t, x, x'/t') \end{aligned} \tag{10.2.10}$$

or

$$H = \dot{x}\frac{\partial L}{\partial \dot{x}} - L. \tag{10.2.11}$$

As observed in Section 8.2, for a Lagrange function of the standard "kinetic energy minus potential energy" form

$$L = \frac{m}{2}\dot{x}^2 - V(x, t), \tag{10.2.12}$$

these are the kinetic momentum

$$p = \frac{\partial L}{\partial \dot{x}} = m\dot{x}, \tag{10.2.13}$$

and the energy

$$H = \dot{x} m \dot{x} - \left[\frac{m}{2}\dot{x}^2 - V(x, t)\right] = \frac{m}{2}\dot{x}^2 + V(x, t), \tag{10.2.14}$$

and we finally understand why there is the minus sign in (10.2.7); it ensures that H is energy, not negative energy.

---

*Sir William Rowan HAMILTON (1805–1865)

In summary, then, the PSA (10.2.5) says

$$\begin{aligned}\delta W_{01} &= (p\delta x - H\delta t)\Big|_0^1 \\ &= (p_1\delta x_1 - H_1\delta t_1) - (p_0\delta x_0 - H_0\delta t_0)\end{aligned} \tag{10.2.15}$$

with p_0, p_1 denoting the momentum values for t_0 and t_1, and likewise H_0, H_1 for the energy of the initial and final configuration.

10.3 Five remarks

10.3.1 *Natural variables*

Here is the first of five brief remarks. When we write the Hamilton function as

$$H = \dot{x}p - L \tag{10.3.1}$$

and then consider the response of H to small changes of its ingredients,

$$\begin{aligned}\delta H &= \delta\dot{x}\,p + \dot{x}\delta p - \delta t\frac{\partial L}{\partial t} - \delta x\frac{\partial L}{\partial x} - \underbrace{\delta\dot{x}\frac{\partial L}{\partial \dot{x}}}_{=\,\delta\dot{x}\,p} \\ &= \dot{x}\delta p - \frac{\partial L}{\partial t}\delta t - \frac{\partial L}{\partial x}\delta x\,,\end{aligned} \tag{10.3.2}$$

we observe that the natural variables of H are x, p, t, not x, $\dot{x}$, t as for the Lagrange function. Accordingly, we should eliminate $\dot{x}$ from $\dot{x}p - L$ in favour of p when stating the Hamilton function. For the standard Lagrange function (10.2.12), this gives

$$\begin{aligned}H(x,p,t) &= \left[\dot{x}p - \frac{m}{2}\dot{x}^2 + V(x,t)\right]_{m\dot{x}\,=\,p} \\ &= \frac{p^2}{2m} + V(x,t)\,,\end{aligned} \tag{10.3.3}$$

the standard form of the Hamilton function.

10.3.2 *The minus sign*

The second remark concerns the minus sign in $p\delta x - H\delta t$, where — to introduce another bit of terminology — one recognizes the momentum p as the *generator* for spatial displacements and the Hamilton function H as the generator for time displacements. These notions carry over to quantum mechanics. Indeed, it is this response of the action which identifies the correct commutation relations for position and momentum. And the minus sign in $p\delta x - H\delta t$ is present, for instance, in the Schrödinger* equation for a position wave function $\psi(x,t)$,

$$H\left(x, \frac{\hbar}{\mathrm{i}}\frac{\partial}{\partial x}, t\right)\psi(x,t) = \mathrm{i}\hbar\frac{\partial}{\partial t}\psi(x,t)\,, \tag{10.3.4}$$

where the differential operator $\frac{\hbar}{\mathrm{i}}\frac{\partial}{\partial x}$ represents the momentum, whereas $\mathrm{i}\hbar\frac{\partial}{\partial t} = -\frac{\hbar}{\mathrm{i}}\frac{\partial}{\partial t}$ relates to the energy. We meet Planck's* constant here (equal to $2\pi\hbar$).

10.3.3 *Legendre transformations*

The third remark comments on the first remark that $H = \dot{x}p - L$ is a function of x, p, and t rather than of x, $\dot{x}$, and t. Such a systematic switch of variables is a so-called *Legendre*† *transformation*. In classical mechanics we encounter it in the transition from the Lagrange function to the Hamilton function, as just seen. Many more examples are met in thermodynamics, where Legendre transformations link the various thermodynamic potentials. A simple example is the switch from the internal energy $U(S,V)$ — which is a function of entropy S and volume V, with the differential

$$\mathrm{d}U = T\mathrm{d}S - p\mathrm{d}V \tag{10.3.5}$$

that identifies the temperature $T = \frac{\partial U}{\partial S}$ and the pressure $p = -\frac{\partial U}{\partial V}$ ($\neq$ momentum) — to the Helmholtz‡ free energy $F = U - TS$ which, as

$$\begin{aligned}\mathrm{d}F &= \mathrm{d}(U - TS)\\ &= T\mathrm{d}S - p\mathrm{d}V - T\mathrm{d}S - S\mathrm{d}T\\ &= -S\mathrm{d}T - p\mathrm{d}V\end{aligned} \tag{10.3.6}$$

tells us, is a function of T and V. There is also the enthalpy $U + pV$, which is a function of entropy and pressure, and the Gibbs§ free energy $U + pV - TS = F + pV$, which is a function of temperature and pressure.

---

*Erwin SCHRÖDINGER (1889–1961) *Max Karl Ernst Ludwig PLANCK (1858–1947)

†Adrien Marie LEGENDRE (1752–1833)

‡Hermann Ludwig Ferdinand von HELMHOLTZ (1821–1894)

§Josiah Willard GIBBS (1839–1903)

10.3.4 *Many coordinates*

We take the small and immediate step from one coordinate x to many,

$$\delta W_{01} = \int_0^1 \mathrm{d}t\, L(t, X, \dot{X}) = \left(\sum_k p_k \delta x_k - H\delta t\right)\Bigg|_0^1 , \tag{10.3.7}$$

where X, as before, stands for the collection of all coordinates and $\dot{X}$ for the collection of their time derivatives. Here we have

$$p_k = \frac{\partial L}{\partial \dot{x}_k} \tag{10.3.8}$$

for the momentum associated with coordinate x_k, and

$$H = \sum_k \dot{x}_k p_k - L = H(X, P, t) \tag{10.3.9}$$

for the Hamilton function that we understand as a function of the coordinates and the momenta $P(p_1, p_2, p_3 \ldots)$, possibly with a parametric time dependence.

10.3.5 *Kinetic and canonical momentum*

The momentum p_k need not be of the simple kinetic "mass times velocity" form, its expression in terms of the velocities depends on the structure of the Lagrange function. In contexts where it deems important to distinguish between kinetic momentum and $\dfrac{\partial L}{\partial \dot{x}_k}$, the latter is often called the *canonical momentum.*

As a first example, let us consider the Lagrange function of (8.1.22) and (8.1.23),

$$\begin{aligned} L &= \frac{m}{2}(\dot{x}^2 + \dot{y}^2) - V\left(\sqrt{x^2 + y^2}\right) \\ &= \frac{m}{2}\left[\dot{s}^2 + (s\dot{\varphi})^2\right] - V(s) . \end{aligned} \tag{10.3.10}$$

In terms of the cartesian coordinates x, y, we have the momenta

$$p_x = m\dot{x} , \quad p_y = m\dot{y} \tag{10.3.11}$$

and get the Hamilton function

$$H(x, y, p_x, p_y) = \frac{1}{2m}(p_x^2 + p_y^2) + V\left(\sqrt{x^2 + y^2}\right) . \tag{10.3.12}$$

If we use the polar-coordinate version, we have the momenta

$$p_s = m\dot{s}\,, \quad p_\varphi = ms^2\dot{\varphi}\,, \tag{10.3.13}$$

and get the Hamilton function

$$\begin{aligned} H(s, p_s, p_\varphi) &= \left(\dot{s}p_s + \dot{\varphi}p_\varphi - L\right)\Big|_{\dot{s} = p_s/m\,,\ \dot{\varphi} = p_\varphi/(ms^2)} \\ &= \frac{1}{2m}\left(p_s^2 + \frac{p_\varphi^2}{s^2}\right) + V(s)\,, \end{aligned} \tag{10.3.14}$$

which does not depend on the cyclic coordinate φ. The momenta p_x, p_y, p_s are of the kinetic "mass times velocity" variety, whereas p_φ is not. Indeed, the physical significance of p_φ is angular momentum, not linear momentum.

Another example is the Lagrange function (8.3.22) of the "oscillator on top of an oscillator" system of (8.3.19),

$$L = \frac{m_1}{2}{\dot{x}_1}^2 + \frac{m_2}{2}(\dot{x}_1 + \dot{x}_2)^2 - k_1x_1^2 - k_2x_2^2\,. \tag{10.3.15}$$

Here, both canonical momenta

$$\begin{aligned} p_1 &= \frac{\partial L}{\partial \dot{x}_1} = m_1\dot{x}_1 + m_2(\dot{x}_1 + \dot{x}_2) = (m_1 + m_2)\dot{x}_1 + m_2\dot{x}_2\,, \\ p_2 &= \frac{\partial L}{\partial \dot{x}_2} = m_2(\dot{x}_1 + \dot{x}_2)\,, \end{aligned} \tag{10.3.16}$$

are sums of kinetic momenta. We solve for the velocities in terms of the momenta,

$$\dot{x}_1 = \frac{1}{m_1}(p_1 - p_2)\,, \quad \dot{x}_2 = \frac{m_1 + m_2}{m_1m_2}p_2 - \frac{1}{m_1}p_1\,, \tag{10.3.17}$$

and use these in

$$\begin{aligned} H(x_1, x_2, p_1, p_2) &= \left(\dot{x}_1p_1 + \dot{x}_2p_2 - L\right)\Big|_{\text{eliminate } \dot{x}_1,\ \dot{x}_2} \\ &= \frac{1}{m_1}(p_1 - p_2)p_1 + \left(\frac{m_1 + m_2}{m_1m_2}p_2 - \frac{1}{m_1}p_1\right)p_2 \\ &\quad - \frac{m_1}{2}\left[\frac{1}{m_1}(p_1 - p_2)\right]^2 - \frac{m_2}{2}\left(\frac{p_2}{m_2}\right)^2 \\ &\quad + k_1x_1^2 + k_2x_2^2 \end{aligned} \tag{10.3.18}$$

to arrive at

$$H = \frac{1}{2m_1}p_1^2 + \frac{m_1 + m_2}{2m_1m_2}p_2^2 - \frac{1}{m_1}p_1p_2 + k_1x_1^2 + k_2x_2^2\,. \tag{10.3.19}$$

10.4 Cyclic coordinates and constants of motion

As we know, there is a conserved quantity, a constant of motion for each cyclic coordinate, namely

$$\frac{\partial L}{\partial \dot{x}_k} = \text{constant} \quad \text{if} \quad \frac{\partial L}{\partial x_k} = 0\,. \tag{10.4.1}$$

We now recognize that the conserved quantity is the canonical momentum associated with the cyclic coordinate,

$$p_k = \text{constant} \quad \text{if} \quad \frac{\partial L}{\partial x_k} = 0\,, \tag{10.4.2}$$

and since

$$\frac{\partial L}{\partial x_k}(t, X, \dot{X}) = -\frac{\partial H}{\partial x_k}(X, P, t) \quad \text{with} \quad p_k = \frac{\partial L}{\partial \dot{x}_k}\,, \tag{10.4.3}$$

this can also be stated in terms of the Hamilton function,

$$p_k = \text{constant} \quad \text{if} \quad \frac{\partial H}{\partial x_k} = 0\,. \tag{10.4.4}$$

It follows that $p_{k0} = p_{k1}$ for the initial and final momentum values of a cyclic coordinate and, therefore, there is no change in the action if $\delta x_{k0} = \delta x_{k1}$ for a cyclic coordinate,

$$(\delta W_{01})_k = p_{k0}\delta x_{k0} - p_{k1}\delta x_{k1} = 0\,. \tag{10.4.5}$$

We can understand this link between cyclic coordinates and conserved momenta by a look at the many-coordinates version of (10.1.10),

$$W_{01} = \int_0^1 \mathrm{d}\alpha\, t'\, L(t, X, X'/t')\,, \tag{10.4.6}$$

where X'/t' stands for all x'_k/t' ratios. We compare W_{01} for given initial and final configurations with W_{01} for configurations that differ solely by a common displacement of a cyclic coordinate, x_k, say. That is

$$\begin{aligned}(t, x_1, x_2, \ldots, x_k, \ldots)_0 &\to (t, x_1, x_2, \ldots, x_k + \epsilon, \ldots)_0\,,\\ (t, x_1, x_2, \ldots, x_k, \ldots)_1 &\to (t, x_1, x_2, \ldots, x_k + \epsilon, \ldots)_1\,,\end{aligned} \tag{10.4.7}$$

and the α functions $t(\alpha)$, $x_1(\alpha)$, ..., $x_k(\alpha)$ ... are also replaced by $t(\alpha)$, $x_1(\alpha)$, ..., $x_k(\alpha) + \epsilon$, ... But since x_k does not actually appear in

$L(t, x, \dot{x})$ and $(x_k(\alpha) + \epsilon)' = x_k'(\alpha)$ is not affected by the constant displacement of $x_k(\alpha)$, there is no change to the value of the action W_{01}: The action is invariant under a displacement of a cyclic coordinate. Put differently, the physical situations described by $(t, x_1, x_2, \ldots, x_k, \ldots)$ and $(t, x_1, x_2, \ldots, x_k + \epsilon, \ldots)$ are completely equivalent.

As an example, consider the situation sketched in (8.3.1), where φ is a cyclic coordinate. Indeed, the actual value of φ is arbitrary, because it refers to the $\varphi = 0$ line on the table, which we can choose anyway we like. A choice must be made to specify the coordinate system, but nothing distinguishes one direction from the other. There is full rotational symmetry around the vertical axis through the hole in the table.

10.4.1 *Energy*

Arguably the most important conservation law is that of energy. We realize that energy is conserved if time is a cyclic coordinate, that is: if there is no difference between

$$(t_0, X_0, \dot{X}_0) \;\rightarrow\; (t_1, X_1, \dot{X}_1) \tag{10.4.8}$$

and

$$(t_0 + \tau, X_0, \dot{X}_0) \;\rightarrow\; (t_1 + \tau, X_1, \dot{X}_1) \tag{10.4.9}$$

with a common displacement τ for the initial and the final configuration,

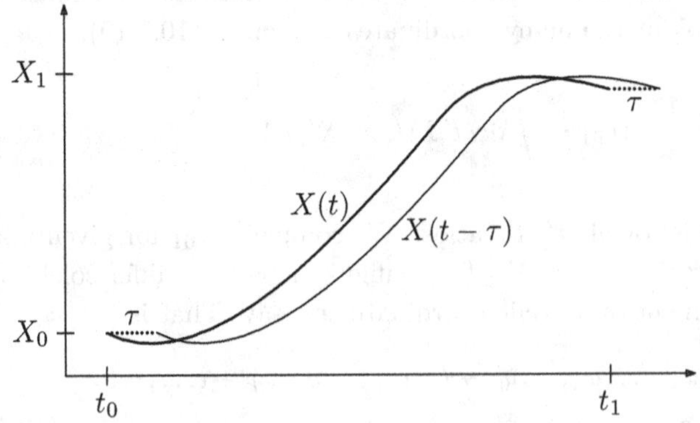

(10.4.10)

This independence of when a particular physical process happens is central for physics and science in general, inasmuch as it enables us to verify experimental findings by repeating the experiment later, many times if necessary. It does not matter *when* we perform an experiment.

When Galilei observed objects rolling down an inclined plane, their motion was not different from such objects today. Should we have doubts about Galilei's observation centuries ago, we could just repeat his experiments and confirm or disprove his records. The possibility of checking the results of past experiments, our own or those of others, is crucial for establishing scientific truth.

10.4.2 *Total momentum*

A system of point masses with conservative line-of-sight forces between them and no external forces acting has the Lagrange function

$$L = \sum_k \frac{1}{2} m_k \boldsymbol{v}_k^2 - \sum_{j<k} V_{(jk)}\big(|\boldsymbol{r}_j - \boldsymbol{r}_k|\big)\,, \tag{10.4.11}$$

a function of all position vectors $\boldsymbol{r}_1, \boldsymbol{r}_2, \ldots$ and all velocity vectors $\boldsymbol{v}_1, \boldsymbol{v}_2, \ldots$, with distance-dependent potential energies, one term for each pair. The energy (4.4.3),

$$E_{\text{tot}} = \sum_k \frac{1}{2} m_k \boldsymbol{v}_k^2 + \sum_{j<k} V_{(jk)}\big(|\boldsymbol{r}_j - \boldsymbol{r}_k|\big)\,, \tag{10.4.12}$$

is conserved as there is no parametric time dependence.

The vectorial canonical momenta,

$$\boldsymbol{p}_k = m_k \boldsymbol{v}_k\,, \tag{10.4.13}$$

are simply the kinetic momenta, and their sum is the total momentum of (4.1.4),

$$\boldsymbol{P}_{\text{tot}} = \sum_k \boldsymbol{p}_k = \sum_k m_k \boldsymbol{v}_k\,. \tag{10.4.14}$$

We know from Section 4.1 that $\boldsymbol{P}_{\text{tot}}$ does not depend on time. How does this momentum conservation result from the PSA?

The PSA reads here

$$\delta W_{01} = \left(\sum_k \boldsymbol{p}_k \cdot \delta \boldsymbol{r}_k - H\delta t\right)\Bigg|_{t_0}^{t_1}\,, \tag{10.4.15}$$

where $W_{01} = \int_{t_0}^{t_1} \mathrm{d}t\, L$ with the Lagrange function (10.4.11), and the Hamilton function is

$$H = \sum_k \frac{\boldsymbol{p}_k^2}{2m_k} + \sum_{j<k} V_{(jk)}\bigl(|\boldsymbol{r}_j - \boldsymbol{r}_k|\bigr)\,. \tag{10.4.16}$$

We apply the PSA to uniform endpoint variations,

$$\delta\boldsymbol{r}_j(t_0) = \delta\boldsymbol{r}_j(t_1) = \boldsymbol{\epsilon} \quad \text{for all } j\,, \tag{10.4.17}$$

as they follow from a uniform displacement of all $\boldsymbol{r}_j(t)$s,

$$\boldsymbol{r}_j(t) \to \boldsymbol{r}_j(t) + \boldsymbol{\epsilon}\,. \tag{10.4.18}$$

This has no effect on the velocities,

$$\boldsymbol{v}_j(t) \to \frac{\mathrm{d}}{\mathrm{d}t}\bigl(\boldsymbol{r}_j(t) + \boldsymbol{\epsilon}\bigr) = \boldsymbol{v}_j(t)\,, \tag{10.4.19}$$

nor on the distances between the point masses,

$$\boldsymbol{r}_j(t) - \boldsymbol{r}_k(t) \to \bigl(\boldsymbol{r}_j(t) + \boldsymbol{\epsilon}\bigr) - \bigl(\boldsymbol{r}_k(t) + \boldsymbol{\epsilon}\bigr) = \boldsymbol{r}_j(t) - \boldsymbol{r}_k(t)\,. \tag{10.4.20}$$

Therefore, the value of the Lagrange function (10.4.11) is not affected by (10.4.18) and so the action is not affected as well, $\delta W_{01} = 0$. Since there is no time shift, $\delta t_0 = \delta t_1 = 0$, we have

$$\delta W_{01} = \sum_k \boldsymbol{p}_k \cdot \delta\boldsymbol{r}_k \Bigr|_{t_0}^{t_1} = \sum_k \boldsymbol{p}_k \cdot \boldsymbol{\epsilon} \Bigr|_{t_0}^{t_1} = \bigl(\boldsymbol{P}_{\mathrm{tot}}(t_1) - \boldsymbol{P}_{\mathrm{tot}}(t_0)\bigr) \cdot \boldsymbol{\epsilon}\,. \tag{10.4.21}$$

This must be true for all $\boldsymbol{\epsilon}$, which requires

$$\boldsymbol{P}_{\mathrm{tot}}(t_1) = \boldsymbol{P}_{\mathrm{tot}}(t_0)\,. \tag{10.4.22}$$

Indeed, momentum is conserved, and we learn here that the *translational invariance* of a system implies the *conservation of its total momentum.*

Translational invariance is also quite important for physics: It does not matter *where* an experiment is performed. You can move the apparatus to a different place and the experiment will show the same results as in the previous location. Or, identical apparatus at different locations give the same results. Clearly, this is important if we wish to repeat an experiment in our lab to check on the results found in another lab.

10.4.3 *Total angular momentum*

The physical situation associated with the Lagrange function (10.4.11) is also invariant under uniform rotations. It is enough to consider the infinitesimal rotation

$$\begin{aligned} \boldsymbol{r}_j &\to \boldsymbol{r}_j + \boldsymbol{\delta\varphi} \times \boldsymbol{r}_j \,, \\ \boldsymbol{v}_j &\to \boldsymbol{v}_j + \boldsymbol{\delta\varphi} \times \boldsymbol{v}_j \,, \end{aligned} \tag{10.4.23}$$

which has no effect on the Lagrange function because the squared velocities $\boldsymbol{v}_j^2$ and the distances $|\boldsymbol{r}_j - \boldsymbol{r}_k|$ between the point masses are not affected,

$$\boldsymbol{v}_j^2 \to \left(\boldsymbol{v}_j + \boldsymbol{\delta\varphi} \times \boldsymbol{v}_j\right)^2 = \boldsymbol{v}_j^2 + \underbrace{2\boldsymbol{v}_j \cdot \left(\boldsymbol{\delta\varphi} \times \boldsymbol{v}_j\right)}_{=0} = \boldsymbol{v}_j^2 \,,$$

$$\begin{aligned} |\boldsymbol{r}_j - \boldsymbol{r}_k| &\to \left|(\boldsymbol{r}_j - \boldsymbol{r}_k) + \boldsymbol{\delta\varphi} \times (\boldsymbol{r}_j - \boldsymbol{r}_k)\right| \\ &= \Bigl((\boldsymbol{r}_j - \boldsymbol{r}_k)^2 + \underbrace{2(\boldsymbol{r}_j - \boldsymbol{r}_k) \cdot \bigl[\boldsymbol{\delta\varphi} \times (\boldsymbol{r}_j - \boldsymbol{r}_k)\bigr]}_{=0}\Bigr)^{1/2} \\ &= |\boldsymbol{r}_j - \boldsymbol{r}_k| \,. \end{aligned} \tag{10.4.24}$$

It follows that there is no change in the action,

$$\begin{aligned} 0 = \delta W_{01} &= \sum_k \boldsymbol{p}_k \cdot (\boldsymbol{\delta\varphi} \times \boldsymbol{r}_k)\Big|_{t_0}^{t_1} = \boldsymbol{\delta\varphi} \cdot \sum_k \boldsymbol{r}_k \times \boldsymbol{p}_k\Big|_{t_0}^{t_1} \\ &= \boldsymbol{\delta\varphi} \cdot \bigl(\boldsymbol{L}_{\mathrm{tot}}(t_1) - \boldsymbol{L}_{\mathrm{tot}}(t_0)\bigr) \,, \end{aligned} \tag{10.4.25}$$

where

$$\boldsymbol{L}_{\mathrm{tot}} = \sum_k \boldsymbol{r}_k \times \boldsymbol{p}_k = \sum_k \boldsymbol{r}_k \times m_k \boldsymbol{v}_k \tag{10.4.26}$$

is the total angular momentum of (4.3.2). It is conserved because (10.4.25) must be true for all $\boldsymbol{\delta\varphi}$,

$$\boldsymbol{L}_{\mathrm{tot}}(t_1) = \boldsymbol{L}_{\mathrm{tot}}(t_0) \,. \tag{10.4.27}$$

The *rotational invariance* of the physical system implies the *conservation of its angular momentum*, and *vice versa*. It does not matter how we orient an experimental apparatus, whether the table on which you have the equipment is aligned north-south or east-west with its long side, say.

When there are external forces, the full rotational invariance may not be the case. For instance, in the presence of the gravitational pull, we have rotational invariance around the vertical axis but not around horizontal axes. Accordingly, the vertical component of $\boldsymbol{L}_{\text{tot}}$ is conserved, while the horizontal components are not.

10.5 Hamilton's equations of motion

In view of the prominent role played by the momenta and the Hamilton function, it is natural to state the action $W_{01} = \int_{t_0}^{t_1} \mathrm{d}t\, L$ itself in terms of them. For this purpose we solve (10.3.9) for L, replace $\mathrm{d}t\, L$ by

$$\mathrm{d}t \left(\sum_k \dot{x}_k p_k - H \right) = \sum_k \mathrm{d}x_k\, p_k - \mathrm{d}t\, H\,, \tag{10.5.1}$$

and write

$$W_{01} = \int_0^1 \left(\sum_k \mathrm{d}x_k\, p_k - \mathrm{d}t\, H \right), \tag{10.5.2}$$

an integral over the path that takes us from the initial configuration to the final configuration

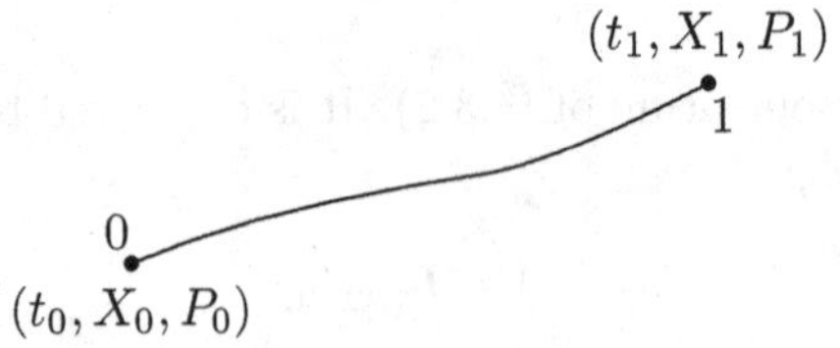

(10.5.3)

where the configurations are still specified by stating their time and the values of all coordinates, just as we did before. The (X, P) space is the so-called *phase space*, and the points in phase space are the *phases*, specified by stating all coordinates and all momenta:

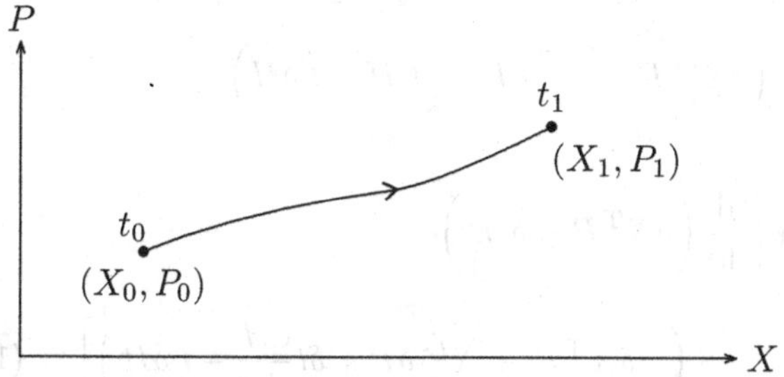

(10.5.4)

A trajectory in phase space takes us from the initial phase (X_0, P_0) at the initial time t_0 to the final phase (X_1, P_1) at the final time t_1. From this point of view, then, the action is given as the time integral

$$W_{01} = \int_{t_0}^{t_1} \mathrm{d}t \left(\sum_k \dot{x}_k p_k - H \right) = \int_{t_0}^{t_1} \mathrm{d}t \left(\dot{X}^{\mathrm{T}} P - H \right), \tag{10.5.5}$$

where the latter version collects all coordinates is the column X, all velocities in the column $\dot{X}$, and all momenta is the column P, so that

$$\sum_k \dot{x}_k p_k = \left(\dot{x}_1 \; \dot{x}_2 \; \ldots \right) \begin{pmatrix} p_1 \\ p_2 \\ \vdots \end{pmatrix} = \dot{X}^{\mathrm{T}} P \tag{10.5.6}$$

is a row-times-column product.

Since we are free to parameterize the action integral any way we like, this reformulation is surely possible, but after the switch from the Lagrange function to the Hamilton function we can no longer invoke the Euler–Lagrange equations that refer to coordinates and velocities. Therefore, let us reconsider the response of the action to small variations of the extremal path, and we do that in the version that uses an independent integration variable α,

$$W_{01} = \int_0^1 \mathrm{d}\alpha \left(X'^{\mathrm{T}} P - t' H \right), \tag{10.5.7}$$

where $X' = \dfrac{\mathrm{d}X}{\mathrm{d}\alpha}$ and $t' = \dfrac{\mathrm{d}t}{\mathrm{d}\alpha}$, exactly as in Section 10.1.

The response of the action to variations of $t(\alpha)$, $X(\alpha)$, and $P(\alpha)$ is then

$$\begin{aligned}\delta W_{01} &= \int_0^1 \mathrm{d}\alpha \left(\delta X'^{\mathrm{T}} P + X'^{\mathrm{T}} \delta P - \delta t' H - t' \delta H\right) \\ &= \int_0^1 \mathrm{d}\alpha \left[\frac{\mathrm{d}}{\mathrm{d}\alpha}\left(\delta X^{\mathrm{T}} P - \delta t H\right)\right. \\ &\qquad \left. + \left(-\delta X^{\mathrm{T}} P' + X'^{\mathrm{T}} \delta P + \delta t \frac{\mathrm{d}H}{\mathrm{d}\alpha} - t' \delta H\right)\right], \end{aligned} \tag{10.5.8}$$

where we have identified a total α derivative. Now,

$$\delta H = \delta t \frac{\partial H}{\partial t} + \sum_k \left(\delta x_k \frac{\partial H}{\partial x_k} + \delta p_k \frac{\partial H}{\partial p_k}\right), \tag{10.5.9}$$

so that

$$\begin{aligned}\delta W_{01} &= \left(\delta X^{\mathrm{T}} P - \delta t H\right)\Big|_0^1 \\ &\quad + \int_0^1 \mathrm{d}\alpha \left[\sum_k \delta x_k \left(-p_k' - t' \frac{\partial H}{\partial x_k}\right)\right. \\ &\qquad + \sum_k \delta p_k \left(x_k' - t' \frac{\partial H}{\partial p_k}\right) \\ &\qquad \left. + \delta t \left(\frac{\mathrm{d}H}{\mathrm{d}\alpha} - t' \frac{\partial H}{\partial t}\right)\right]. \end{aligned} \tag{10.5.10}$$

The first term is the familiar contribution from the variation of the initial and final configurations, and the PSA requires that that is all there is, implying that the terms multiplying δx_k, δp_k, and δt vanish for all values of α,

$$\begin{aligned} p_k' &= -t' \frac{\partial H}{\partial x_k}, \\ x_k' &= t' \frac{\partial H}{\partial p_k}, \\ \frac{\mathrm{d}}{\mathrm{d}\alpha} H &= t' \frac{\partial H}{\partial t}. \end{aligned} \tag{10.5.11}$$

After getting rid of the α parameterization with the aid of (10.1.8) we then have

$$\begin{aligned} \frac{\mathrm{d}}{\mathrm{d}t}p_k &= -\frac{\partial H}{\partial x_k}\,, \\ \frac{\mathrm{d}}{\mathrm{d}t}x_k &= \frac{\partial H}{\partial p_k}\,, \\ \frac{\mathrm{d}}{\mathrm{d}t}H &= \frac{\partial H}{\partial t}\,, \end{aligned} \tag{10.5.12}$$

which are *Hamilton's equations of motion*, the classical precursor of Heisenberg's* equations of motion in quantum mechanics. The first equation states that the force is the negative position-derivative of the energy, and the second equation says that the momentum-derivative of the energy is the velocity. The third equation repeats what we learned earlier: Energy is conserved if there is no parametric time dependence in the Hamilton function.

In this context, it is important to fully appreciate the difference between the total time derivative $\dfrac{\mathrm{d}}{\mathrm{d}t}$ and the parametric time derivative $\dfrac{\partial}{\partial t}$. For an arbitrary phase-space function $F\bigl(X(t), P(t), t\bigr)$, we have

$$\underbrace{\frac{\mathrm{d}}{\mathrm{d}t}F\bigl(X(t), P(t), t\bigr)}_{\text{all}} = \underbrace{\frac{\partial}{\partial t}F\bigl(X(t), P(t), t\bigr)}_{\text{only}} + \sum_k \left(\frac{\mathrm{d}x_k}{\mathrm{d}t}\frac{\partial F}{\partial x_k} + \frac{\mathrm{d}p_k}{\mathrm{d}t}\frac{\partial F}{\partial p_k}\right). \tag{10.5.13}$$

This is quite analogous to the difference between $\dfrac{\mathrm{d}}{\mathrm{d}x}$ and $\dfrac{\partial}{\partial x}$ in (7.2.12) and (7.2.14).

Unless it is necessary to avoid a misunderstanding we shall not, as a rule, exhibit the time dependence of $X(t)$ and $P(t)$ explicitly, so that (10.5.13) appears as

$$\frac{\mathrm{d}}{\mathrm{d}t}F = \frac{\partial}{\partial t}F + \sum_k \left(\dot{x}_k\frac{\partial F}{\partial x_k} + \dot{p}_k\frac{\partial F}{\partial p_k}\right), \tag{10.5.14}$$

and it is understood that F is a function of all $x_k(t)$s, all $p_k(t)$s, as well as time t itself.

When we apply this to the Hamilton function and make use of the first and the second equation in (10.5.12), then

$$\frac{\mathrm{d}}{\mathrm{d}t}H = \frac{\partial}{\partial t}H + \sum_k \bigl[\dot{x}_k(-\dot{p}_k) + \dot{p}_k\dot{x}_k\bigr] = \frac{\partial}{\partial t}H \tag{10.5.15}$$

---

*Werner HEISENBERG (1901–1976)

simply reproduces the third equation. It follows that this third equation is a consequence of the other two and not really an independent statement.

For an arbitrary function $F(X,P,t)$, we combine Hamilton's equations of motion with (10.5.14) and then have

$$\frac{\mathrm{d}}{\mathrm{d}t}F = \frac{\partial}{\partial t}F + \sum_k \left(\frac{\partial F}{\partial x_k} \underbrace{\frac{\partial H}{\partial p_k}}_{=\dot{x}_k} - \frac{\partial F}{\partial p_k} \underbrace{\frac{\partial H}{\partial x_k}}_{=-\dot{p}_k} \right) \tag{10.5.16}$$

or

$$\frac{\mathrm{d}}{\mathrm{d}t}F = \frac{\partial}{\partial t}F + \{F,H\} \tag{10.5.17}$$

upon introducing the *Poisson* bracket* that is defined by

$$\{F,G\} = \sum_k \left(\frac{\partial F}{\partial x_k}\frac{\partial G}{\partial p_k} - \frac{\partial F}{\partial p_k}\frac{\partial G}{\partial x_k} \right) \tag{10.5.18}$$

for any pair of functions $F(X,P,t)$ and $G(X,P,t)$. In (10.5.17), we have Hamilton's equation of motion in its general form. What we had earlier in (10.5.12) are particular examples,

$$\begin{aligned} F = x_k :\ & \frac{\mathrm{d}}{\mathrm{d}t}x_k = \{x_k, H\} = \frac{\partial H}{\partial p_k}\,, \\ F = p_k :\ & \frac{\mathrm{d}}{\mathrm{d}t}p_k = \{p_k, H\} = -\frac{\partial H}{\partial x_k}\,, \\ F = H :\ & \frac{\mathrm{d}}{\mathrm{d}t}H = \frac{\partial H}{\partial t} + \{H,H\} = \frac{\partial}{\partial t}H\,. \end{aligned} \tag{10.5.19}$$

The last equality realizes that $\{G,G\} = 0$ for any G because of the antisymmetry of the Poisson bracket,

$$\{F,G\} = -\{G,F\}\,, \tag{10.5.20}$$

which is an immediate consequence of its definition.

10.6 Poisson bracket

Important properties of the Poisson bracket (10.5.18) are its *linearity* in both arguments,

$$\{\alpha_1 F_1 + \alpha_2 F_2, G\} = \alpha_1\{F_1,G\} + \alpha_2\{F_2,G\}\,, \tag{10.6.1}$$

*Siméon Denise POISSON (1781–1840)

where α_1, α_2 are numbers (that is: they are *not* functions of the coordinates and momenta), and likewise for G,

$$\{F, \alpha_1 G_1 + \alpha_2 G_2\} = \alpha_1\{F, G_1\} + \alpha_2\{F, G_2\}, \tag{10.6.2}$$

and its *product rule*

$$\begin{aligned}\{F, G_1 G_2\} &= \{F, G_1\}G_2 + G_1\{F, G_2\}, \\ \{F_1 F_2, G\} &= F_1\{F_2, G\} + \{F_1, G\}F_2,\end{aligned} \tag{10.6.3}$$

which is a consequence of the product rule of differentiation. Taken together these properties tell us that the mappings

$$G \to \{F, G\} \quad \text{for given } F \tag{10.6.4}$$

and

$$F \to \{F, G\} \quad \text{for given } G \tag{10.6.5}$$

are *derivations*, if we wish to use mathematical terminology.

Another property of the Poisson bracket with quite some importance as well is the *Jacobi identity*

$$\{\{A, B\}, C\} + \{\{B, C\}, A\} + \{\{C, A\}, B\} = 0, \tag{10.6.6}$$

a very close relative of the Jacobi identity (1.1.50) for double-vector products. As an exercise, verify (10.6.6) yourself.

Here are a few examples for the position vector $\boldsymbol{r}$ and the momentum vector $\boldsymbol{p}$ of a point mass, as well as its angular momentum vector $\boldsymbol{l} = \boldsymbol{r} \times \boldsymbol{p}$. With numerical vectors $\boldsymbol{a}$, $\boldsymbol{b}$ we have first

$$\{\boldsymbol{a}\cdot\boldsymbol{r}, \boldsymbol{b}\cdot\boldsymbol{p}\} = \sum_k \Bigg(\underbrace{\frac{\partial \boldsymbol{a}\cdot\boldsymbol{r}}{\partial x_k}}_{=a_k}\underbrace{\frac{\partial \boldsymbol{b}\cdot\boldsymbol{p}}{\partial p_k}}_{=b_k} - \underbrace{\frac{\partial \boldsymbol{a}\cdot\boldsymbol{r}}{\partial p_k}}_{=0}\underbrace{\frac{\partial \boldsymbol{b}\cdot\boldsymbol{p}}{\partial x_k}}_{=0}\Bigg) = \boldsymbol{a}\cdot\boldsymbol{b}, \tag{10.6.7}$$

as well as $\{\boldsymbol{a}\cdot\boldsymbol{r}, \boldsymbol{b}\cdot\boldsymbol{r}\} = 0$ and $\{\boldsymbol{a}\cdot\boldsymbol{p}, \boldsymbol{b}\cdot\boldsymbol{p}\} = 0$. Then we find

$$\begin{aligned}\{\boldsymbol{a}\cdot\boldsymbol{r}, \boldsymbol{b}\cdot\boldsymbol{l}\} &= \{\boldsymbol{a}\cdot\boldsymbol{r}, (\boldsymbol{b}\times\boldsymbol{r})\cdot\boldsymbol{p}\} \\ &= \boldsymbol{a}\cdot(\boldsymbol{b}\times\boldsymbol{r}) = (\boldsymbol{a}\times\boldsymbol{b})\cdot\boldsymbol{r}, \\ \{\boldsymbol{a}\cdot\boldsymbol{p}, \boldsymbol{b}\cdot\boldsymbol{l}\} &= \{\boldsymbol{a}\cdot\boldsymbol{p}, \boldsymbol{r}\cdot(\boldsymbol{p}\times\boldsymbol{b})\} \\ &= -\boldsymbol{a}\cdot(\boldsymbol{p}\times\boldsymbol{b}) = (\boldsymbol{a}\times\boldsymbol{b})\cdot\boldsymbol{p}\end{aligned} \tag{10.6.8}$$

and

$$\begin{aligned}\{\boldsymbol{a}\cdot\boldsymbol{l},\boldsymbol{b}\cdot\boldsymbol{l}\} &= (\boldsymbol{a}\times\boldsymbol{r})\cdot\{\boldsymbol{p},\boldsymbol{b}\cdot\boldsymbol{l}\}+(\boldsymbol{p}\times\boldsymbol{a})\cdot\{\boldsymbol{r},\boldsymbol{b}\cdot\boldsymbol{l}\}\\ &= \big[(\boldsymbol{a}\times\boldsymbol{r})\times\boldsymbol{b}\big]\cdot\boldsymbol{p}+\big[(\boldsymbol{p}\times\boldsymbol{a})\times\boldsymbol{b}\big]\cdot\boldsymbol{r}\\ &= \boldsymbol{a}\cdot\boldsymbol{b}\,\boldsymbol{r}\cdot\boldsymbol{p}-\boldsymbol{b}\cdot\boldsymbol{r}\,\boldsymbol{a}\cdot\boldsymbol{p}+\boldsymbol{a}\cdot\boldsymbol{r}\,\boldsymbol{b}\cdot\boldsymbol{p}-\boldsymbol{a}\cdot\boldsymbol{b}\,\boldsymbol{r}\cdot\boldsymbol{p}\\ &= (\boldsymbol{a}\times\boldsymbol{b})\cdot(\boldsymbol{r}\times\boldsymbol{p})\end{aligned} \tag{10.6.9}$$

or

$$\{\boldsymbol{a}\cdot\boldsymbol{l},\boldsymbol{b}\cdot\boldsymbol{l}\} = (\boldsymbol{a}\times\boldsymbol{b})\cdot\boldsymbol{l}\,, \tag{10.6.10}$$

where the sum rule and the product rule for the Poisson bracket have been used repeatedly. The linearity of the Poisson bracket, see (10.6.1) and (10.6.2), then implies that

$$\{\boldsymbol{a}\cdot\boldsymbol{f},\boldsymbol{b}\cdot\boldsymbol{l}\} = (\boldsymbol{a}\times\boldsymbol{b})\cdot\boldsymbol{f} = \{\boldsymbol{a}\cdot\boldsymbol{l},\boldsymbol{b}\cdot\boldsymbol{f}\} \tag{10.6.11}$$

for any vector $\boldsymbol{f}$ that is a weighted sum of $\boldsymbol{r}$, $\boldsymbol{p}$, and $\boldsymbol{l}$, that is: $\boldsymbol{f} = \alpha\boldsymbol{r}+\beta\boldsymbol{p}+\gamma\boldsymbol{l}$ with numerical coefficients α, β, and γ. More generally, the coefficients could be rotationally invariant combinations of $\boldsymbol{r}$, $\boldsymbol{p}$, and $\boldsymbol{l}$ — such as $\boldsymbol{r}^2$ or $\boldsymbol{r}\cdot\boldsymbol{p}$, say, or functions of them — because such combinations have vanishing Poisson brackets with $\boldsymbol{l}$, as exemplified by $\{\boldsymbol{r}\cdot\boldsymbol{p},\boldsymbol{l}\} = 0$.

10.7 Conservation laws and symmetries. Noether's theorem

If $F(X,P)$ is a constant of motion, perhaps a component of angular momentum or a component of the axis vector of Exercise 82, then

$$0 = \frac{\mathrm{d}F}{\mathrm{d}t} = \{F,H\}\,. \tag{10.7.1}$$

There is now another way of reading this statement, namely

$$\begin{aligned}0 = \epsilon\{F,H\} &= \epsilon\sum_k\left(\frac{\partial F}{\partial x_k}\frac{\partial H}{\partial p_k}-\frac{\partial F}{\partial p_k}\frac{\partial H}{\partial x_k}\right)\\ &= \sum_k\left(\delta x_k\frac{\partial H}{\partial x_k}+\delta p_k\frac{\partial H}{\partial p_k}\right) = \delta H\end{aligned} \tag{10.7.2}$$

with

$$\delta x_k = -\epsilon\frac{\partial F}{\partial p_k}\,,\qquad \delta p_k = \epsilon\frac{\partial F}{\partial x_k}\,, \tag{10.7.3}$$

where ϵ is an infinitesimal increment. It follows that H does not change if the coordinates and momenta are infinitesimally varied as stated above, with increments deriving from the conserved quantity F.

The fundamental observation is that symmetries of the Hamilton function — invariance properties of the Hamilton function if you like — are intimately related to conservation laws. This connection between symmetries and conservation laws is known as *Noether's* Theorem.* We saw examples thereof earlier, in the context of conserved angular momentum and rotational invariance: Angular momentum is conserved if the physical situation is rotationally invariant, and *vice versa:* There must be rotational invariance if angular momentum is conserved.

10.8 One-dimensional motion; two-dimensional phase space

Let us now consider one-dimensional motion with the standard Hamilton function

$$H = \frac{1}{2m}p^2 + V(x)\,. \tag{10.8.1}$$

Hamilton's equations of motion are

$$\begin{aligned}\frac{\mathrm{d}}{\mathrm{d}t}x &= \frac{\partial H}{\partial p} = \frac{1}{m}p\,,\\ \frac{\mathrm{d}}{\mathrm{d}t}p &= -\frac{\partial H}{\partial x} = -\frac{\partial V}{\partial x}(x) = F(x)\end{aligned} \tag{10.8.2}$$

with the force $F(x)$. Phase space points (x,p) with $p = m\dot{x} > 0$ will move to the right between t and $t + \mathrm{d}t$, those with $p = m\dot{x} < 0$ move to the left:

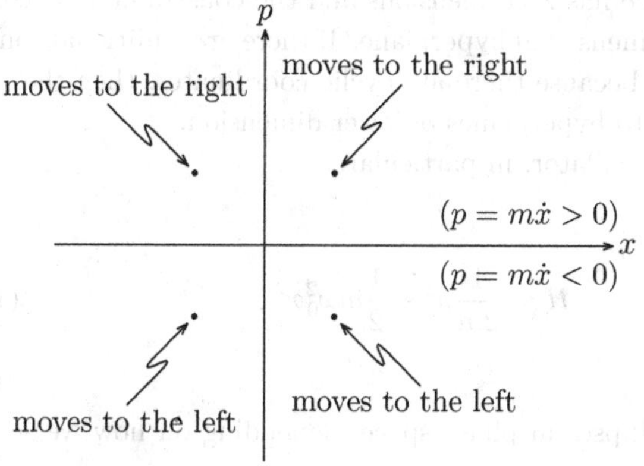

(10.8.3)

*Emmy NOETHER (1882–1935)

It follows that there are possible patterns such as depicted here:

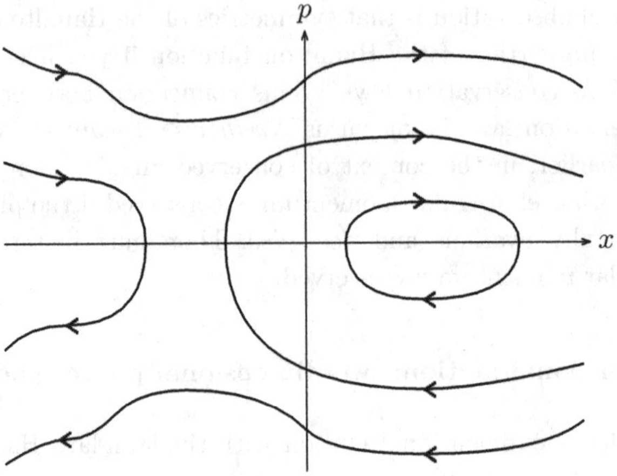

(10.8.4)

Each trajectory in phase space is a solution of the equations of motion, and different trajectories do not intersect because there are unique values of $\dot{x}$ and $\dot{p}$ for a given phase (x, p). The initial conditions select the actual trajectory from the possible ones.

The crossing of the x axis ($p = 0$) is always with a vertical tangent because $\dot{x} = 0$ there. If the Hamilton function has no parametric time dependence, the phase-space trajectories trace the lines of constant H in the phase space, here the two-dimensional plane. For motion in n spatial dimensions, the phase space has $2n$ dimensions and the constraint $H =$ constant defines a $(2n-1)$-dimensional hyperplane. If there are additional constants of motion, typically because there are cyclic coordinates, then the motion is further restricted to hyperplanes of lower dimension.

For a harmonic oscillator, in particular,

$$H = \frac{1}{2m}p^2 + \frac{1}{2}m\omega_0^2 x^2 \,, \tag{10.8.5}$$

we have circles or ellipses in phase space, depending on how we scale the axes of the drawing:

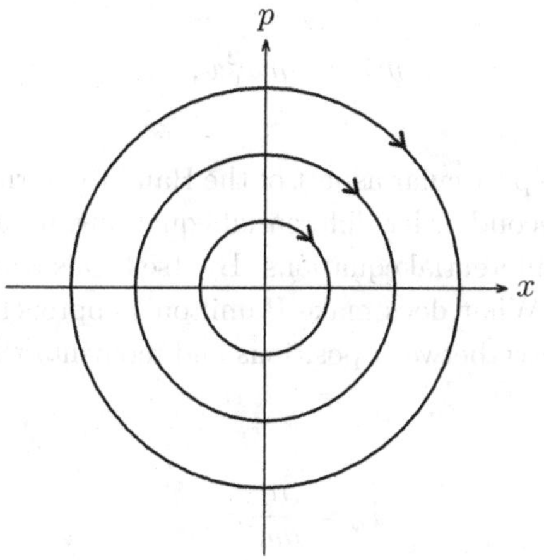

(10.8.6)

The initial conditions select one circle and then all phases $\bigl(x(t), p(t)\bigr)$ are on that circle. This is easy to see also from the explicit solution of Hamilton's equations of motion for the harmonic oscillator,

$$\frac{\mathrm{d}}{\mathrm{d}t}x = \frac{1}{m}p, \quad \frac{\mathrm{d}}{\mathrm{d}t}p = -m\omega_0^2 x, \tag{10.8.7}$$

or

$$\frac{\mathrm{d}}{\mathrm{d}t}(m\omega_0 x + \mathrm{i}p) = \omega_0 p - \mathrm{i}m\omega_0^2 x = -\mathrm{i}\omega_0(m\omega_0 x + \mathrm{i}p) \tag{10.8.8}$$

with the solution

$$m\omega_0 x(t) + \mathrm{i}p(t) = e^{-\mathrm{i}\omega_0 t}(m\omega_0 x_0 + \mathrm{i}p_0). \tag{10.8.9}$$

Not accidentally, this procedure is reminiscent of the "second method" in Section 1.1.8. If we now identify the phase-space point (x, p) with point $z = m\omega_0 x + \mathrm{i}p$ in the complex plane, then $|z(t)| = |z_0|$, which is just the statement that all $z(t)$ are on the circle with radius $|z_0|$.

Upon disentangling the real and imaginary parts, we get

$$\begin{aligned} x(t) &= x_0 \cos(\omega_0 t) + \frac{p_0}{m\omega_0}\sin(\omega_0 t), \\ p(t) &= p_0 \cos(\omega_0 t) - m\omega_0 x_0 \sin(\omega_0 t), \end{aligned} \tag{10.8.10}$$

which are utterly familiar. Earlier, in Section 2.2.4, we wrote $p(t) = m\dot{x}(t)$ and $p_0 = m\dot{x}_0$, but this is really the only difference. The earlier version was obtained from Newton's equation of motion

$$m\ddot{x} = -m\omega_0^2 x\,, \tag{10.8.11}$$

and so we notice one particular aspect of the Hamilton formulation: Rather than dealing with second-order differential equations, we are working with pairs of first-order differential equations. By itself, this transition would be hardly remarkable. What does make Hamilton's approach powerful is the fundamental symmetry between positions and momenta that is manifested in the equation pair

$$\begin{aligned} \dot{x}_k &= \frac{\partial H}{\partial p_k}\,, \\ \dot{p}_k &= -\frac{\partial H}{\partial x_k}\,. \end{aligned} \tag{10.8.12}$$

This symmetry in the equations of motion is an actual structure of phase space itself. We get an important glimpse of it by the following consideration.

10.9 Phase-space density. Liouville's theorem

Another aspect of the structural symmetry between coordinates and momenta in the Hamilton formulation of the equations of motion is revealed upon considering an ensemble of point masses (or other objects) rather than a single one, such as one would do if one studied the behavior of a substantial amount of a gas or a liquid, where the attempt of describing the individual constituents is simply hopeless. Rather than having a single point in phase space, we then have very many of them, all following the same equation of motion, so many indeed that it makes good sense to speak of a *phase-space density* $\rho(X, P, t)$ describing the situation. The physical meaning of $\rho(X, P, t)$ is this: At time t, the number of representative point masses in the phase-space volume $\mathrm{d}x_1\mathrm{d}x_2\ldots\mathrm{d}p_1\mathrm{d}p_2\ldots = (\mathrm{d}X)(\mathrm{d}P)$ is the product of the phase-space density and the volume element,

$$(\mathrm{d}X)(\mathrm{d}P)\,\rho(X, P, t)\,. \tag{10.9.1}$$

What can we say about the evolution of $\rho(X,P,t)$, how does it change in time? We consider the x_k, p_k section of phase space:

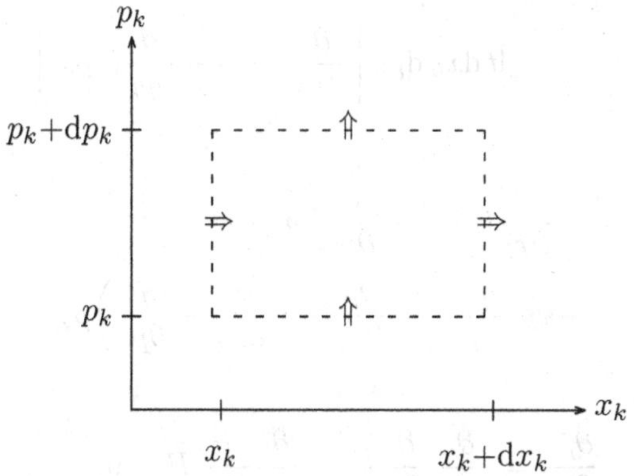

(10.9.2)

The infinitesimal rectangular area, with extensions $\mathrm{d}x_k$ and $\mathrm{d}p_k$, contains $\mathrm{d}x_k\,\mathrm{d}p_k\,\rho(X,P,t)$ objects at time t (per volume element of the other degrees of freedom, the other x,p pairs), and this amount changes because there is flux into the volume and out of it, as indicated in the figure.

The two contributions of in-flux are

$$\begin{aligned} \Rightarrow_{\text{in}} &= \mathrm{d}t\,\mathrm{d}p_k\,\rho\dot{x}_k\,, \\ \Uparrow_{\text{in}} &= \mathrm{d}t\,\mathrm{d}x_k\,\rho\dot{p}_k\,, \end{aligned} \tag{10.9.3}$$

and the two out-flux contributions are

$$\begin{aligned} \Rightarrow_{\text{out}} &= \mathrm{d}t\,\mathrm{d}p_k\left[\rho\dot{x}_k + \mathrm{d}x_k\,\frac{\partial}{\partial x_k}(\rho\dot{x}_k)\right] \\ &= \Rightarrow_{\text{in}} + \mathrm{d}t\,\mathrm{d}p_k\,\mathrm{d}x_k\,\frac{\partial}{\partial x_k}(\rho\dot{x}_k)\,, \\ \Uparrow_{\text{out}} &= \mathrm{d}t\,\mathrm{d}x_k\left[\rho\dot{p}_k + \mathrm{d}p_k\,\frac{\partial}{\partial p_k}(\rho\dot{p}_k)\right] \\ &= \Uparrow_{\text{in}} + \mathrm{d}t\,\mathrm{d}x_k\,\mathrm{d}p_k\,\frac{\partial}{\partial p_k}(\rho\dot{p}_k)\,, \end{aligned} \tag{10.9.4}$$

where the differences of $\rho\dot{x}_k$ and $\rho\dot{p}_k$ at the two respective sides are accounted for. The total change is then

$$\begin{aligned} \mathrm{d}t\,\mathrm{d}x_k\,\mathrm{d}p_k \left(\frac{\partial \rho}{\partial t}\right)_k &= \text{(in-flux)} - \text{(out-flux)} \\ &= -\mathrm{d}t\,\mathrm{d}x_k\,\mathrm{d}p_k \left[\frac{\partial}{\partial x_k}(\rho\dot{x}_k) + \frac{\partial}{\partial p_k}(\rho\dot{p}_k)\right], \end{aligned} \tag{10.9.5}$$

giving

$$\begin{aligned} \left(\frac{\partial \rho}{\partial t}\right)_k &= -\frac{\partial}{\partial x_k}(\rho\dot{x}_k) - \frac{\partial}{\partial p_k}(\rho\dot{p}_k) \\ &= -\dot{x}_k\frac{\partial \rho}{\partial x_k} - \dot{p}_k\frac{\partial \rho}{\partial p_k} - \left(\frac{\partial \dot{x}_k}{\partial x_k} + \frac{\partial \dot{p}_k}{\partial p_k}\right)\rho\,. \end{aligned} \tag{10.9.6}$$

In view of

$$\frac{\partial \dot{x}_k}{\partial x_k} + \frac{\partial \dot{p}_k}{\partial p_k} = \frac{\partial}{\partial x_k}\frac{\partial}{\partial p_k}H - \frac{\partial}{\partial p_k}\frac{\partial}{\partial x_k}H = 0\,, \tag{10.9.7}$$

an immediate consequence of Hamilton's equations of motion, this means

$$\left(\frac{\partial}{\partial t}\rho\right)_k = -\dot{x}_k\frac{\partial \rho}{\partial x_k} - \dot{p}_k\frac{\partial \rho}{\partial p_k} \tag{10.9.8}$$

for the contribution to $\frac{\partial \rho}{\partial t}$ from the x_k, p_k sector. Summation over all sectors then establishes

$$\frac{\partial}{\partial t}\rho = -\sum_k \left(\dot{x}_k\frac{\partial \rho}{\partial x_k} + \dot{p}_k\frac{\partial \rho}{\partial p_k}\right). \tag{10.9.9}$$

We recognize that this is simply the statement

$$\frac{\mathrm{d}}{\mathrm{d}t}\rho = \frac{\partial}{\partial t}\rho + \sum_k \left(\dot{x}_k\frac{\partial \rho}{\partial x_k} + \dot{p}_k\frac{\partial \rho}{\partial p_k}\right) = 0 \tag{10.9.10}$$

that the total time derivative of the phase-space density is zero — the phase-space density is constant in time. This is *Liouville's* Theorem*, which is of great importance in statistical physics. The statement

$$\frac{\partial}{\partial t}\rho = \{H, \rho\} \tag{10.9.11}$$

is known as the *Liouville equation*, where we recognize a Poisson bracket on the right-hand side of (10.9.9).

---

*Joseph LIOUVILLE (1809–1882)

The picture is this:

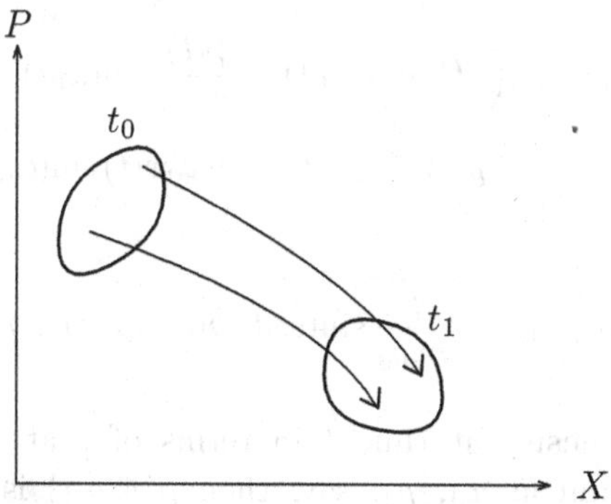

(10.9.12)

At time t_0 a collection of representative point masses are enclosed by a certain surface. Each point of the surface moves in accordance with the equations of motion to form a new surface at time t_1. This new surface will enclose exactly the same collection of representative point masses as the $t = t_0$ surface; none can leave or enter the volume because the phase-space trajectories cannot cross. As we are moving with the ensemble, the density stays the same, but of course the distribution as a whole changes: There is a parametric time dependence of $\rho\bigl(X(t), P(t), t\bigr)$ since each point mass follows its trajectory.

We then have two special cases of the general Hamilton's equation of motion (10.5.17), namely (i) if F is a phase space variable, $F = x_k$ or $F = p_k$, then $\dfrac{\partial F}{\partial t} = 0$ and there is only the dynamical time derivative $\{F, H\}$; and (ii) if $F = \rho$ is the phase-space density, then the total time derivative vanishes, $\dfrac{\mathrm{d}F}{\mathrm{d}t} = 0$, and the dynamical time dependence compensates fully for the parametric time dependence. That is:

$$\frac{\mathrm{d}}{\mathrm{d}t}F = \{F, H\} \quad \text{if} \quad F = x_k \quad \text{or} \quad F = p_k\,,$$
$$\frac{\partial}{\partial t}F = \{H, F\} \quad \text{if} \quad F = \rho\,. \tag{10.9.13}$$

The constancy in time of the phase-space density ρ,

$$\rho\bigl(X(t), P(t), t\bigr) = \rho\bigl(X(0), P(0), 0\bigr) = \rho_0\bigl(X(0), P(0)\bigr)\,, \tag{10.9.14}$$

tells us how to find $\rho\bigl(X(t), P(t), t\bigr)$ from the initial density ρ_0: Solve the equations of motion and express $X(0)$ and $P(0)$ in terms of $X(t)$, $P(t)$, and t. For the example of the one-dimensional harmonic oscillator of (10.8.5)–(10.8.10), this is

$$\rho\big(x(t), p(t), t\big) = \rho_0\Big(x(t)\cos(\omega_0 t) - \frac{p(t)}{m\omega_0}\sin(\omega_0 t),\\ p(t)\cos(\omega_0 t) + m\omega_0 x(t)\sin(\omega_0 t)\Big)\,, \qquad (10.9.15)$$

so that

$$\rho(x, p, t) = \rho_0\Big(x\cos(\omega_0 t) - \frac{p}{m\omega_0}\sin(\omega_0 t), p\cos(\omega_0 t) + m\omega_0 x\sin(\omega_0 t)\Big) \qquad (10.9.16)$$

is the phase-space density at time t in terms of that at $t = 0$. Thus, if ρ_0 has its maximum at x_0, p_0, say, then $\rho(x, p, t)$ is maximal where $x = x_0\cos(\omega_0 t) + \frac{p_0}{m\omega_0}\sin(\omega_0 t)$ and $p = p_0\cos(\omega_0 t) - m\omega_0 x_0\sin(\omega_0 t)$. The location of the maximum just follows the phase space trajectory of a (fictitious) point mass that is at x_0, p_0 at time $t = 0$.

10.10 Velocity-dependent forces and Schwinger's action

The standard "kinetic minus potential energy" form of the Lagrange function or the corresponding "kinetic plus potential energy" form of the Hamilton function — we recall that they read

$$L(t, \boldsymbol{r}, \dot{\boldsymbol{r}}) = \frac{m}{2}\dot{\boldsymbol{r}}^2 - V(\boldsymbol{r}, t)\,,\\ H(\boldsymbol{r}, \boldsymbol{p}, t) = \frac{1}{2m}\boldsymbol{p}^2 + V(\boldsymbol{r}, t) \qquad (10.10.1)$$

for a single point mass — yield equations of motion that do not contain velocity-dependent forces, such as the Lorentz force

$$\boldsymbol{F} = q\boldsymbol{E}(\boldsymbol{r}, t) + \frac{q}{c}\boldsymbol{v}\times\boldsymbol{B}(\boldsymbol{r}, t) \qquad (10.10.2)$$

for a point mass that carries electric charge q and moves in the electric field $\boldsymbol{E}(\boldsymbol{r}, t)$ and the magnetic field $\boldsymbol{B}(\boldsymbol{r}, t)$; the symbol c denotes the speed of light. For the incorporation of velocity-dependent forces, it is expedient to use Schwinger's* version of the action,

$$W_{01} = \int_0^1 \big[\mathrm{d}\boldsymbol{r}\cdot\boldsymbol{p} - \mathrm{d}t\, H(\boldsymbol{r}, \boldsymbol{v}, \boldsymbol{p}, t)\big]\,, \qquad (10.10.3)$$

*Julian SCHWINGER (1918–1994)

where we have the velocity $\boldsymbol{v}$ as an independent variable in the Hamilton function

$$H(\boldsymbol{r},\boldsymbol{v},\boldsymbol{p},t)=\boldsymbol{p}\cdot\boldsymbol{v}-\frac{m}{2}\boldsymbol{v}^2+V(\boldsymbol{r},t)\,. \tag{10.10.4}$$

The response of the action (10.10.3) to infinitesimal variations of $\boldsymbol{r}(t)$, $\boldsymbol{v}(t)$, and $\boldsymbol{p}(t)$ is

$$\begin{aligned}\delta W_{01}&=\int_0^1\bigl[\mathrm{d}\delta\boldsymbol{r}\cdot\boldsymbol{p}+\mathrm{d}\boldsymbol{r}\cdot\delta\boldsymbol{p}-\mathrm{d}\delta t\,H-\mathrm{d}t\,\delta H\bigr]\\ &=\int_0^1\biggl[\mathrm{d}(\delta\boldsymbol{r}\cdot\boldsymbol{p}-\delta t\,H)-\delta\boldsymbol{r}\cdot\mathrm{d}\boldsymbol{p}+\delta t\,\mathrm{d}H+\delta\boldsymbol{p}\cdot\mathrm{d}\boldsymbol{r}\\ &\qquad-\mathrm{d}t\biggl(\delta\boldsymbol{r}\cdot\frac{\partial H}{\partial\boldsymbol{r}}+\delta\boldsymbol{v}\cdot\frac{\partial H}{\partial\boldsymbol{v}}+\delta\boldsymbol{p}\cdot\frac{\partial H}{\partial\boldsymbol{p}}+\delta t\frac{\partial H}{\partial t}\biggr)\biggr]\\ &=(\delta\boldsymbol{r}\cdot\boldsymbol{p}-\delta t\,H)\Big|_0^1\\ &\quad+\int_0^1\biggl[\delta\boldsymbol{r}\cdot\biggl(-\mathrm{d}\boldsymbol{p}-\mathrm{d}t\frac{\partial H}{\partial\boldsymbol{r}}\biggr)+\delta\boldsymbol{v}\cdot\biggl(-\mathrm{d}t\frac{\partial H}{\partial\boldsymbol{v}}\biggr)\\ &\qquad+\delta\boldsymbol{p}\cdot\biggl(\mathrm{d}\boldsymbol{r}-\mathrm{d}t\frac{\partial H}{\partial\boldsymbol{p}}\biggr)+\delta t\biggl(\mathrm{d}H-\mathrm{d}t\frac{\partial H}{\partial t}\biggr)\biggr]\,.\end{aligned} \tag{10.10.5}$$

Accordingly, since the PSA requires that

$$\delta W_{01}=(\delta\boldsymbol{r}\cdot\boldsymbol{p}-\delta t\,H)\Big|_0^1\,, \tag{10.10.6}$$

which is (10.3.7) in the present context, it follows that

$$\begin{aligned}\delta\boldsymbol{r}:&\qquad \frac{\mathrm{d}}{\mathrm{d}t}\boldsymbol{p}=-\frac{\partial H}{\partial\boldsymbol{r}}=-\boldsymbol{\nabla}V(\boldsymbol{r},t)\,,\\ \delta\boldsymbol{v}:&\qquad 0=\frac{\partial H}{\partial\boldsymbol{v}}=\boldsymbol{p}-m\boldsymbol{v}\,,\\ \delta\boldsymbol{p}:&\qquad \frac{\mathrm{d}}{\mathrm{d}t}\boldsymbol{r}=\frac{\partial H}{\partial\boldsymbol{p}}=\boldsymbol{v}\,,\\ \delta t:&\qquad \frac{\mathrm{d}H}{\mathrm{d}t}=\frac{\partial H}{\partial t}\,.\end{aligned} \tag{10.10.7}$$

There are equations of motion for $\boldsymbol{r}$ and $\boldsymbol{p}$, the variables paired in $\mathrm{d}\boldsymbol{r}\cdot\boldsymbol{p} = \mathrm{d}t\,\dot{\boldsymbol{r}}\cdot\boldsymbol{p}$, whereas the variation $\delta\boldsymbol{v}$ of the velocity yields $\boldsymbol{p} = m\boldsymbol{v}$, which is not an equation of motion but an equation of constraint.

If we use this constraint to accept $\boldsymbol{v} = \dfrac{1}{m}\boldsymbol{p}$ as the definition of $\boldsymbol{v}$, then Schwinger's Hamilton function becomes

$$H(\boldsymbol{r},\boldsymbol{v},\boldsymbol{p},t) = \frac{1}{2m}\boldsymbol{p}^2 + V(\boldsymbol{r},t)\,, \tag{10.10.8}$$

which is Hamilton's original version. Alternatively, we can accept $\boldsymbol{v} = \dfrac{\mathrm{d}}{\mathrm{d}t}\boldsymbol{r}$ as the definition of $\boldsymbol{v}$ and $\boldsymbol{p} = m\boldsymbol{v} = m\dot{\boldsymbol{r}}$ as the definition of $\boldsymbol{p}$, and then the Lagrange function acquires Lagrange's form,

$$\begin{aligned} L &= \left[\boldsymbol{p}\cdot\left(\frac{\mathrm{d}\boldsymbol{r}}{\mathrm{d}t} - \boldsymbol{v}\right) + \frac{1}{2}m\boldsymbol{v}^2 - V(\boldsymbol{r},t)\right]\Bigg|_{\boldsymbol{v}=\dot{\boldsymbol{r}},\ \boldsymbol{p}=m\dot{\boldsymbol{r}}} \\ &= \frac{m}{2}\dot{\boldsymbol{r}}^2 - V(\boldsymbol{r},t)\,. \end{aligned} \tag{10.10.9}$$

Clearly, Schwinger's formulation of the PSA is intermediate between the viewpoints of Lagrange and Hamilton and contains both as special cases.

Let us now subtract a term $\boldsymbol{v}\cdot\boldsymbol{f}(\boldsymbol{r},t)$ with some vector function $\boldsymbol{f}$ of $\boldsymbol{r}$ and t from the Hamilton function in (10.10.4),

$$H(\boldsymbol{r},\boldsymbol{v},\boldsymbol{p},t) = \boldsymbol{p}\cdot\boldsymbol{v} - \frac{m}{2}\boldsymbol{v}^2 + V(\boldsymbol{r},t) - \boldsymbol{v}\cdot\boldsymbol{f}(\boldsymbol{r},t)\,. \tag{10.10.10}$$

This modification is of no consequence for the $\delta\boldsymbol{p}$ contribution to the variation δW_{01} of the action, so that

$$\frac{\mathrm{d}}{\mathrm{d}t}\boldsymbol{r} = \boldsymbol{v} \tag{10.10.11}$$

remains valid (we would not like to give up *that*), whereas the $\delta\boldsymbol{r}$ and $\delta\boldsymbol{v}$ contributions now imply

$$\begin{aligned} \frac{\mathrm{d}}{\mathrm{d}t}\boldsymbol{p} &= -\frac{\partial H}{\partial\boldsymbol{r}} = -\boldsymbol{\nabla}V + \boldsymbol{\nabla}(\boldsymbol{v}\cdot\boldsymbol{f})\,, \\ 0 &= \frac{\partial H}{\partial\boldsymbol{v}} = \boldsymbol{p} - m\boldsymbol{v} - \boldsymbol{f}\,. \end{aligned} \tag{10.10.12}$$

The resulting Newton's equation of motion, $\dfrac{\mathrm{d}}{\mathrm{d}t}(m\boldsymbol{v}) = \boldsymbol{F}$, has the force

$$\boldsymbol{F} = \frac{\mathrm{d}}{\mathrm{d}t}(\boldsymbol{p}-\boldsymbol{f}) = -\boldsymbol{\nabla}V + \boldsymbol{\nabla}(\boldsymbol{v}\cdot\boldsymbol{f}) - \frac{\mathrm{d}}{\mathrm{d}t}\boldsymbol{f}\,, \tag{10.10.13}$$

with the familiar negative gradient of the potential energy V and the unfamiliar velocity-dependent force

$$\begin{aligned} \boldsymbol{\nabla}(\boldsymbol{v}\cdot\boldsymbol{f}) - \frac{\mathrm{d}}{\mathrm{d}t}\boldsymbol{f} &= \boldsymbol{\nabla}(\boldsymbol{v}\cdot\boldsymbol{f}) - \boldsymbol{v}\cdot\boldsymbol{\nabla}\boldsymbol{f} - \frac{\partial}{\partial t}\boldsymbol{f} \\ &= \boldsymbol{v}\times(\boldsymbol{\nabla}\times\boldsymbol{f}) - \frac{\partial}{\partial t}\boldsymbol{f}\,; \end{aligned} \tag{10.10.14}$$

the last step makes use of the double-vector product identity (1.1.36) for $\boldsymbol{a}=\boldsymbol{v}$, $\boldsymbol{b}=\boldsymbol{\nabla}$, and $\boldsymbol{c}=\boldsymbol{f}$. Together, then, we have

$$\boldsymbol{F} = -\boldsymbol{\nabla}V - \frac{\partial}{\partial t}\boldsymbol{f} + \boldsymbol{v}\times(\boldsymbol{\nabla}\times\boldsymbol{f})\,, \tag{10.10.15}$$

which is the Lorentz force (10.10.2) if we choose V and $\boldsymbol{f}$ such that

$$q\boldsymbol{E} = -\boldsymbol{\nabla}V - \frac{\partial}{\partial t}\boldsymbol{f} \quad\text{and}\quad \frac{q}{c}\boldsymbol{B} = \boldsymbol{\nabla}\times\boldsymbol{f}\,. \tag{10.10.16}$$

We compare this with the usual expressions that involve the scalar potential $\Phi(\boldsymbol{r},t)$ and the vector potential $\boldsymbol{A}(\boldsymbol{r},t)$,

$$\boldsymbol{E} = -\boldsymbol{\nabla}\Phi - \frac{1}{c}\frac{\partial}{\partial t}\boldsymbol{A}\,, \qquad \boldsymbol{B} = \boldsymbol{\nabla}\times\boldsymbol{A}\,, \tag{10.10.17}$$

and conclude that the choices $V = q\Phi$ and $\boldsymbol{f} = \frac{q}{c}\boldsymbol{A}$ are appropriate.

For these, then, we have

$$H(\boldsymbol{r},\boldsymbol{v},\boldsymbol{p},t) = \boldsymbol{p}\cdot\boldsymbol{v} - \frac{m}{2}\boldsymbol{v}^2 + q\Phi - \frac{q}{c}\boldsymbol{v}\cdot\boldsymbol{A} \tag{10.10.18}$$

for Schwinger's intermediate version of the Hamilton function, and putting

$$\boldsymbol{v} = \frac{1}{m}\left(\boldsymbol{p} - \frac{q}{c}\boldsymbol{A}\right), \tag{10.10.19}$$

as required by (10.10.12), gives Hamilton's version

$$H(\boldsymbol{r},\boldsymbol{p},t) = \frac{1}{2m}\left(\boldsymbol{p} - \frac{q}{c}\boldsymbol{A}\right)^2 + q\Phi\,. \tag{10.10.20}$$

We note that the first term on the right is just the kinetic energy $\frac{m}{2}\boldsymbol{v}^2$ and that there is a term $\propto \boldsymbol{A}^2$ in this H, although we only have the linear term

$-\frac{q}{c}\boldsymbol{v}\cdot\boldsymbol{A}$ in (10.10.18). The corresponding Lagrange function,

$$L = \left[\boldsymbol{p}\cdot\left(\frac{\mathrm{d}\boldsymbol{r}}{\mathrm{d}t} - \boldsymbol{v}\right) + \frac{1}{2}m\boldsymbol{v}^2 - q\Phi + \frac{q}{c}\boldsymbol{v}\cdot\boldsymbol{A}\right]\Bigg|_{\boldsymbol{v}=\dot{\boldsymbol{r}},\ \boldsymbol{p}=m\dot{\boldsymbol{r}}+(q/c)\boldsymbol{A}}$$
$$= \frac{m}{2}\dot{\boldsymbol{r}}^2 - q\Phi + \frac{q}{c}\dot{\boldsymbol{r}}\cdot\boldsymbol{A}\,. \qquad (10.10.21)$$

does not contain any $\boldsymbol{A}^2$ term either.

The generalization of these expressions to a system of charged point masses is immediate. We have

$$H = \sum_k \left(\boldsymbol{p}_k\cdot\boldsymbol{v}_k - \frac{1}{2}m_k\boldsymbol{v}_k^2 + q_k\Phi(\boldsymbol{r}_k,t) - \frac{q_k}{c}\boldsymbol{v}_k\cdot\boldsymbol{A}(\boldsymbol{r}_k,t)\right) \qquad (10.10.22)$$

for Schwinger's Hamilton function,

$$L = \sum_k \left(\frac{1}{2}m_k\dot{\boldsymbol{r}}_k^2 - q_k\Phi(\boldsymbol{r}_k,t) + \frac{q_k}{c}\dot{\boldsymbol{r}}_k\cdot\boldsymbol{A}(\boldsymbol{r}_k,t)\right) \qquad (10.10.23)$$

for the Lagrange function, and

$$H = \sum_k \left[\frac{1}{2m_k}\left(\boldsymbol{p}_k - \frac{q_k}{c}\boldsymbol{A}(\boldsymbol{r}_k,t)\right)^2 + q_k\Phi(\boldsymbol{r}_k,t)\right] \qquad (10.10.24)$$

for the standard Hamilton function. These apply when charged masses move in external electric and magnetic fields, not in the fields that the charged point masses produce themselves, as described by Maxwell's equations. The complete treatment that accounts for both aspects is given in the *Lectures on Classical Electrodynamics*.

10.11 An excursion into the quantum realm

Here are a few remarks on the link between classical mechanics in Hamilton form and quantum mechanics. For simplicity, we restrict ourselves to a one-dimensional system described by the cartesian x, p pair of position and momentum. In quantum mechanics (QM) they are the position and momentum operators with the Heisenberg commutation relation

$$[\mathrm{QM}] \qquad \frac{1}{\mathrm{i}\hbar}[x,p] = 1\,, \qquad (10.11.1)$$

for which

$$\text{[CM]} \qquad \{x, p\} = 1 \qquad (10.11.2)$$

is the analog in classical mechanics (CM). Heisenberg's equation of motion

$$\text{[QM]} \qquad \frac{\mathrm{d}}{\mathrm{d}t}F = \frac{\partial}{\partial t}F + \frac{1}{\mathrm{i}\hbar}[F, H] \qquad (10.11.3)$$

is fully analogous to Hamilton's equation of motion (10.5.17),

$$\text{[CM]} \qquad \frac{\mathrm{d}}{\mathrm{d}t}F = \frac{\partial}{\partial t}F + \{F, H\}, \qquad (10.11.4)$$

where F and H are corresponding quantities, functions of the phase-space variables $x(t)$, $p(t)$ in classical mechanics, but of the position and momentum operators in quantum mechanics. In particular, we have

$$\frac{\mathrm{d}x}{\mathrm{d}t} = \left\{ \begin{array}{ll} \text{[QM]} & \frac{1}{\mathrm{i}\hbar}[x, H] \\ \text{[CM]} & \{x, H\} \end{array} \right\} = \frac{\partial H}{\partial p} \qquad (10.11.5)$$

and

$$\frac{\mathrm{d}p}{\mathrm{d}t} = \left\{ \begin{array}{ll} \text{[QM]} & \frac{1}{\mathrm{i}\hbar}[p, H] \\ \text{[CM]} & \{p, H\} \end{array} \right\} = -\frac{\partial H}{\partial x} \qquad (10.11.6)$$

for the time evolution of $x(t)$ and $p(t)$. The Liouville equation (10.9.11) of classical mechanics

$$\text{[CM]} \qquad \frac{\partial}{\partial t}\rho + \{\rho, H\} = 0 \qquad (10.11.7)$$

becomes the von Neumann* equation

$$\text{[QM]} \qquad \frac{\partial}{\partial t}\rho + \frac{1}{\mathrm{i}\hbar}[\rho, H] = 0 \qquad (10.11.8)$$

in quantum mechanics, both stating that the total time derivative vanishes,

$$\text{[CM \&QM]} \qquad \frac{\mathrm{d}}{\mathrm{d}t}\rho = 0\,. \qquad (10.11.9)$$

What is the phase-space density $\rho\bigl(x(t), p(t), t\bigr)$ in classical mechanics becomes the statistical operator (*vulgo* the density matrix or the density operator or the state operator or ...) in quantum mechanics.

---

*John VON NEUMANN (1903–1957)

For two quantum-mechanical operators $F(x,p)$ and $G(x,p)$, with x and p referring to the same time (but it does not matter which common time), the commutator $\frac{1}{\mathrm{i}\hbar}[F,G]$ is not equal to their Poisson bracket $\{F,G\}$, as it is the case for $F=x$, $G=p$. Instead one has the general expression

$$\text{[QM]} \qquad \frac{1}{\mathrm{i}\hbar}[F,G]=\frac{1}{\hbar}\Bigl(F\tan\Bigl(\frac{1}{2}\hbar\Lambda\Bigr)G-G\tan\Bigl(\frac{1}{2}\hbar\Lambda\Bigr)F\Bigr) \tag{10.11.10}$$

where

$$\Lambda=\underset{\leftarrow}{\frac{\partial}{\partial x}}\,\underset{\rightarrow}{\frac{\partial}{\partial p}}-\underset{\leftarrow}{\frac{\partial}{\partial p}}\,\underset{\rightarrow}{\frac{\partial}{\partial x}} \tag{10.11.11}$$

is the two-sided differential operator of the Poisson bracket,

$$\begin{aligned}\{F,G\}&=\frac{\partial F}{\partial x}\frac{\partial G}{\partial p}-\frac{\partial F}{\partial p}\frac{\partial G}{\partial x}\\&=F\left(\underset{\leftarrow}{\frac{\partial}{\partial x}}\,\underset{\rightarrow}{\frac{\partial}{\partial p}}-\underset{\leftarrow}{\frac{\partial}{\partial p}}\,\underset{\rightarrow}{\frac{\partial}{\partial x}}\right)G=F\Lambda G\,,\end{aligned} \tag{10.11.12}$$

where the subarrows indicate whether the function on the left or the function on the right is differentiated. The power series of the tangent function,

$$\tan\phi=\phi+\frac{1}{3}\phi^3+\frac{2}{15}\phi^5+\cdots\,, \tag{10.11.13}$$

gives a corresponding series for the commutator,

$$\text{[QM]} \qquad \begin{aligned}\frac{1}{\mathrm{i}\hbar}[F,G]&=\frac{1}{2}(\{F,G\}-\{G,F\})\\&\quad+\frac{\hbar^2}{24}(F\Lambda^3G-G\Lambda^3F)\\&\quad+\frac{\hbar^4}{240}(F\Lambda^5G-G\Lambda^5F)+\cdots\,;\end{aligned} \tag{10.11.14}$$

all the terms of order $\hbar^2$, $\hbar^4$, ... vanish if either F or G is at most quadratic in x and p. In particular, we have

$$\text{[QM]} \qquad \frac{1}{\mathrm{i}\hbar}[x,F]=\frac{\partial F}{\partial p}\,,\quad \frac{1}{\mathrm{i}\hbar}[F,p]=\frac{\partial F}{\partial x}\,, \tag{10.11.15}$$

which were used in (10.11.5) and (10.11.6). These statements about the relation between quantum mechanics and classical mechanics carry over to higher-dimensional systems, with the obvious generalization of Λ to a sum of terms like the ones in (10.11.11), one term for each x,p pair.

Chapter 11

Rigid Bodies

11.1 Inertia dyadic. Steiner's theorem. Principal axes

The description of the physical system as being composed of point masses ceases to be useful when we are dealing with rigid bodies. Yes, we can think of them as being composed of very many very small mass elements that can be treated as point masses, but then we need to take into account a vast number of constraints that enforce the rigidity of the rigid body. A different course of action is called for, quite similar to the situation of the gravitational forces exerted by an extended body in Chapter 6.

We consider a massive body with mass density $\rho(\boldsymbol{r}')$ that is rotating about a fixed axis with angular velocity ω. The rotation axis is the line $u\boldsymbol{e}$ with u any real number and $\boldsymbol{\omega} = \omega\boldsymbol{e}$ the angular velocity vector, as depicted here:

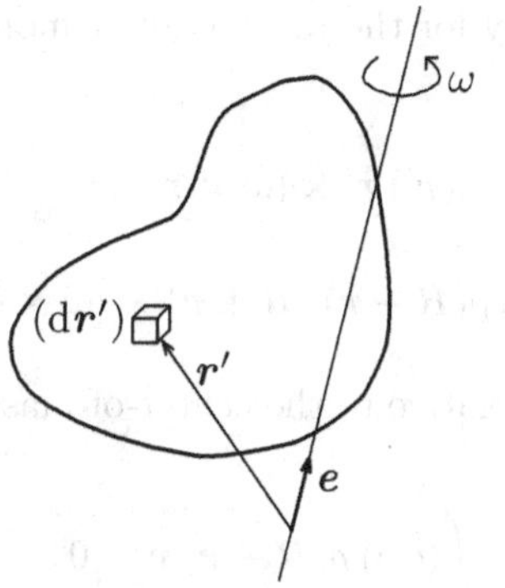

(11.1.1)

A mass element $(\mathrm{d}\boldsymbol{r}')\rho(\boldsymbol{r}')$ at $\boldsymbol{r}'$ moves with the velocity

$$\boldsymbol{v}' = \boldsymbol{\omega} \times \boldsymbol{r}' , \tag{11.1.2}$$

which is (1.1.44) with $\boldsymbol{\delta\varphi} = \boldsymbol{\omega}\,\delta t$ applied to $\boldsymbol{r} = \boldsymbol{r}'$ and $\delta\boldsymbol{r} = \boldsymbol{v}'\delta t$. The component of $\boldsymbol{r}'$ that is parallel to $\boldsymbol{\omega}$ does not contribute to $\boldsymbol{v}'$ and, therefore, it does not matter which reference point on the axis we identify with $\boldsymbol{r}' = 0$.

The total momentum of the rigid body is then given by the sum over all mass elements,

$$\boldsymbol{P} = \int (\mathrm{d}\boldsymbol{r}')\,\rho(\boldsymbol{r}')\,\boldsymbol{\omega} \times \boldsymbol{r}' . \tag{11.1.3}$$

With the total mass M as in (6.1.2),

$$M = \int (\mathrm{d}\boldsymbol{r}')\,\rho(\boldsymbol{r}') , \tag{11.1.4}$$

we recognize the integral of (6.2.2),

$$\int (\mathrm{d}\boldsymbol{r}')\,\rho(\boldsymbol{r}')\,\boldsymbol{r}' = M\boldsymbol{R} , \tag{11.1.5}$$

that identifies the center-of-mass position $\boldsymbol{R}$, so that

$$\boldsymbol{P} = M\boldsymbol{\omega} \times \boldsymbol{R} . \tag{11.1.6}$$

It is as if all of the mass were concentrated in the center-of mass, at least as far as the total momentum is concerned. This is, of course, a special case of (4.5.8) with the center-of-mass velocity

$$\boldsymbol{V} = \boldsymbol{\omega} \times \boldsymbol{R} . \tag{11.1.7}$$

We are dropping the subscript of $\boldsymbol{P}_{\mathrm{tot}}$ here and write simply $\boldsymbol{P}$ for the total momentum, and similarly for the total angular momentum.

For that we get

$$\begin{aligned} \boldsymbol{L} &= \int (\mathrm{d}\boldsymbol{r}')\,\rho(\boldsymbol{r}')\,\boldsymbol{r}' \times (\boldsymbol{\omega} \times \boldsymbol{r}') \\ &= \int (\mathrm{d}\boldsymbol{r})\,\rho(\boldsymbol{R} + \boldsymbol{r})\,(\boldsymbol{R} + \boldsymbol{r}) \times \left[\boldsymbol{\omega} \times (\boldsymbol{R} + \boldsymbol{r})\right] , \end{aligned} \tag{11.1.8}$$

where $\boldsymbol{r}$ is the position relative to the center-of-mass in the second version. Since

$$\int (\mathrm{d}\boldsymbol{r})\,\rho(\boldsymbol{R} + \boldsymbol{r})\,\boldsymbol{r} = 0 , \tag{11.1.9}$$

the terms in $(\boldsymbol{R}+\boldsymbol{r})\times\left[\boldsymbol{\omega}\times(\boldsymbol{R}+\boldsymbol{r})\right]$ that are linear in $\boldsymbol{r}$ do not contribute to the integral, and the replacement

$$(\boldsymbol{R}+\boldsymbol{r})\times\left[\boldsymbol{\omega}\times(\boldsymbol{R}+\boldsymbol{r})\right]\to\boldsymbol{R}\times(\boldsymbol{\omega}\times\boldsymbol{R})+\boldsymbol{r}\times(\boldsymbol{\omega}\times\boldsymbol{r}) \tag{11.1.10}$$

is permissible. This takes us to

$$\begin{aligned}\boldsymbol{L}&=M\boldsymbol{R}\times(\boldsymbol{\omega}\times\boldsymbol{R})+\int(\mathrm{d}\boldsymbol{r})\,\rho_{\mathrm{CM}}(\boldsymbol{r})\,\boldsymbol{r}\times(\boldsymbol{\omega}\times\boldsymbol{r})\\&=\boldsymbol{R}\times\boldsymbol{P}+\int(\mathrm{d}\boldsymbol{r})\,\rho_{\mathrm{CM}}(\boldsymbol{r})\,(r^2\mathbf{1}-\boldsymbol{r}\,\boldsymbol{r})\cdot\boldsymbol{\omega}\,,\end{aligned} \tag{11.1.11}$$

where $\rho_{\mathrm{CM}}(\boldsymbol{r})=\rho(\boldsymbol{R}+\boldsymbol{r})$ is the mass density for the center-of-mass at $\boldsymbol{r}=0$, and we wrote the double vector product as a dyadic dot-multiplied by the angular velocity vector,

$$\boldsymbol{r}\times(\boldsymbol{\omega}\times\boldsymbol{r})=(r^2\mathbf{1}-\boldsymbol{r}\,\boldsymbol{r})\cdot\boldsymbol{\omega}\,. \tag{11.1.12}$$

Upon introducing the *inertia dyadic* I in accordance with

$$\mathsf{I}=\int(\mathrm{d}\boldsymbol{r})\,\rho_{\mathrm{CM}}(\boldsymbol{r})\,(r^2\mathbf{1}-\boldsymbol{r}\,\boldsymbol{r})\,, \tag{11.1.13}$$

we have

$$\boldsymbol{L}=\boldsymbol{R}\times\boldsymbol{P}+\mathsf{I}\cdot\boldsymbol{\omega} \tag{11.1.14}$$

for the total angular momentum. The first term is as if all the mass were concentrated in the center-of-mass, and the second term accounts for the actual distribution of the mass around the center-of-mass.

In passing we note the similarity between the inertia dyadic of (11.1.13) and the quadrupole moment dyadic of (6.2.10); see Exercise 150. This similarity is not an accidental coincidence but rather a manifestation of the equivalence of inertial mass and gravitating mass, established by Eötvös's* precision measurements and the key ingredient of Einstein's† general relativity.

Since

$$\boldsymbol{R}\times\boldsymbol{P}=M\boldsymbol{R}\times(\boldsymbol{\omega}\times\boldsymbol{R})=M(R^2\mathbf{1}-\boldsymbol{R}\,\boldsymbol{R})\cdot\boldsymbol{\omega}\,, \tag{11.1.15}$$

we can also write

$$\boldsymbol{L}=\mathsf{I}_{\boldsymbol{R}}\cdot\boldsymbol{\omega} \tag{11.1.16}$$

---

*Loránd Eötvös (1849–1919) †Albert Einstein (1879–1955)

with

$$\mathsf{I}_R = \mathsf{I} + M(R^2\mathbf{1} - \boldsymbol{R}\,\boldsymbol{R})\,. \tag{11.1.17}$$

Harking back to the first line in (11.1.8), we observe that

$$\mathsf{I}_R = \int (\mathrm{d}\boldsymbol{r})\,\rho(\boldsymbol{r})\,(r^2\mathbf{1} - \boldsymbol{r}\,\boldsymbol{r}) \tag{11.1.18}$$

which has the same structure as (11.1.13), only that $\rho_{\mathrm{CM}}(\boldsymbol{r})$ is replaced by $\rho(\boldsymbol{r}) = \rho_{\mathrm{CM}}(\boldsymbol{r} - \boldsymbol{R})$.

The relation (11.1.17), which expresses I_R in terms of the center-of-mass inertia dyadic I and the inertia dyadic of a point mass M at the center-of-mass, is *Steiner's** *Theorem*, also known as the *Huygens*† *–Steiner Theorem*, or as the *parallel-axis theorem*. It is quite a useful insight because one often has rotation about an axis that does not go through the center-of-mass.

The total kinetic energy is

$$\begin{aligned} E_{\mathrm{kin}} &= \frac{1}{2}\int (\mathrm{d}\boldsymbol{r}')\,\rho(\boldsymbol{r}')\,(\boldsymbol{\omega}\times\boldsymbol{r}')^2 \\ &= \frac{1}{2}\int (\mathrm{d}\boldsymbol{r}')\,\rho(\boldsymbol{r}')\,(\boldsymbol{\omega}\times\boldsymbol{r}')\cdot(\boldsymbol{\omega}\times\boldsymbol{r}') \\ &= \frac{1}{2}\int (\mathrm{d}\boldsymbol{r}')\,\rho(\boldsymbol{r}')\,\boldsymbol{\omega}\cdot\big[\boldsymbol{r}'\times(\boldsymbol{\omega}\times\boldsymbol{r}')\big]\,, \end{aligned} \tag{11.1.19}$$

where we recognize the integral of (11.1.8), so that

$$E_{\mathrm{kin}} = \frac{1}{2}\boldsymbol{\omega}\cdot\boldsymbol{L} = \frac{1}{2}\boldsymbol{\omega}\cdot\mathsf{I}_R\cdot\boldsymbol{\omega}\,. \tag{11.1.20}$$

With Steiner's theorem (11.1.17) and the center-of-mass velocity (11.1.7), we have

$$E_{\mathrm{kin}} = \frac{1}{2}M\,\boldsymbol{V}^2 + \frac{1}{2}\boldsymbol{\omega}\cdot\mathsf{I}\cdot\boldsymbol{\omega}\,, \tag{11.1.21}$$

which tells us that the total kinetic energy is the sum of the kinetic energy of the center-of-mass motion and the kinetic energy of the rotation around the center-of-mass.

Upon solving (11.1.16) for $\boldsymbol{\omega}$, that is: $\boldsymbol{\omega} = \mathsf{I}_R^{-1}\cdot\boldsymbol{L}$, the kinetic energy (11.1.20) can be expressed as

$$E_{\mathrm{kin}} = \frac{1}{2}\boldsymbol{L}\cdot\mathsf{I}_R^{-1}\cdot\boldsymbol{L}\,. \tag{11.1.22}$$

---

*Jakob STEINER (1796–1863) †Christiaan HUYGENS (1629–1695)

Clearly, (11.1.20) is the analog of the point-mass expression $E_{\text{kin}} = \frac{1}{2}m\boldsymbol{v}^2$ and (11.1.22) is the analog of $E_{\text{kin}} = \boldsymbol{p}^2/(2m)$. The velocity version in (11.1.20) would be part of a Lagrange function, and its response to small changes of the angular velocity,

$$\delta\Bigl(E_{\text{kin}} = \frac{1}{2}\boldsymbol{\omega}\cdot\mathbf{I}_{\boldsymbol{R}}\cdot\boldsymbol{\omega}\Bigr) = \delta\boldsymbol{\omega}\cdot\mathbf{I}_{\boldsymbol{R}}\cdot\boldsymbol{\omega} = \delta\boldsymbol{\omega}\cdot\boldsymbol{L}\,, \tag{11.1.23}$$

is of the familiar $\delta\boldsymbol{v}\cdot\boldsymbol{p}$ kind. Likewise, the momentum version (11.1.22) could be part of a Hamilton function, and its response to small changes of the angular momentum,

$$\delta\Bigl(E_{\text{kin}} = \frac{1}{2}\boldsymbol{L}\cdot\mathbf{I}_{\boldsymbol{R}}^{-1}\cdot\boldsymbol{L}\Bigr) = \delta\boldsymbol{L}\cdot\mathbf{I}_{\boldsymbol{R}}^{-1}\cdot\boldsymbol{L} = \delta\boldsymbol{L}\cdot\boldsymbol{\omega}\,, \tag{11.1.24}$$

is of the familiar $\delta\boldsymbol{p}\cdot\boldsymbol{v}$ kind.

Both for the inertia dyadic $\mathbf{I}$ for rotations about axes through the center-of-mass, and for the extended inertia dyadic $\mathbf{I}_{\boldsymbol{R}}$ for other axes, the angular momentum (11.1.16) is not parallel to the angular velocity $\boldsymbol{\omega}$, as a rule. But since $\mathbf{I}$ is a positive symmetric dyadic, it has a set of three orthogonal eigenvectors $\boldsymbol{e}_1$, $\boldsymbol{e}_2$, $\boldsymbol{e}_3$, which we can choose to be right-handed in this order. That is

$$\mathbf{I}\cdot\boldsymbol{e}_j = I_j\boldsymbol{e}_j \quad \text{with} \quad I_1 \geq I_2 \geq I_3 \geq 0\,, \tag{11.1.25}$$

if we order the eigenvalues such that I_1 is the largest and I_3 is the smallest among them. These eigenvalues are the *principal moments of inertia* of the body, and the directions specified by the eigenvectors $\boldsymbol{e}_j$ are its *principal axes.* The principal axes are such that for rotations about them, $\boldsymbol{L}$ and $\boldsymbol{\omega}$ are parallel,

$$\boldsymbol{\omega} = \omega\boldsymbol{e}_j: \quad \boldsymbol{L} = \mathbf{I}\cdot\boldsymbol{\omega} = I_j\boldsymbol{\omega}\,. \tag{11.1.26}$$

We can, therefore, adopt a coordinate system that fits to the rotating body by specifying cartesian coordinates that refer to the principal axes. The inertia dyadic is then represented by a diagonal 3×3 matrix,

$$\mathbf{I} = \sum_{j=1}^{3} I_j \underbrace{\boldsymbol{e}_j\boldsymbol{e}_j}_{\substack{\text{dyadic}\\ \text{product}}} = \underbrace{\bigl(\boldsymbol{e}_1\ \boldsymbol{e}_2\ \boldsymbol{e}_3\bigr)}_{\substack{\text{row of}\\ \text{columns}}} \begin{pmatrix} I_1 & 0 & 0\\ 0 & I_2 & 0\\ 0 & 0 & I_3\end{pmatrix} \underbrace{\begin{pmatrix}\boldsymbol{e}_1^{\mathrm{T}}\\ \boldsymbol{e}_2^{\mathrm{T}}\\ \boldsymbol{e}_3^{\mathrm{T}}\end{pmatrix}}_{\substack{\text{column}\\ \text{of rows}}} \mathrel{\widehat{=}} \begin{pmatrix} I_1 & 0 & 0\\ 0 & I_2 & 0\\ 0 & 0 & I_3\end{pmatrix}. \tag{11.1.27}$$

11.2 Euler's equation of motion

11.2.1 *The general case*

When a rigid body is rotating, the unit vectors of its principal axes change direction in accordance with

$$\frac{\mathrm{d}}{\mathrm{d}t}\boldsymbol{e}_j = \boldsymbol{\omega}\times\boldsymbol{e}_j = -\boldsymbol{e}_j\times\boldsymbol{\omega}\,, \tag{11.2.1}$$

whereas the principal moments of inertia are constant in time — the size, shape, and internal mass distribution of the *rigid* body do not change. Then, the inertia dyadic is also time-dependent and its time derivative is an immediate consequence of (11.2.1),

$$\begin{aligned}\frac{\mathrm{d}}{\mathrm{d}t}\mathbf{I} &= \frac{\mathrm{d}}{\mathrm{d}t}\sum_{j=1}^{3}\boldsymbol{e}_j I_j\boldsymbol{e}_j = \sum_{j=1}^{3}\Big(\boldsymbol{\omega}\times\boldsymbol{e}_j I_j\boldsymbol{e}_j - \boldsymbol{e}_j I_j\boldsymbol{e}_j\times\boldsymbol{\omega}\Big)\\ &= \boldsymbol{\omega}\times\mathbf{I} - \mathbf{I}\times\boldsymbol{\omega}\,.\end{aligned} \tag{11.2.2}$$

This is an ingredient in deriving the equation of motion for $\boldsymbol{\omega}(t)$ from the fundamental statement

$$\frac{\mathrm{d}}{\mathrm{d}t}\boldsymbol{L} = \boldsymbol{\tau} \tag{11.2.3}$$

that equates the time derivative of the angular momentum $\boldsymbol{L} = \mathbf{I}\cdot\boldsymbol{\omega}$ with the torque $\boldsymbol{\tau}$ that is acting on the rigid body; recall (4.5.4) in this context.

An application of the product rule to $\mathbf{I}\cdot\boldsymbol{\omega}$ gives

$$\begin{aligned}\frac{\mathrm{d}}{\mathrm{d}t}\mathbf{I}\cdot\boldsymbol{\omega} &= \dot{\mathbf{I}}\cdot\boldsymbol{\omega} + \mathbf{I}\cdot\dot{\boldsymbol{\omega}}\\ &= \big(\boldsymbol{\omega}\times\mathbf{I} - \mathbf{I}\times\boldsymbol{\omega}\big)\cdot\boldsymbol{\omega} + \mathbf{I}\cdot\dot{\boldsymbol{\omega}}\\ &= \boldsymbol{\omega}\times\mathbf{I}\cdot\boldsymbol{\omega} + \mathbf{I}\cdot\dot{\boldsymbol{\omega}}\,,\end{aligned} \tag{11.2.4}$$

where the term $\big(\mathbf{I}\times\boldsymbol{\omega}\big)\cdot\boldsymbol{\omega} = \mathbf{I}\cdot(\boldsymbol{\omega}\times\boldsymbol{\omega}) = 0$ does not contribute, and so we arrive at

$$\mathbf{I}\cdot\dot{\boldsymbol{\omega}} + \boldsymbol{\omega}\times\mathbf{I}\cdot\boldsymbol{\omega} = \boldsymbol{\tau}\,, \tag{11.2.5}$$

which is known as *Euler's equation.* Together with (11.2.2), we have a pair of coupled first-order differential equations for $\boldsymbol{\omega}(t)$ and $\mathbf{I}(t)$. Whereas both equations are linear in $\mathbf{I}$, (11.2.5) is nonlinear in $\boldsymbol{\omega}$, which is one of the reasons why these equations are notoriously difficult to solve.

Some insight can be gained by expressing $\boldsymbol{\omega}(t)$ in the time-dependent basis specified by the unit vectors of the principal axes,

$$\boldsymbol{\omega}(t) = \sum_{j=1}^{3} \boldsymbol{e}_j(t)\,\omega_j(t)\,. \tag{11.2.6}$$

Then we have

$$\dot{\boldsymbol{\omega}} = \sum_{j=1}^{3} \boldsymbol{e}_j\dot{\omega}_j + \underbrace{\sum_{j=1}^{3} \boldsymbol{\omega}\times\boldsymbol{e}_j\omega_j}_{=\boldsymbol{\omega}\times\boldsymbol{\omega}=0} = \sum_{j=1}^{3} \boldsymbol{e}_j\dot{\omega}_j \tag{11.2.7}$$

and

$$\mathbf{I}\cdot\boldsymbol{\omega} = \sum_{j=1}^{3} \boldsymbol{e}_j I_j\omega_j\,, \qquad \mathbf{I}\cdot\dot{\boldsymbol{\omega}} = \sum_{j=1}^{3} \boldsymbol{e}_j I_j\dot{\omega}_j \tag{11.2.8}$$

as well as

$$\begin{aligned}\boldsymbol{\omega}\times\mathbf{I}\cdot\boldsymbol{\omega} &= \sum_{j,k=1}^{3} \omega_j\boldsymbol{e}_j\times\boldsymbol{e}_k I_k\omega_k\\ &= \sum_{j,k,l=1}^{3} \boldsymbol{e}_l\epsilon_{jkl}\omega_j I_k\omega_k\end{aligned} \tag{11.2.9}$$

with the epsilon symbol of (1.1.24), and find

$$I_l\dot{\omega}_l + \sum_{j,k=1}^{3} \epsilon_{jkl}\omega_j I_k\omega_k = \tau_l = \boldsymbol{\tau}\cdot\boldsymbol{e}_l \tag{11.2.10}$$

for the $\boldsymbol{e}_l$ component of (11.2.5). More explicitly, these are the three *Euler's equations*,

$$\begin{aligned}I_1\dot{\omega}_1 &= (I_2 - I_3)\omega_2\omega_3 + \tau_1\,,\\ I_2\dot{\omega}_2 &= (I_3 - I_1)\omega_3\omega_1 + \tau_2\,,\\ I_3\dot{\omega}_3 &= (I_1 - I_2)\omega_1\omega_2 + \tau_3\,.\end{aligned} \tag{11.2.11}$$

We have to keep in mind that they refer to the body-fixed system of coordinates, defined by the time-dependent unit vectors $\boldsymbol{e}_1(t)$, $\boldsymbol{e}_2(t)$, and $\boldsymbol{e}_3(t)$. The usefulness of (11.2.11) is often limited because the torque is usually known with reference to an external, fixed coordinate system — the laboratory system — and the components $\tau_l = \boldsymbol{\tau}\cdot\boldsymbol{e}_l$ in the body-fixed system cannot be stated before the time-dependent unit vectors $\boldsymbol{e}_j(t)$ have been found by solving the equation pair (11.2.2) and (11.2.5).

11.2.2 *No torque acting*

This is of no concern when the rigid body is tumbling without a torque acting, $\boldsymbol{\tau} = 0$, and

$$\begin{aligned} I_1\dot{\omega}_1 &= (I_2 - I_3)\omega_2\omega_3\,, \\ I_2\dot{\omega}_2 &= (I_3 - I_1)\omega_3\omega_1\,, \\ I_3\dot{\omega}_3 &= (I_1 - I_2)\omega_1\omega_2 \end{aligned} \tag{11.2.12}$$

is a closed system for the three components $\omega_1(t)$, $\omega_2(t)$, and $\omega_3(t)$. The general solution of this trio of coupled nonlinear differential equations cannot be expressed in terms of elementary functions.

(a) Two equal moments of inertia

We shall not discuss these general solutions and consider first the particular symmetric situation in which two of the principal moments of inertia are equal, $I_2 = I_3$, say. Then

$$I_1\dot{\omega}_1 = 0\,, \tag{11.2.13}$$

so that

$$\omega_1(t) = \omega_1(0) \tag{11.2.14}$$

and

$$\dot{\omega}_2 = -\Omega\omega_3\,, \qquad \dot{\omega}_3 = \Omega\omega_2 \tag{11.2.15}$$

with

$$\Omega = \frac{I_1 - I_2}{I_2}\omega_1(0)\,. \tag{11.2.16}$$

The equation pair (11.2.15), which we first met in Section 1.1.8, is familiar and is solved by

$$\begin{aligned} \omega_2(t) &= \omega_2(0)\cos(\Omega t) - \omega_3(0)\sin(\Omega t)\,, \\ \omega_3(t) &= \omega_3(0)\cos(\Omega t) + \omega_2(0)\sin(\Omega t)\,. \end{aligned} \tag{11.2.17}$$

With $\boldsymbol{\omega}(t)$ now at hand with reference to the $\boldsymbol{e}_j(t)$s, one combines (11.2.1) and (11.2.6) and so determines the time-dependent unit vectors $\boldsymbol{e}_j(t)$. The details are an exercise for the reader (see Appendix C).

(b) Rotation about a principal axis

When all three principal moments of inertia are different, the solution of (11.2.12) is immediate if only one of the initial values $\omega_1(0)$, $\omega_2(0)$, and $\omega_3(0)$ is nonzero; all $\omega_j(t)$s are then constant in time. The rigid body just rotates about one of its principal axes.

To investigate the stability of this situation, we consider a large $\omega_1(t)$ and very small $\omega_2(t)$ and $\omega_3(t)$. In the pair of equations for the small components,

$$\dot{\omega}_2 = -\frac{I_1 - I_3}{I_2}\omega_1\omega_3\,,$$
$$\dot{\omega}_3 = \frac{I_1 - I_2}{I_3}\omega_1\omega_2\,, \tag{11.2.18}$$

we regard $\omega_1(t)$ as a given, even if currently unknown, function of time and note that

$$\frac{I_1 - I_2}{I_3}\omega_2(t)^2 + \frac{I_1 - I_3}{I_2}\omega_3(t)^2 = \frac{I_1 - I_2}{I_3}\omega_2(0)^2 + \frac{I_1 - I_3}{I_2}\omega_3(0)^2 \tag{11.2.19}$$

is a constant of motion. When

$$\operatorname{sgn}(I_1 - I_2) = \operatorname{sgn}(I_1 - I_3) \tag{11.2.20}$$

we have a weighted sum of $\omega_2(t)^2$ and $\omega_3(t)^2$ in (11.2.19) — as in $(\cos\phi)^2 + (\sin\phi)^2 = 1$ — so that $\omega_2(t)$ and $\omega_3(t)$ must remain small for all times if they are small initially. By contrast, if

$$\operatorname{sgn}(I_1 - I_2) = -\operatorname{sgn}(I_1 - I_3) \tag{11.2.21}$$

the left-hand side of (11.2.19) is a weighted difference of $\omega_2(t)^2$ and $\omega_3(t)^2$ — as in $(\cosh\theta)^2 - (\sinh\theta)^2 = 1$ — and nothing prevents $\omega_2(t)$ and $\omega_3(t)$ from growing from initially small values to sizeable values later.

The favorable condition (11.2.20) is obeyed, when I_1 is the largest ($I_1 > I_2$ and $I_1 > I_3$) or the smallest ($I_1 < I_2$ and $I_1 < I_3$) principal moment of inertia, whereas the unfavorable condition (11.2.21) holds if I_1 is the middle value ($I_2 > I_1 > I_3$ or $I_2 < I_1 < I_3$). It follows that rotation about the axis of the largest or the smallest principal moment of inertia is stable as small perturbations do not grow, whereas rotation about the axis with the middle principal moment of inertia is not stable.

More specifically, we have

$$\omega_2(t) = \omega_2(0)\cos\bigl(\phi(t)\bigr) - \sqrt{\frac{(I_1-I_3)I_3}{(I_1-I_2)I_2}}\,\omega_3(0)\sin\bigl(\phi(t)\bigr)\,,$$

$$\omega_3(t) = \omega_3(0)\cos\bigl(\phi(t)\bigr) + \sqrt{\frac{(I_1-I_2)I_2}{(I_1-I_3)I_3}}\,\omega_2(0)\sin\bigl(\phi(t)\bigr) \tag{11.2.22}$$

with

$$\phi(t) = \sqrt{\frac{(I_1-I_2)(I_1-I_3)}{I_2I_3}}\int_0^t \mathrm{d}t'\,\omega_1(t') \tag{11.2.23}$$

for $I_1 > I_2$ and $I_1 > I_3$ when (11.2.20) is the case. If $I_1 < I_2$ and $I_1 < I_3$, the signs of the $\sin\bigl(\phi(t)\bigr)$ terms are reversed, and for $I_2 = I_3$ we recover (11.2.17). Indeed, the small components $\omega_2(t)$ and $\omega_3(t)$ stay small here, and the right-hand side in

$$\dot{\omega}_1 = \frac{I_2-I_3}{I_1}\omega_2\omega_3 \tag{11.2.24}$$

is second-order small for all times and oscillatory.

Likewise, if (11.2.21) is the case we have

$$\omega_2(t) = \omega_2(0)\cosh\bigl(\theta(t)\bigr) - \sqrt{\frac{(I_1-I_3)I_3}{(I_2-I_1)I_2}}\,\omega_3(0)\sinh\bigl(\theta(t)\bigr)\,,$$

$$\omega_3(t) = \omega_3(0)\cosh\bigl(\theta(t)\bigr) - \sqrt{\frac{(I_2-I_1)I_2}{(I_1-I_3)I_3}}\,\omega_2(0)\sinh\bigl(\theta(t)\bigr) \tag{11.2.25}$$

with

$$\theta(t) = \sqrt{\frac{(I_2-I_1)(I_1-I_3)}{I_2I_3}}\int_0^t \mathrm{d}t'\,\omega_1(t') \tag{11.2.26}$$

for $I_2 > I_1 > I_3$, and reversed signs for the $\sinh\bigl(\theta(t)\bigr)$ terms if $I_2 < I_1 < I_3$. Here, the initially-small components $\omega_2(t)$ and $\omega_3(t)$ will not stay small forever.

11.3 Examples

11.3.1 *Physical pendulum*

As a first example, we consider a physical pendulum, a massive body that can rotate about a horizontal axis that does not go through the center-of-mass, as depicted here:

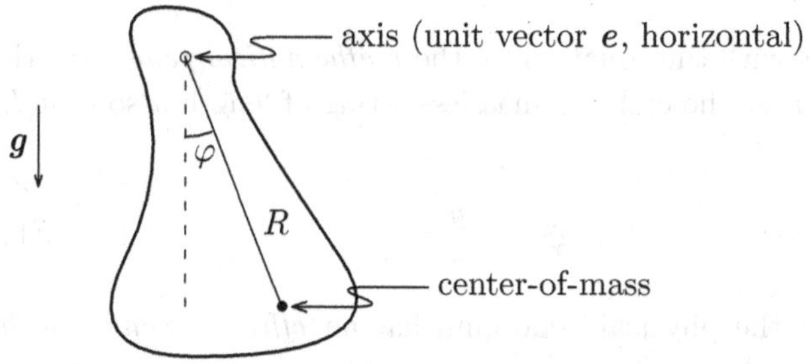

(11.3.1)

With the inertia dyadic $\mathbf{I}$, we have the relevant *moment of inertia* for rotation about the axis specified by unit vector $\boldsymbol{e}$ given by

$$I_e = \boldsymbol{e} \cdot \mathbf{I} \cdot \boldsymbol{e}\,, \tag{11.3.2}$$

so that $\frac{1}{2}I_e\omega^2 = \frac{1}{2}\boldsymbol{\omega} \cdot \mathbf{I} \cdot \boldsymbol{\omega}$ is the rotational kinetic energy for angular velocity vector $\boldsymbol{\omega} = \omega\boldsymbol{e} = \dot{\varphi}\boldsymbol{e}$. This is for rotation about an axis through the center-of-mass but the actual rotation is about an axis at distance R from the center-of-mass. The kinetic energy associated with the motion of the center-of-mass, the first summand in (11.1.21), is then $\frac{1}{2}MR^2\omega^2$. Accordingly, we have

$$E_{\text{kin}} = \frac{1}{2}(I_e + MR^2)\omega^2 = \frac{1}{2}(I_e + MR^2)\dot{\varphi}^2 \tag{11.3.3}$$

for the total kinetic energy. Combined with the potential energy of the center-of-mass at height $-R\cos\varphi$,

$$E_{\text{pot}} = -MgR\cos\varphi\,, \tag{11.3.4}$$

this gives the Lagrange function

$$L = \frac{1}{2}(I_e + MR^2)\dot{\varphi}^2 + MgR\cos\varphi\,. \tag{11.3.5}$$

The Euler–Lagrange equation for $\varphi(t)$ then reads

$$\frac{\mathrm{d}}{\mathrm{d}t}\frac{\partial L}{\partial \dot{\varphi}} = \frac{\partial L}{\partial \varphi}: \quad (I_e + MR^2)\ddot{\varphi} = -MgR\sin\varphi\,, \tag{11.3.6}$$

and for small-amplitude oscillations around $\varphi = 0$ we have

$$\ddot{\varphi} = -\frac{MgR}{I_e + MR^2}\varphi\,. \tag{11.3.7}$$

We compare this with the equation for the *mathematical pendulum*, that is: the point mass m at the end of a massless string of length a so that $I_e = 0$ and $R = a$,

$$\ddot{\varphi} = -\frac{g}{a}\varphi\,, \tag{11.3.8}$$

and so find that the physical pendulum has an *effective pendulum length* given by

$$a_{\mathrm{eff}} = R + \frac{1}{MR}I_e = R + \frac{1}{MR}\boldsymbol{e}\cdot\mathbf{I}\cdot\boldsymbol{e}\,. \tag{11.3.9}$$

The resulting period of the small-amplitude oscillations is

$$T = 2\pi\sqrt{\frac{a_{\mathrm{eff}}}{g}}\,. \tag{11.3.10}$$

11.3.2 *Thin rod*

Next, let us consider a thin homogenous rod of length l:

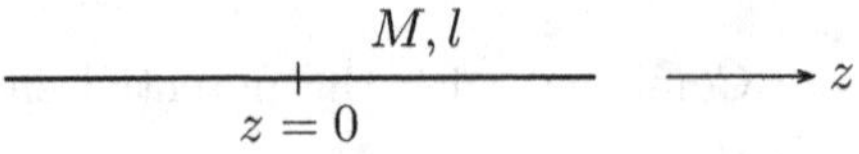

(11.3.11)

and we choose the coordinates such that the middle of the rod is at $x = y = z = 0$ and the rod is aligned with the z axis. Then the mass density is

$$\rho(\boldsymbol{r}) = \frac{M}{l}\delta(x)\delta(y)\eta\big(\tfrac{1}{2}l - |z|\big)\,, \tag{11.3.12}$$

where the step function selects $-\frac{1}{2}l < z < \frac{1}{2}l$ and the product of the two delta functions accounts for the rod being thin. The inertia dyadic is

$$\begin{aligned}\mathsf{I} &= \int (\mathrm{d}\boldsymbol{r})\,\rho(\boldsymbol{r})\,(r^2\mathbf{1} - \boldsymbol{r}\,\boldsymbol{r}) \\ &= \frac{M}{l}\int_{-\frac{1}{2}l}^{\frac{1}{2}l} \mathrm{d}z\, z^2(\mathbf{1} - \boldsymbol{e}_z\,\boldsymbol{e}_z) = \frac{Ml^2}{12}(\mathbf{1} - \boldsymbol{e}_z\,\boldsymbol{e}_z) \\ &= \frac{Ml^2}{12}(\boldsymbol{e}_x\,\boldsymbol{e}_x + \boldsymbol{e}_y\,\boldsymbol{e}_y) \mathrel{\widehat{=}} \frac{Ml^2}{12}\begin{pmatrix}1&0&0\\0&1&0\\0&0&0\end{pmatrix},\end{aligned} \tag{11.3.13}$$

where the 3×3 matrix refers to the cartesian coordinates used. There is no moment of inertia for rotation about the z axis,

$$\boldsymbol{e}_z \cdot \mathsf{I} \cdot \boldsymbol{e}_z = 0\,, \tag{11.3.14}$$

and the moments of inertia for rotation about the x axis and the y axis are equal,

$$\boldsymbol{e}_x \cdot \mathsf{I} \cdot \boldsymbol{e}_x = \boldsymbol{e}_y \cdot \mathsf{I} \cdot \boldsymbol{e}_y = \frac{Ml^2}{12}\,; \tag{11.3.15}$$

any axis perpendicular to the z axis has the same moment of inertia.

Such a thin rod is now suspended at one end and free to rotate about this endpoint:

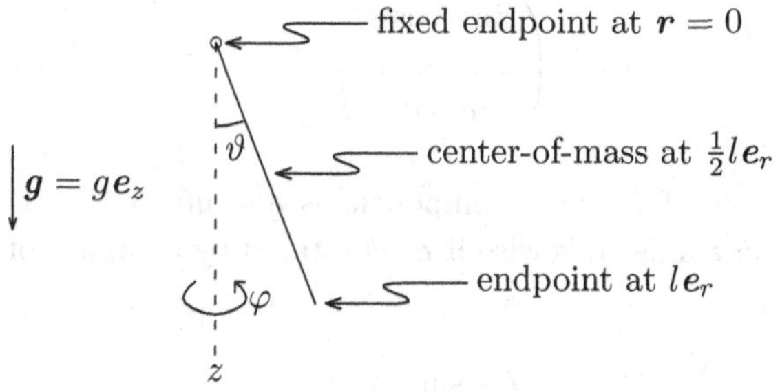

(11.3.16)

The rod is aligned with $\boldsymbol{e}_r$, the unit vector in radial direction. For rotations about the center-of-mass we now have the inertia dyadic

$$\mathsf{I} = \frac{Ml^2}{12}(\mathbf{1} - \boldsymbol{e}_r\,\boldsymbol{e}_r)\,, \tag{11.3.17}$$

and for rotations about an endpoint it is

$$\mathbf{I}_{\text{end}} = \mathbf{I} + M\left(\frac{l}{2}\right)^2 (\mathbf{1} - \boldsymbol{e}_r\,\boldsymbol{e}_r) = \frac{Ml^2}{3}(\mathbf{1} - \boldsymbol{e}_r\,\boldsymbol{e}_r)\,. \tag{11.3.18}$$

With both the polar angle ϑ and azimuth φ depending on time we have an angular velocity vector $\boldsymbol{\omega}$ which is such that

$$\boldsymbol{\omega} \times l\boldsymbol{e}_r = l\dot{\vartheta}\frac{\partial}{\partial\vartheta}\boldsymbol{e}_r + l\dot{\varphi}\frac{\partial}{\partial\varphi}\boldsymbol{e}_r \tag{11.3.19}$$

for the velocity of the free endpoint of the rod. Recalling (1.1.134) and (1.1.135), we observe that

$$\begin{aligned}\frac{\partial}{\partial\vartheta}\boldsymbol{e}_r &= \boldsymbol{e}_\vartheta = \boldsymbol{e}_\varphi \times \boldsymbol{e}_r\,,\\ \frac{\partial}{\partial\varphi}\boldsymbol{e}_r &= \sin\vartheta\,\boldsymbol{e}_\varphi = -\sin\vartheta\,\boldsymbol{e}_\vartheta \times \boldsymbol{e}_r\\ &= (\cos\vartheta\,\boldsymbol{e}_r - \sin\vartheta\,\boldsymbol{e}_\vartheta) \times \boldsymbol{e}_r = \boldsymbol{e}_z \times \boldsymbol{e}_r\,,\end{aligned} \tag{11.3.20}$$

and conclude that

$$\boldsymbol{\omega} = \dot{\vartheta}\boldsymbol{e}_\varphi + \dot{\varphi}\boldsymbol{e}_z\,. \tag{11.3.21}$$

Indeed, when

$$\boldsymbol{e}_r \mathrel{\widehat{=}} \begin{pmatrix}\sin\vartheta\cos\varphi\\ \sin\vartheta\sin\varphi\\ \cos\vartheta\end{pmatrix} \tag{11.3.22}$$

is rotated by a change of φ, the z component is not affected, so that the rotation is about the z axis. Likewise if $\boldsymbol{e}_r$ is rotated by a change of ϑ, the rotation axis is

$$\boldsymbol{e}_\varphi \mathrel{\widehat{=}} \begin{pmatrix}-\sin\varphi\\ \cos\varphi\\ 0\end{pmatrix}, \tag{11.3.23}$$

which is $\boldsymbol{e}_y$ for $\varphi = 0$, and for $\varphi \neq 0$ it is the unit vector that one gets upon rotating $\boldsymbol{e}_y$ about the z axis by angle φ; see Exercise 157 in this context.

The kinetic energy is then

$$\begin{aligned} E_{\text{kin}} &= \frac{1}{2}\boldsymbol{\omega} \cdot \mathsf{I}_{\text{end}} \cdot \boldsymbol{\omega} \\ &= \frac{Ml^2}{6}\boldsymbol{\omega} \cdot (\mathbf{1} - \boldsymbol{e}_r\, \boldsymbol{e}_r) \cdot \boldsymbol{\omega} \\ &= \frac{Ml^2}{6}(\boldsymbol{\omega} \times \boldsymbol{e}_r)^2 \,, \end{aligned} \tag{11.3.24}$$

and with

$$\boldsymbol{\omega} \times \boldsymbol{e}_r = \dot{\vartheta}\boldsymbol{e}_\vartheta + \dot{\varphi}\sin\vartheta\, \boldsymbol{e}_\varphi \tag{11.3.25}$$

we obtain

$$E_{\text{kin}} = \frac{Ml^2}{6}\left[\dot{\vartheta}^2 + (\dot{\varphi}\sin\vartheta)^2\right]. \tag{11.3.26}$$

We combine this with the potential energy

$$E_{\text{pot}} = -\frac{1}{2}Mgl\cos\vartheta \tag{11.3.27}$$

to arrive at the Lagrange function

$$L = \frac{Ml^2}{6}\left[\dot{\vartheta}^2 + (\dot{\varphi}\sin\vartheta)^2\right] + \frac{Mgl}{2}\cos\vartheta\,. \tag{11.3.28}$$

We compare this with the point-mass-at-the-end-of-a-massless-string Lagrange function in (9.3.9),

$$L = \frac{m}{2}a^2\left[\dot{\vartheta}^2 + (\dot{\varphi}\sin\vartheta)^2\right] + mga\cos\vartheta \tag{11.3.29}$$

and note that the replacements

$$\frac{1}{3}Ml^2 \leftrightarrow ma^2\,, \qquad \frac{1}{2}Ml \leftrightarrow ma \tag{11.3.30}$$

turn the two Lagrange functions into each other. Therefore, the rod is equivalent to a point-mass pendulum with an effective length of

$$a_{\text{eff}} = \frac{\frac{1}{3}Ml^2}{\frac{1}{2}Ml} = \frac{2}{3}l \tag{11.3.31}$$

and an effective mass of

$$m_{\text{eff}} = \frac{(\frac{1}{2}Ml)^2}{\frac{1}{3}Ml^2} = \frac{3}{4}M\,, \tag{11.3.32}$$

so that the period of small-amplitude oscillations is

$$T = 2\pi\sqrt{\frac{a_{\text{eff}}}{g}} = 2\pi\sqrt{\frac{2l}{3g}}\,. \tag{11.3.33}$$

11.3.3 *Symmetric top*

If the rod is not so thin, we have a nonzero moment of inertia also for rotation about the third axis. We shall assume that the body has rotational symmetry around this third axis, the so-called *figure axis.* One then speaks of a *symmetric top*, for which the inertia dyadic has the form

$$\mathsf{I} = I_1(\mathbf{1} - \boldsymbol{e}\,\boldsymbol{e}) + I_2\boldsymbol{e}\,\boldsymbol{e}\,, \tag{11.3.34}$$

where $\boldsymbol{e}$ is the unit vector of the figure axis. One point on the figure axis is fixed, indicated here as the tip of the conical body:

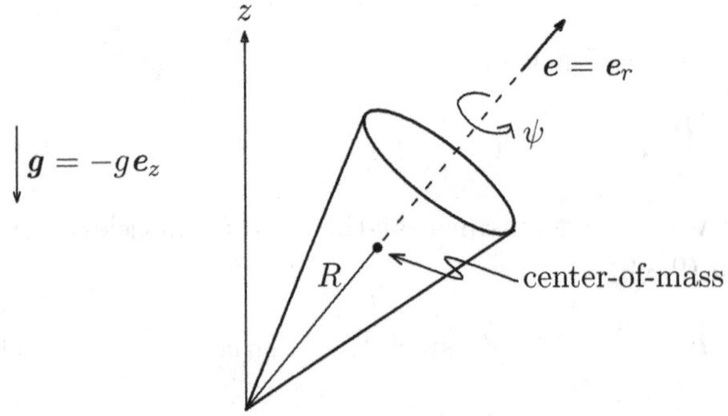

(11.3.35)

The center-of-mass is at distance R from that fixed point.

We use spherical coordinates with the z axis pointing up, so that $\boldsymbol{e} = \boldsymbol{e}_r$, the center-of-mass position vector is $R\boldsymbol{e}_r$, and the inertia dyadic of the fixed tip is

$$\begin{aligned}\mathsf{I}_{\text{tip}} &= \mathsf{I} + MR^2(\mathbf{1} - \boldsymbol{e}_r\,\boldsymbol{e}_r)\\ &= (I_1 + MR^2)(\mathbf{1} - \boldsymbol{e}_r\,\boldsymbol{e}_r) + I_2\boldsymbol{e}_r\,\boldsymbol{e}_r\,,\end{aligned} \tag{11.3.36}$$

as follows from Steiner's theorem (11.1.17). The angular velocity vector has one more component now: the rotation about the figure axis, for which angle ψ is the conventional symbol. Thus, rather than (11.3.21), we now have the angular velocity vector

$$\boldsymbol{\omega} = \dot{\vartheta}\boldsymbol{e}_\varphi + \dot{\varphi}\boldsymbol{e}_z + \dot{\psi}\boldsymbol{e}_r\,. \tag{11.3.37}$$

In this context, the angles φ, ϑ, ψ are often called *Euler angles* (see Exercise 157). You can think of getting the current orientation of the top as follows: Initially the top has its figure axis aligned with the z axis and $\psi = 0$. Then you rotate the top by φ about the z axis, followed by a rotation by ϑ about the new y axis, followed by a rotation about the new-new z axis by angle ψ. In the case of the symmetric top, the initial φ rotation and the final ψ rotation do not actually alter the physical situation, as they would for a body of arbitrary shape. We should, therefore, expect that φ and ψ are cyclic coordinates of the symmetric top.

With the inertia dyadic of (11.3.36) and the angular velocity of (11.3.37), the kinetic energy is

$$\begin{aligned} E_{\text{kin}} &= \frac{1}{2}\boldsymbol{\omega}\cdot\mathsf{I}_{\text{tip}}\cdot\boldsymbol{\omega} \\ &= \frac{1}{2}(I_1 + MR^2)(\boldsymbol{\omega}\times\boldsymbol{e}_r)^2 + \frac{1}{2}I_2(\boldsymbol{\omega}\cdot\boldsymbol{e}_r)^2 \end{aligned} \tag{11.3.38}$$

with $\boldsymbol{\omega}\times\boldsymbol{e}_r$ given in (11.3.25) and

$$\boldsymbol{\omega}\cdot\boldsymbol{e}_r = \dot{\varphi}\cos\vartheta + \dot{\psi}\,. \tag{11.3.39}$$

This gives the Lagrange function

$$L = \frac{1}{2}(I_1 + MR^2)\left[\dot{\vartheta}^2 + (\dot{\varphi}\sin\vartheta)^2\right] + \frac{1}{2}I_2(\dot{\varphi}\cos\vartheta + \dot{\psi})^2 - MgR\cos\vartheta\,, \tag{11.3.40}$$

where the last term subtracts the potential energy. As expected, φ and ψ are cyclic coordinates; the respective momenta

$$\begin{aligned} p_\varphi &= \frac{\partial L}{\partial\dot{\varphi}} = (I_1 + MR^2)\dot{\varphi}(\sin\vartheta)^2 + I_2(\dot{\varphi}\cos\vartheta + \dot{\psi})\cos\vartheta\,, \\ p_\psi &= \frac{\partial L}{\partial\dot{\psi}} = I_2(\dot{\varphi}\cos\vartheta + \dot{\psi}) \end{aligned} \tag{11.3.41}$$

are constants of motion, with their values determined by the initial conditions.

With no parametric time dependence in the Lagrange function, another constant of motion is the energy

$$E = \frac{1}{2}(I_1 + MR^2)\left[\dot{\vartheta}^2 + (\dot{\varphi}\sin\vartheta)^2\right] + \frac{1}{2}I_2(\dot{\varphi}\cos\vartheta + \dot{\psi})^2 + MgR\cos\vartheta\,, \tag{11.3.42}$$

from which we eliminate $\dot{\varphi}$ and $\dot{\psi}$ in favor of p_φ and p_ψ. This takes us to

$$E = \frac{1}{2}(I_1 + MR^2)\dot{\vartheta}^2 + V_{\mathrm{eff}}(\vartheta) \tag{11.3.43}$$

with

$$V_{\mathrm{eff}}(\vartheta) = \frac{1}{2(I_1 + MR^2)}\left(\frac{p_\varphi - p_\psi\cos\vartheta}{\sin\vartheta}\right)^2 + \frac{p_\psi^2}{2I_2} + MgR\cos\vartheta\,. \tag{11.3.44}$$

This effective potential energy is very large for $\sin\vartheta \ll 1$, that is: for $\vartheta \gtrsim 0$ and $\vartheta \lesssim \pi$, when the figure axis is almost vertical with the center-of-mass either above or below the fixed point, and has a minimum at some ϑ_0 in between:

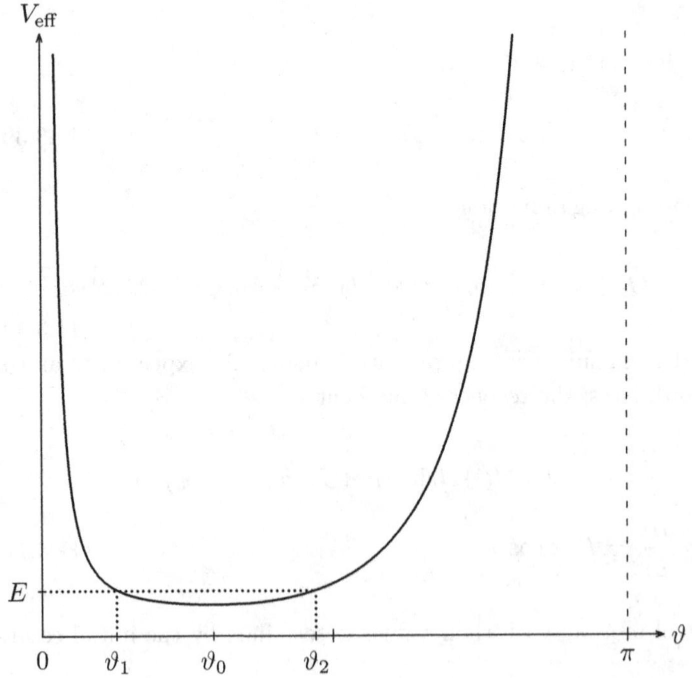

(11.3.45)

The ϑ motion is then periodic between ϑ_1 and ϑ_2, meaning that the figure axis has a varying angle with the vertical axis. This motion is called *nutation* ("nodding"). It is accompanied by the φ motion, for which

$$\dot{\varphi} = \frac{p_\varphi - p_\psi \cos\vartheta}{(I_1 + MR^2)(\sin\vartheta)^2}, \tag{11.3.46}$$

the *precession* of the figure axis about the vertical axis. Depending on the values of ϑ_1 and ϑ_2, $\dot{\varphi}$ can have the same sign at all times, or it can be positive for some fraction of the nutation period and negative for other times. Finally, there is the *spinning* of the top, its rotation about its figure axis with angular velocity

$$\begin{aligned} \dot{\psi} &= \frac{p_\psi}{I_2} - \dot{\varphi}\cos\vartheta \\ &= p_\psi\left(\frac{1}{I_2} + \frac{(\cot\vartheta)^2}{I_1 + MR^2}\right) - \frac{p_\varphi \cos\vartheta}{(I_1 + MR^2)(\sin\vartheta)^2}. \end{aligned} \tag{11.3.47}$$

Clearly, the motion of the spinning symmetric top is quite complicated, but we have a good idea of its characteristic features.

The important aspect of the motion is the direction of the figure axis as the spinning top undergoes precession and nutation. We imagine a sphere centered at the fixed tip of the top, and mark the direction of the figure axis as a point on this sphere. Then, there are trajectories of three kinds. First, when $\dot{\varphi}$ does not change sign, we have regular precession:

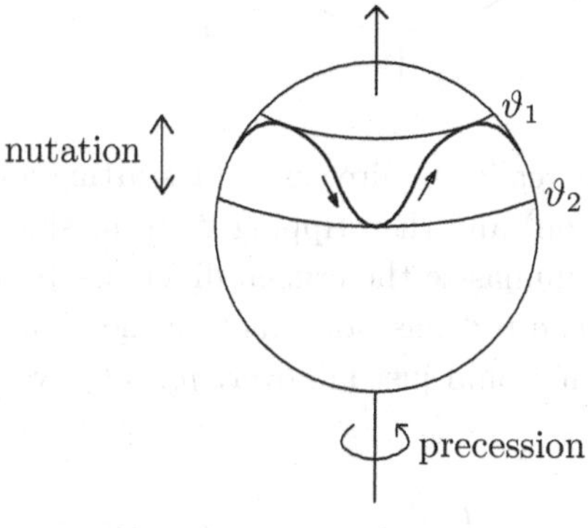

(11.3.48)

with the top precessing forward all the time if $\dot{\varphi} > 0$ as in the figure, or backward if $\dot{\varphi} < 0$. Second, when $\dot{\varphi}$ changes sign, the precession is forward at some times and backward at others, and a typical trajectory looks like this:

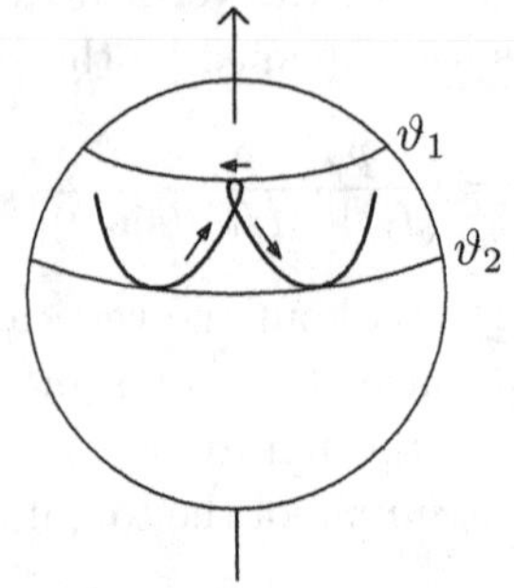

(11.3.49)

where we depict the situation of $\dot{\varphi} > 0$ for $\vartheta = \vartheta_2$ and $\dot{\varphi} < 0$ for $\vartheta = \vartheta_1$, which is the case if $p_\psi \cos\vartheta_2 < p_\varphi < p_\psi \cos\vartheta_1$; see (11.3.46). Third, if $p_\varphi = p_\psi \cos\vartheta_1$, we have $\dot{\varphi} = 0$ for ϑ_1 and $\dot{\varphi}$ of one sign for $\vartheta_1 < \vartheta \leq \vartheta_2$, positive here:

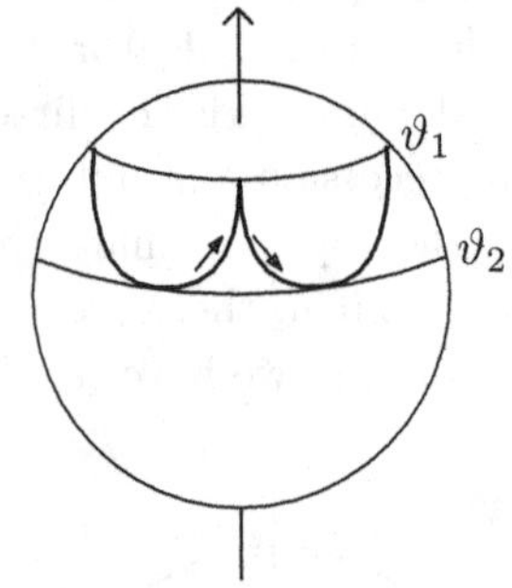

(11.3.50)

Motion of this third kind is realized quite easily by starting a top vertically (as you would with a toy top) and then tipping it to the side.

We should not fail to emphasize the crucial difference between the rod and the symmetric top. The rod has no ψ motion, no rotation about its figure axis and, therefore, it would just fall over: $p_\psi = 0$ gives the effective potential energy

$$V_{\text{eff}}(\vartheta) = \frac{1}{2(I_1 + MR^2)}\left(\frac{p_\varphi}{\sin\vartheta}\right)^2 + MgR\cos\vartheta\,, \tag{11.3.51}$$

which has its minimum at ϑ_0 with

$$\frac{p_\varphi^2}{I_1 + MR^2}\cos\vartheta_0 = -MgR(\sin\vartheta_0)^4\,, \tag{11.3.52}$$

so that $\cos\vartheta_0 < 0$ and $\vartheta_0 > \frac{\pi}{2}$, below the horizontal plane with the fixed point.

The spinning top can stay above the horizontal plane — the situation $\vartheta_1 < \vartheta_0 < \vartheta_2 < \frac{\pi}{2}$ is possible, and rather typical; it is the case in (11.3.45) — because the torque $R\boldsymbol{e}_r \times (-Mg\boldsymbol{e}_z)$ is horizontal and the implied change of the angular momentum, $\frac{\mathrm{d}}{\mathrm{d}t}\boldsymbol{L} = MgR\sin\vartheta\,\boldsymbol{e}_\varphi$, makes this vector precess. Then, if the lion's share of the angular momentum comes from the fast ψ rotation about the figure axis, the figure axis and the angular momentum vector are (almost) aligned, so that the figure axis itself has to precess.

This is an example of stabilization achieved by fast rotation, a mechanical device with many technical applications. Perhaps the best known is the spinning projectile shot from a rifle, which has grooves and lands, as compared to the not-spinning bullet you get from a gun with a smooth bore. Shooting with a rifle is much more precise because the large angular momentum of the spinning projectile makes it less responsive to forces that would otherwise deflect it.

Chapter 12

Earth-Bound Laboratories

12.1 Coriolis force, centrifugal force

Imagine you are at the north pole (N) and set off a pendulum that oscillates in the plane defined by the direction to Singapore (S):

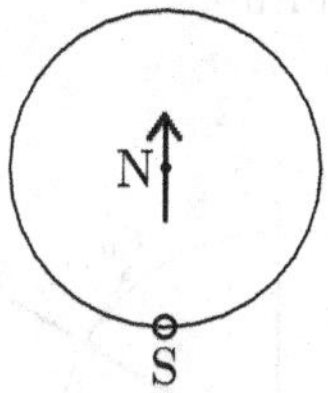

(12.1.1)

Six hours later, the situation is this:

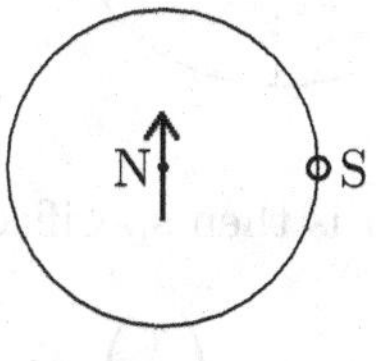

(12.1.2)

as looked from outer space, but you are perceiving the situation as this:

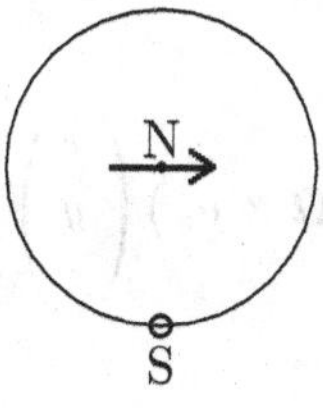

(12.1.3)

You report that the plane of the pendulum oscillation has precessed clockwise by 90°. This is how you see it because you take your laboratory as reference, and this laboratory is fixed to the surface of the earth and rotating with it about the south-to-north axis through the earth with a period of 1 sidereal day = 23 h 56 min 4 s = 86 164 s. The sidereal day matters here, rather than the 24 h period of the solar day which is a bit longer because the earth proceeds along its Kepler ellipse. Per year, we have one more sidereal day than there are solar days.

While possible, it is often inconvenient to describe such experiments in terms of the non-rotating reference frame of outer space. We take all our data by apparatus that are at fixed locations relative to the laboratory. Therefore, it will be expedient to have an efficient method that accounts for the consequences of the earth rotation as we observe them in earth-bound laboratories.

Consider a surface point with position vector $\boldsymbol{r}_0$ from the center of the earth and a local surface-bound coordinate system with unit vectors $\boldsymbol{e}_x$, $\boldsymbol{e}_y$, $\boldsymbol{e}_z$ that, say, point east, north, and up:

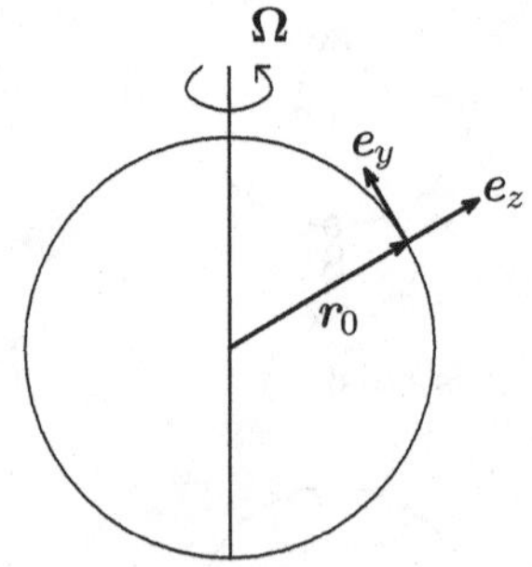

(12.1.4)

The position of a point mass m is then specified by the vector

$$\boldsymbol{r}_0 + \begin{pmatrix} \boldsymbol{e}_x & \boldsymbol{e}_y & \boldsymbol{e}_z \end{pmatrix} \begin{pmatrix} x \\ y \\ z \end{pmatrix} = \boldsymbol{r}_0 + \boldsymbol{r} \tag{12.1.5}$$

and its velocity is given by

$$\begin{aligned} &\boldsymbol{\Omega} \times \boldsymbol{r}_0 + \begin{pmatrix} \boldsymbol{\Omega} \times \boldsymbol{e}_x & \boldsymbol{\Omega} \times \boldsymbol{e}_y & \boldsymbol{\Omega} \times \boldsymbol{e}_z \end{pmatrix} \begin{pmatrix} x \\ y \\ z \end{pmatrix} + \begin{pmatrix} \boldsymbol{e}_x & \boldsymbol{e}_y & \boldsymbol{e}_z \end{pmatrix} \begin{pmatrix} \dot{x} \\ \dot{y} \\ \dot{z} \end{pmatrix} \\ &= \boldsymbol{\Omega} \times \boldsymbol{r}_0 + \boldsymbol{\Omega} \times \boldsymbol{r} + \boldsymbol{v}\,, \end{aligned} \tag{12.1.6}$$

where $\boldsymbol{r}$ is the *local* position vector and $\boldsymbol{v}$ is the *local* velocity vector,

$$\boldsymbol{v} = \begin{pmatrix} \boldsymbol{e}_x & \boldsymbol{e}_y & \boldsymbol{e}_z \end{pmatrix} \begin{pmatrix} \dot{x} \\ \dot{y} \\ \dot{z} \end{pmatrix} = \boldsymbol{e}_x \dot{x} + \boldsymbol{e}_y \dot{y} + \boldsymbol{e}_z \dot{z} \,, \tag{12.1.7}$$

that accounts solely for the time dependence of the coordinates $x(t)$, $y(t)$, $z(t)$ but not for the time dependence of the unit vectors which change direction because the earth rotates with angular velocity vector $\boldsymbol{\Omega}$, where

$$\Omega = |\boldsymbol{\Omega}| = \frac{2\pi}{1\,\text{sidereal day}} = 2\pi \times \frac{1}{86\,164\,\text{s}} \,. \tag{12.1.8}$$

The effect of the earth rotation is contained in the terms $\boldsymbol{\Omega} \times \boldsymbol{r}_0 + \boldsymbol{\Omega} \times \boldsymbol{r}$ in (12.1.6).

With potential energy $V(\boldsymbol{r})$, a function of the local coordinates x, y, z, we have then the Lagrange function

$$L = \frac{m}{2}(\boldsymbol{\Omega} \times \boldsymbol{r}_0 + \boldsymbol{\Omega} \times \boldsymbol{r} + \boldsymbol{v})^2 - V(\boldsymbol{r}) \tag{12.1.9}$$

from which we get the momentum

$$\boldsymbol{p} = m(\boldsymbol{\Omega} \times \boldsymbol{r}_0 + \boldsymbol{\Omega} \times \boldsymbol{r} + \boldsymbol{v}) \tag{12.1.10}$$

and the force

$$\boldsymbol{\nabla} L = -m\boldsymbol{\Omega} \times (\boldsymbol{\Omega} \times \boldsymbol{r}_0 + \boldsymbol{\Omega} \times \boldsymbol{r} + \boldsymbol{v}) - \boldsymbol{\nabla} V(\boldsymbol{r}) \,. \tag{12.1.11}$$

These establish the equation of motion

$$\begin{aligned} \frac{\mathrm{d}}{\mathrm{d}t}\big[m(\boldsymbol{\Omega} \times \boldsymbol{r}_0 + \boldsymbol{\Omega} \times \boldsymbol{r} + \boldsymbol{v})\big] &= m\boldsymbol{\Omega} \times \boldsymbol{v} + m\dot{\boldsymbol{v}} \\ &= -m\boldsymbol{\Omega} \times (\boldsymbol{\Omega} \times \boldsymbol{r}_0 + \boldsymbol{\Omega} \times \boldsymbol{r} + \boldsymbol{v}) - \boldsymbol{\nabla} V(\boldsymbol{r}) \,, \end{aligned} \tag{12.1.12}$$

where $\frac{\mathrm{d}}{\mathrm{d}t}\boldsymbol{r}_0 = 0$, $\frac{\mathrm{d}}{\mathrm{d}t}\boldsymbol{r} = \boldsymbol{v}$ because the variables of the Lagrange function are $\boldsymbol{r}(t)$ and $\boldsymbol{v}(t)$, the position and velocity vectors of the local surface-bound coordinate system, in which $\boldsymbol{r}_0$ and $\boldsymbol{\Omega}$ are constant.

We rearrange the terms and arrive at

$$m\dot{\boldsymbol{v}} = -2m\boldsymbol{\Omega} \times \boldsymbol{v} - m\boldsymbol{\Omega} \times (\boldsymbol{\Omega} \times \boldsymbol{r}) + \boldsymbol{F} \,, \tag{12.1.13}$$

where

$$\boldsymbol{F} = -\boldsymbol{\nabla} V(\boldsymbol{r}) - m\boldsymbol{\Omega} \times (\boldsymbol{\Omega} \times \boldsymbol{r}_0) \tag{12.1.14}$$

includes a constant mass-times-acceleration term in addition to the gradient of the potential energy. Now, part of $-\boldsymbol{\nabla}V(\boldsymbol{r})$ is the gravitational force,

$$-\boldsymbol{\nabla}V(\boldsymbol{r}) = -GMm\frac{\boldsymbol{r}_0}{r_0^3} + \boldsymbol{F}_{\text{other}}\,, \tag{12.1.15}$$

with the gravitational constant G and the earth mass M, and $\boldsymbol{F}_{\text{other}}$ accounts for all other forces (springs, charges in electric and magnetic fields, ...), possibly including frictional forces, and also including the difference between $-GMm\frac{\boldsymbol{r}_0}{r_0^3}$ and $-GMm\frac{\boldsymbol{r}_0+\boldsymbol{r}}{|\boldsymbol{r}_0+\boldsymbol{r}|^3}$ if it should be relevant, as is the case in the situation depicted in (6.3.18). Together with the $\boldsymbol{r}$-independent term we then have

$$\boldsymbol{F} = m\boldsymbol{g} + \boldsymbol{F}_{\text{other}} \tag{12.1.16}$$

where

$$\boldsymbol{g} = -GM\frac{\boldsymbol{r}_0}{r_0^3} - \boldsymbol{\Omega}\times(\boldsymbol{\Omega}\times\boldsymbol{r}_0) \tag{12.1.17}$$

is the constant effective gravitational acceleration at reference position $\boldsymbol{r}_0$. It deviates slightly from $-GM\frac{\boldsymbol{r}_0}{r_0^3}$ because the term $-\boldsymbol{\Omega}\times(\boldsymbol{\Omega}\times\boldsymbol{r}_0)$ — the outward-pointing centrifugal acceleration — is about three-tenths of a percent ($\sim 3\times 10^{-3}$) of the whole. The relative size of the centrifugal acceleration is grossly exaggerated in this sketch:

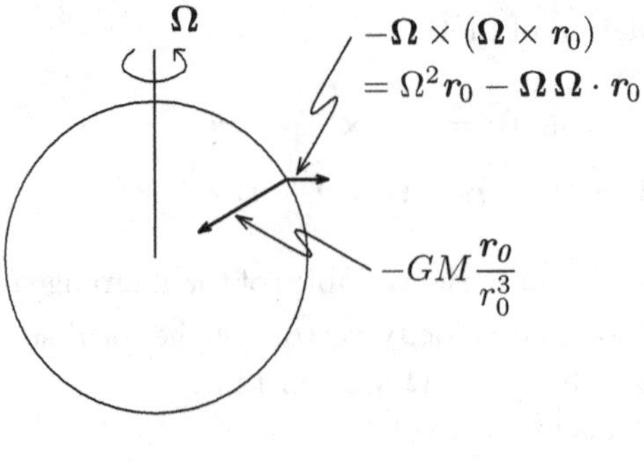

(12.1.18)

This $\boldsymbol{g}$, composed of the genuine gravitational pull and the centrifugal acceleration, is what we experience as downward acceleration. The direction of $-\boldsymbol{g}$ is "up", and the surface of calm ocean water (no waves) is perpendicular to $\boldsymbol{g}$, not to $-GM\frac{\boldsymbol{r}_0}{r_0^3} \propto \boldsymbol{r}_0$.

In summary, the equation of motion for $\boldsymbol{r}(t)$ that refers to an earth-bound frame is

$$m\dot{\boldsymbol{v}} = -2m\boldsymbol{\Omega} \times \boldsymbol{v} - m\boldsymbol{\Omega} \times (\boldsymbol{\Omega} \times \boldsymbol{r}) + m\boldsymbol{g} + \boldsymbol{F}_{\text{other}} . \qquad (12.1.19)$$

We have dealt with $m\boldsymbol{g} + \boldsymbol{F}_{\text{other}}$ in many examples, and now want to understand the effects of the two additional terms,

$$\begin{aligned} &\text{the velocity-dependent Coriolis* force } -2m\boldsymbol{\Omega} \times \boldsymbol{v} \\ &\text{and the conservative centrifugal force } -m\boldsymbol{\Omega} \times (\boldsymbol{\Omega} \times \boldsymbol{r}) . \end{aligned} \qquad (12.1.20)$$

The Coriolis force is reminiscent of the Lorentz force by the magnetic field on a charged particle; it is perpendicular to $\boldsymbol{v}$ and gives rise to a deflection without, however, transferring energy to the moving mass. On the northern hemisphere, $\boldsymbol{\Omega}$ has a vertical-up component so that objects moving horizontally (trains on a track, cruising airplanes) are deflected to the right. The deflection is to the left on the southern hemisphere where $\boldsymbol{\Omega}$ has a vertical-down component, and there is no deflection at the equator where $\boldsymbol{\Omega}$ is horizontal and points north. In Singapore, the Coriolis force affects falling objects, for example, which are deflected to the east.

The deflection by the Coriolis force is important on the large scale of weather phenomena such as the clockwise and counter clockwise motion of big air masses north or south of the equator respectively:

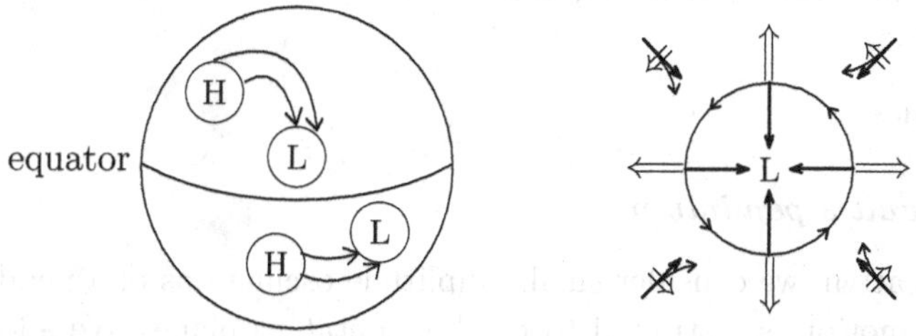

(12.1.21)

Air moves from high-pressure zones (H) to low-pressure zones (L) with the path bent clockwise on the north and counter clockwise on the south, so that we get the picture on the left for big air masses that move on global distances. On the smaller scale of a cyclone (Pacific typhoon or Atlantic hurricane), the northern-hemisphere situation is sketched on the right: The pressure gradient gives a force toward the center (single arrows $\longrightarrow$), while the Coriolis force (double arrows $\Rightarrow$, perpendicular to the velocity) deflects to the right. Far away, the net effect is a deflection from the direct path

*Gaspard-Gustave DE CORIOLIS (1792–1843)

to the center; near by, the air masses move in circles around the center counter clockwise, with the Coriolis force pointing away from the center and partially compensating for the pressure gradient, so that a net centripetal force results.

What about tornados? They have a much smaller extension, and the Coriolis force is not so important for them. Although most tornados (in the north) rotate counter clockwise, clockwise rotating tornados occur also.

A rule of thumb is provided by the *Rossby* number*. At latitude λ ($\lambda = \frac{1}{2}\pi$: north pole, $\lambda = 0$: equator, $\lambda = -\frac{1}{2}\pi$: south pole), the Rossby number is the ratio of a characteristic velocity and twice the product of $\Omega \sin\lambda$ and a characteristic length. Tropical cyclones have small Rossby numbers (Coriolis force rather important) whereas tornados have large Rossby numbers (Coriolis force not so important). Very close to the equator, we have $\sin\lambda \cong 0$ and all Rossby numbers are large, which is one of the reasons why cyclones form at some distance from, but not at, the equator.

Whereas the Coriolis force has these quite noticeable consequences, the centrifugal force is usually negligibly small because it involves two factors of the small angular velocity Ω of the rotating earth. For $r = 1\,\mathrm{m}$, we have

$$\Omega^2 r = \left(\frac{2\pi}{86\,164}\right)^2 \frac{\mathrm{m}}{\mathrm{s}^2} = 5.3 \times 10^{-9}\,\frac{\mathrm{m}}{\mathrm{s}^2}\,, \qquad (12.1.22)$$

which is really tiny compared with $g = 9.8\,\mathrm{m/s}^2$.

12.2 Examples

12.2.1 *Foucault's pendulum*

As a first application, we consider small-amplitude oscillations of a pendulum, for which motion is restricted to the horizontal xy plane. We adopt the conventions of (12.1.4), and use a coordinate system where $\boldsymbol{e}_x$ points east, $\boldsymbol{e}_y$ points north, and $\boldsymbol{e}_z$ points up, so that we have $\boldsymbol{g} = -g\boldsymbol{e}_z$ for the gravitational acceleration. For a pendulum with effective length a (point mass m at the end of a string of length a, thin rod of length $\frac{3}{2}a$, or a physical pendulum of some kind), we have

$$\boldsymbol{F}_{\mathrm{other}} = -m\frac{g}{a}(x\boldsymbol{e}_x + y\boldsymbol{e}_y) = -m\omega_0^2(x\boldsymbol{e}_x + y\boldsymbol{e}_y)\,, \qquad (12.2.1)$$

the harmonic restoring force of the small-amplitude oscillations, as obtained from the potential energy in the Lagrange function (9.3.16), for instance.

*Carl-Gustav Arvid ROSSBY (1898–1957)

At latitude λ, the angular velocity vector of the earth is

$$\boldsymbol{\Omega} = \Omega(\boldsymbol{e}_y \cos\lambda + \boldsymbol{e}_z \sin\lambda)\,; \tag{12.2.2}$$

at the poles it only has the vertical z component, at the equator there is only the horizontal north component. For the small-amplitude motion of the pendulum we need the horizontal components of

$$\begin{aligned} -2m\boldsymbol{\Omega} \times \boldsymbol{v} &= -2m\Omega(\boldsymbol{e}_y \cos\lambda + \boldsymbol{e}_z \sin\lambda) \times (\boldsymbol{e}_x \dot{x} + \boldsymbol{e}_y \dot{y}) \\ &= -2m\Omega(-\boldsymbol{e}_z \dot{x}\cos\lambda + \boldsymbol{e}_y \dot{x}\sin\lambda - \boldsymbol{e}_x \dot{y}\sin\lambda)\,, \end{aligned} \tag{12.2.3}$$

which are

$$2m\Omega\sin\lambda\,(\boldsymbol{e}_x\dot{y} - \boldsymbol{e}_y\dot{x}) = 2m\Omega_\lambda(\boldsymbol{e}_x\dot{y} - \boldsymbol{e}_y\dot{x})\,, \tag{12.2.4}$$

where $\Omega_\lambda = \Omega\sin\lambda = \boldsymbol{e}_z \cdot \boldsymbol{\Omega}$ is the latitude-dependent vertical component of $\boldsymbol{\Omega}$.

The equation of motion is then

$$m(\boldsymbol{e}_x\ddot{x} + \boldsymbol{e}_y\ddot{y}) = 2m\Omega_\lambda(\boldsymbol{e}_x\dot{y} - \boldsymbol{e}_y\dot{x}) - m\omega_0^2(x\boldsymbol{e}_x + y\boldsymbol{e}_y)\,, \tag{12.2.5}$$

which is the pair

$$\begin{aligned} \ddot{x} &= 2\Omega_\lambda\dot{y} - \omega_0^2 x\,, \\ \ddot{y} &= -2\Omega_\lambda\dot{x} - \omega_0^2 y \end{aligned} \tag{12.2.6}$$

of coupled, linear, second-order differential equations. We write them more compactly as one equation for the complex function $u(t) = x(t) + \mathrm{i}y(t)$,

$$\ddot{u} = -2\mathrm{i}\Omega_\lambda\dot{u} - \omega_0^2 u\,. \tag{12.2.7}$$

This has the structure of the equation of motion (2.2.59) for a harmonic oscillator with Newtonian friction, with the friction constant γ replaced by the imaginary $2\mathrm{i}\Omega_\lambda$.

Recalling the lessons of Section 2.2.5, we expect that $\mathrm{e}^{\mathrm{i}\Omega_\lambda t}u(t)$ obeys the equation of motion of a harmonic oscillator with angular frequency $\omega_\lambda = \sqrt{\omega_0^2 + \Omega_\lambda^2}$. This is indeed the case,

$$\begin{aligned} \left(\frac{\mathrm{d}}{\mathrm{d}t}\right)^2 \left[\mathrm{e}^{\mathrm{i}\Omega_\lambda t}u(t)\right] &= (\ddot{u} + 2\mathrm{i}\Omega_\lambda\dot{u} - \Omega_\lambda^2 u)\,\mathrm{e}^{\mathrm{i}\Omega_\lambda t} \\ &= -(\omega_0^2 + \Omega_\lambda^2)\left[\mathrm{e}^{\mathrm{i}\Omega_\lambda t}u(t)\right] \\ &= -\omega_\lambda^2\left[\mathrm{e}^{\mathrm{i}\Omega_\lambda t}u(t)\right]\,, \end{aligned} \tag{12.2.8}$$

so that

$$u(t) = e^{-i\Omega_\lambda t}\left[A\cos(\omega_\lambda t) + \frac{B}{\omega_\lambda}\sin(\omega_\lambda t)\right], \tag{12.2.9}$$

where the values of

$$A = e^{i\Omega_\lambda t}u(t)\Big|_{t=0} = u(0) \tag{12.2.10}$$

and

$$B = \frac{d}{dt}\left[e^{i\Omega_\lambda t}u(t)\right]\Big|_{t=0} = \dot{u}(0) + i\Omega_\lambda u(0) \tag{12.2.11}$$

are determined by the initial coordinate values $u(0) = x_0 + iy_0$ and the initial values of their time derivatives, $\dot{u}(0) = \dot{x}_0 + i\dot{y}_0$.

We arrived at (12.2.9) by neglecting the tiny centrifugal force, but then it is only consistent to also neglect the second-order difference between ω_0 and ω_λ,

$$\begin{aligned}\omega_\lambda = \sqrt{\omega_0^2 + \Omega_\lambda^2} &= \omega_0\left[1 + \frac{1}{2}\left(\frac{\Omega_\lambda}{\omega_0}\right)^2 + \cdots\right] \\ &= \omega_0 + (\text{second-order terms}),\end{aligned} \tag{12.2.12}$$

which means that

$$\begin{aligned}x(t) + iy(t) = e^{-i\Omega_\lambda t}\bigg[&(x_0 + iy_0)\left(\cos(\omega_0 t) + i\frac{\Omega_\lambda}{\omega_0}\sin(\omega_0 t)\right) \\ &+ (\dot{x}_0 + i\dot{y}_0)\frac{\sin(\omega_0 t)}{\omega_0}\bigg]\end{aligned} \tag{12.2.13}$$

is sufficiently accurate. Let us read this as a two-factor solution. If we only had the ω_0 part, it would just be a harmonic oscillation with period $\frac{2\pi}{\omega_0}$, which is of the order of seconds. The prefactor alone would give us

$$x(t) + iy(t) = e^{-i\Omega_\lambda t}\left[x(0) + iy(0)\right], \tag{12.2.14}$$

or

$$\begin{pmatrix} x(t) \\ y(t) \end{pmatrix} = \begin{pmatrix} \cos(\Omega_\lambda t) & \sin(\Omega_\lambda t) \\ -\sin(\Omega_\lambda t) & \cos(\Omega_\lambda t) \end{pmatrix}\begin{pmatrix} x(0) \\ y(0) \end{pmatrix}. \tag{12.2.15}$$

This is a rotation about the vertical z axis with period $\left|\frac{2\pi}{\Omega_\lambda}\right| = \frac{2\pi}{\Omega|\sin\lambda|}$, clockwise for $\sin\lambda > 0$ but counter clockwise for $\sin\lambda < 0$, which is a day at the poles, longer away from the poles, many days at low latitude.

The motion as a whole is, therefore, that of a swinging pendulum with period $2\pi/\omega_0 = 2\pi\sqrt{a/g}$, which precesses with period (1 sidereal day) $\times$ $|\sin\lambda|$. This precession can actually be observed rather easily if you are not too close to the equator, usually by a long pendulum ($a \sim 20\,\mathrm{m}$) for which the precession is noticeable over the stretch of a few hours. The *Deutsches Museum* in Munich, for example, has such a *Foucault* Pendulum*, named after the French physicist who came up with this clever laboratory demonstration of the earth rotation in 1851.

In Singapore, or other places near the equator, $\boldsymbol{\Omega}$ is pointing north with a negligible vertical component. The Coriolis force then vanishes if the velocity is north or south, and is vertical for east-west travel. It increases your weight when traveling from east to west, and decreases your weight in the opposite direction — the so-called *Eötvös effect.* At $20\,\mathrm{m/s} = 72\,\mathrm{km/h}$, the acceleration is

$$2 \times \frac{2\pi}{86\,164} \times 20\,\frac{\mathrm{m}}{\mathrm{s}^2} = 3 \times 10^{-3}\,\frac{\mathrm{m}}{\mathrm{s}^2}\,, \tag{12.2.16}$$

so that a 50 kg person experiences a weight difference between westward and eastward travel that corresponds to an effective mass difference of about 30 g. This is surely measurable but it is hardly enough for a weight-loss business that exploits the Coriolis force.

12.2.2 *Deflection of a falling mass*

As another example, we consider a point mass m, released from rest at height h above ground and falling until it hits the ground under the combined influence of the gravitational pull (including the centrifugal-force correction) as well as the Coriolis force and the centrifugal force. When ignoring frictional forces, the equation of motion is (12.1.19) with $\boldsymbol{F}_{\text{other}} = 0$,

$$\dot{\boldsymbol{v}}(t) = \boldsymbol{g} - 2\boldsymbol{\Omega} \times \boldsymbol{v}(t) - \boldsymbol{\Omega} \times \big(\boldsymbol{\Omega} \times \boldsymbol{r}(t)\big)\,, \tag{12.2.17}$$

where the Coriolis force is a correction of first-order in $\boldsymbol{\Omega}$ and the centrifugal force is a second-order correction. Here, too, we have $\boldsymbol{g} = -g\boldsymbol{e}_z$ and $\boldsymbol{\Omega}$ as given in (12.2.2), and the initial conditions are those of a mass that is dropped from height h, that is: $\boldsymbol{r}(0) = h\boldsymbol{e}_z$ and $\dot{\boldsymbol{r}}(0) = \boldsymbol{v}(0) = 0$.

The zeroth-order solution

$$\boldsymbol{r}_0(t) = \boldsymbol{r}(0) + \frac{1}{2}\boldsymbol{g}t^2\,, \quad \boldsymbol{v}_0(t) = \boldsymbol{g}t \tag{12.2.18}$$

---

*Jean Bernard Léon FOUCAULT (1819–1868)

ignores the Coriolis force and the centrifugal force altogether. We take them into account by a systematic procedure that begins by first converting the differential equation with its initial conditions into an equivalent integral equation.

We have the velocity

$$\begin{aligned} \boldsymbol{v}(t) &= \int_0^t \mathrm{d}t'\, \dot{\boldsymbol{v}}(t') \\ &= \boldsymbol{g}t - \int_0^t \mathrm{d}t' \left[2\boldsymbol{\Omega} \times \boldsymbol{v}(t') + \boldsymbol{\Omega} \times \left(\boldsymbol{\Omega} \times \boldsymbol{r}(t')\right)\right], \end{aligned} \tag{12.2.19}$$

where $\boldsymbol{g}t$ is the zeroth-order solution and the integral accounts for the $\boldsymbol{\Omega}$-dependent corrections. Another integration gives

$$\begin{aligned} \boldsymbol{r}(t) &= \boldsymbol{r}(0) + \int_0^t \mathrm{d}t''\, \boldsymbol{v}(t'') \\ &= \boldsymbol{r}(0) + \frac{1}{2}\boldsymbol{g}t^2 - \int_0^t \mathrm{d}t'' \int_0^{t''} \mathrm{d}t' \left[2\boldsymbol{\Omega} \times \boldsymbol{v}(t') + \boldsymbol{\Omega} \times \left(\boldsymbol{\Omega} \times \boldsymbol{r}(t')\right)\right] \\ &= \boldsymbol{r}_0(t) - \int_0^t \mathrm{d}t' \int_{t'}^t \mathrm{d}t'' \left[2\boldsymbol{\Omega} \times \boldsymbol{v}(t') + \boldsymbol{\Omega} \times \left(\boldsymbol{\Omega} \times \boldsymbol{r}(t')\right)\right], \end{aligned} \tag{12.2.20}$$

where the two double integrations are identical because both cover the same range of $0 \le t' \le t'' \le t$, only that in one version we integrate over t' for fixed t'', followed by the t'' integration, whereas in the other version we integrate over t'' for fixed t', followed by the t' integration. The latter form has the advantage that the t'' integral can be done, with the outcome

$$\boldsymbol{r}(t) = \boldsymbol{r}_0(t) - \int_0^t \mathrm{d}t'\,(t-t')\left[2\boldsymbol{\Omega} \times \boldsymbol{v}(t') + \boldsymbol{\Omega} \times \left(\boldsymbol{\Omega} \times \boldsymbol{r}(t')\right)\right]. \tag{12.2.21}$$

This integral equation incorporates the initial conditions and is fully equivalent to the differential equation plus the initial conditions.

We use the integral equation for an iteration procedure to generate a hierarchy of approximate solutions that are of first, second, third, ... order in $\boldsymbol{\Omega}$. The nth-order solution is obtained from the $(n-1)$th-order solution and the $(n-2)$th-order solution in accordance with

$$\boldsymbol{r}_n(t) = \boldsymbol{r}_0(t) - \int_0^t \mathrm{d}t'\,(t-t')\big[2\boldsymbol{\Omega}\times\boldsymbol{v}_{n-1}(t') + \boldsymbol{\Omega}\times\big(\boldsymbol{\Omega}\times\boldsymbol{r}_{n-2}(t')\big)\big]\,. \quad (12.2.22)$$

An iteration of this kind is particularly useful if one only needs a low-order approximation.

Starting from the zeroth-order solution (12.2.18) we obtain the first-order solution

$$\boldsymbol{r}_1(t) = \boldsymbol{r}_0(t) - \int_0^t \mathrm{d}t'\,(t-t')\,2\boldsymbol{\Omega}\times\boldsymbol{v}_0(t') = \boldsymbol{r}_0(t) - \frac{1}{3}\boldsymbol{\Omega}\times\boldsymbol{g}t^3\,,$$

$$\boldsymbol{v}_1(t) = \boldsymbol{v}_0(t) - \int_0^t \mathrm{d}t'\,2\boldsymbol{\Omega}\times\boldsymbol{v}_0(t') = \boldsymbol{v}_0(t) - \boldsymbol{\Omega}\times\boldsymbol{g}t^2\,, \quad (12.2.23)$$

and then the second-order solution

$$\begin{aligned}\boldsymbol{r}_2(t) &= \boldsymbol{r}_0(t) - \int_0^t \mathrm{d}t'\,(t-t')\big[2\boldsymbol{\Omega}\times\boldsymbol{v}_1(t') + \boldsymbol{\Omega}\times\big(\boldsymbol{\Omega}\times\boldsymbol{r}_0(t')\big)\big]\\ &= \boldsymbol{r}_1(t) - \int_0^t \mathrm{d}t'\,(t-t')\big[2\boldsymbol{\Omega}\times\big(\boldsymbol{v}_1(t') - \boldsymbol{v}_0(t')\big) + \boldsymbol{\Omega}\times\big(\boldsymbol{\Omega}\times\boldsymbol{r}_0(t')\big)\big]\,.\end{aligned} \quad (12.2.24)$$

With $\boldsymbol{v}_1(t') - \boldsymbol{v}_0(t') = -\boldsymbol{\Omega}\times\boldsymbol{g}t'^2$ and $\boldsymbol{r}_0(t') = \boldsymbol{r}(0) + \frac{1}{2}\boldsymbol{g}t'^2$ this is

$$\begin{aligned}\boldsymbol{r}_2(t) &= \boldsymbol{r}_1(t) - \int_0^t \mathrm{d}t'\,(t-t')\Big[-\frac{3}{2}\boldsymbol{\Omega}\times(\boldsymbol{\Omega}\times\boldsymbol{g})t'^2 + \boldsymbol{\Omega}\times\big(\boldsymbol{\Omega}\times\boldsymbol{r}(0)\big)\Big]\\ &= \boldsymbol{r}(0) + \frac{1}{2}\boldsymbol{g}t^2 - \frac{1}{3}\boldsymbol{\Omega}\times\boldsymbol{g}t^3\\ &\quad + \frac{1}{8}\boldsymbol{\Omega}\times(\boldsymbol{\Omega}\times\boldsymbol{g})t^4 - \frac{1}{2}\boldsymbol{\Omega}\times\big(\boldsymbol{\Omega}\times\boldsymbol{r}(0)\big)t^2\,.\end{aligned} \quad (12.2.25)$$

Since the terms of second order are quite small already, it is not worth the trouble to go beyond this level of accuracy and calculate approximations of third or higher order.

For $\boldsymbol{r}(0) = h\boldsymbol{e}_z$, $\boldsymbol{g} = -g\boldsymbol{e}_z$, and $\boldsymbol{\Omega}$ as in (12.2.2), the various vector products in (12.2.25) are

$$\begin{aligned} \boldsymbol{\Omega} \times \boldsymbol{g} &= -g\boldsymbol{\Omega} \times \boldsymbol{e}_z \\ \text{and} \quad \boldsymbol{\Omega} \times (\boldsymbol{\Omega} \times \boldsymbol{g}) &= -g\boldsymbol{\Omega} \times (\boldsymbol{\Omega} \times \boldsymbol{e}_z)\,, \\ \boldsymbol{\Omega} \times \bigl(\boldsymbol{\Omega} \times \boldsymbol{r}(0)\bigr) &= h\boldsymbol{\Omega} \times (\boldsymbol{\Omega} \times \boldsymbol{e}_z)\,, \end{aligned} \tag{12.2.26}$$

where

$$\begin{aligned} \boldsymbol{\Omega} \times \boldsymbol{e}_z &= \boldsymbol{e}_x \Omega \cos\lambda\,, \\ \boldsymbol{\Omega} \times (\boldsymbol{\Omega} \times \boldsymbol{e}_z) &= \boldsymbol{e}_y \Omega^2 \cos\lambda \sin\lambda - \boldsymbol{e}_z (\Omega \cos\lambda)^2\,. \end{aligned} \tag{12.2.27}$$

In summary, then, we have the second-order approximation

$$\boldsymbol{r}(t) \cong \boldsymbol{e}_x x(t) + \boldsymbol{e}_y y(t) + \boldsymbol{e}_z z(t) \tag{12.2.28}$$

with

$$\begin{aligned} x(t) &= \frac{1}{3}\Omega g t^3 \cos\lambda\,, \\ y(t) &= -\frac{1}{16}(\Omega t)^2 (4h + gt^2) \sin(2\lambda)\,, \\ z(t) &= h - \frac{1}{2}gt^2 + \frac{1}{8}(\Omega t \cos\lambda)^2 (4h + gt^2)\,. \end{aligned} \tag{12.2.29}$$

Here, $z(t)$ is the height above ground at time t, so that the mass hits the ground at time T that is given by $z(T) = 0$. But rather than hitting at the plum-line point below the position of release at $(x, y, z) = (0, 0, h)$, there is a deflection east by $x(T)$ and a deflection by the distance $|y(T)|$ toward the equator, south for $\lambda > 0$ and north for $\lambda < 0$.

The duration T of the fall is

$$T = \sqrt{\frac{2h}{g}} + \text{corrections of order } \Omega^2\,, \tag{12.2.30}$$

and the corrections do not matter for the east deflection

$$x(T) = \frac{2}{3}(\Omega T \cos\lambda)\frac{1}{2}gT^2 = \frac{2\Omega h}{3}\sqrt{\frac{2h}{g}} \cos\lambda \tag{12.2.31}$$

and for the south deflection

$$-y(T) = \frac{1}{16}(\Omega T)^2 (4h + gT^2) \sin(2\lambda) = \frac{3(\Omega h)^2}{4g} \sin(2\lambda)\,. \tag{12.2.32}$$

To have an idea of the size of these deflections, let us evaluate the expressions at latitude $\lambda = 45°$, where $\cos\lambda = 1/\sqrt{2}$ and $\sin(2\lambda) = 1$, and height $h = 200\,\text{m}$. For $g = 9.8\,\text{m/s}^2$, we obtain

$$x(T) = 44\,\text{mm} \quad \text{and} \quad -y(T) = 0.016\,\text{mm}\,. \tag{12.2.33}$$

Such experiments have actually been conducted, with objects dropped into deep mine shafts. The observed east deflections are smaller than this calculated value, with such reduction originating in the friction of the air drag. But it is hard to believe the reports of observing also the tiny south deflection, in particular since the objects themselves are much larger than the few micrometers in question.

12.2.3 *Gyrocompass*

Our final example is the gyrocompass, first considered by Foucault and eventually turned into a practical device by Anschütz-Kaempfe.* A gyrocompass is a symmetric top, such as the one in Section 11.3.3, but mounted differently: The top is supported at its center-of-mass and the figure axis is always horizontal. With the center-of-mass at a fixed position, there is no potential-energy difference to be taken into account, as the weight of the top does not give rise to a torque. But in the earth-bound laboratory, there are Coriolis forces on the rotating mass elements of the top, and the vertical component of the resulting torque can change the orientation of the figure axis in the horizontal plane to which it is confined.

We return to (12.1.6), regard $\boldsymbol{r}_0$ as the vector from the center of the earth to the center-of-mass of the spinning top, and note that the local velocity $\boldsymbol{v}$ equals $\boldsymbol{\omega} \times \boldsymbol{r}$, where $\boldsymbol{\omega}$ is the angular velocity vector of the top with respect to the laboratory system. Accordingly, the kinetic energy is

$$\begin{aligned} E_{\text{kin}} &= \frac{1}{2}\int (\mathrm{d}\boldsymbol{r})\,\rho_{\text{CM}}(\boldsymbol{r})\big(\boldsymbol{\Omega} \times \boldsymbol{r}_0 + (\boldsymbol{\omega} + \boldsymbol{\Omega}) \times \boldsymbol{r}\big)^2 \\ &= \frac{1}{2}M(\boldsymbol{\Omega} \times \boldsymbol{r}_0)^2 + \frac{1}{2}(\boldsymbol{\omega} + \boldsymbol{\Omega}) \cdot \mathsf{I} \cdot (\boldsymbol{\omega} + \boldsymbol{\Omega})\,, \end{aligned} \tag{12.2.34}$$

the analog of (11.1.21) with the center-of-mass velocity $\boldsymbol{V} = \boldsymbol{\Omega} \times \boldsymbol{r}_0$ and the total angular momentum vector $\boldsymbol{\omega} + \boldsymbol{\Omega}$. The kinetic energy of the center-of-mass motion is constant here, so we do not need to account for it in the Lagrange function

$$L = \frac{1}{2}(\boldsymbol{\omega} + \boldsymbol{\Omega}) \cdot \mathsf{I} \cdot (\boldsymbol{\omega} + \boldsymbol{\Omega})\,. \tag{12.2.35}$$

*Hermann Franz Joseph Hubertus Maria ANSCHÜTZ-KAEMPFE (1872–1931)

It is expedient to use cylindrical coordinates s, φ, z for the top with the figure axis aligned with the unit vector $\boldsymbol{e}_s$, so that

$$\mathsf{I} = I_1(\mathbf{1} - \boldsymbol{e}_s\,\boldsymbol{e}_s) + I_2\boldsymbol{e}_s\,\boldsymbol{e}_s \tag{12.2.36}$$

is the appropriate version of (11.3.34) and

$$\boldsymbol{\Omega} = (\boldsymbol{e}_s \sin\varphi + \boldsymbol{e}_\varphi \cos\varphi)\Omega\cos\lambda + \boldsymbol{e}_z\Omega\sin\lambda \tag{12.2.37}$$

that of (12.2.2), and we have

$$\boldsymbol{\omega} = \boldsymbol{e}_s\dot{\psi} + \boldsymbol{e}_z\dot{\varphi} \tag{12.2.38}$$

for the angular velocity vector. Owing to the confinement to the horizontal plane, the component $\propto \dot{\vartheta}$ in (11.3.37) vanishes ($\dot{\vartheta} = 0$) here.

Since the Lagrange function

$$\begin{aligned} L &= \frac{1}{2}I_1\big[\boldsymbol{e}_s \times (\boldsymbol{\omega} + \boldsymbol{\Omega})\big]^2 + \frac{1}{2}I_2\big[\boldsymbol{e}_s \cdot (\boldsymbol{\omega} + \boldsymbol{\Omega})\big]^2 \\ &= \frac{1}{2}I_1(\Omega\cos\lambda\cos\varphi)^2 + \frac{1}{2}I_1(\dot{\varphi} + \Omega\sin\lambda)^2 \\ &\quad + \frac{1}{2}I_2(\dot{\psi} + \Omega\cos\lambda\sin\varphi)^2 \end{aligned} \tag{12.2.39}$$

depends on φ, $\dot{\varphi}$, and $\dot{\psi}$ but not on ψ, the coordinate ψ is cyclic and its canonical momentum is a constant of motion,

$$p_\psi = \frac{\partial L}{\partial\dot{\psi}} = I_2(\dot{\psi} + \Omega\cos\lambda\sin\varphi) = \text{constant}, \tag{12.2.40}$$

and the equation of motion for $\varphi(t)$ reads

$$\frac{\mathrm{d}}{\mathrm{d}t}\frac{\partial L}{\partial\dot{\varphi}} = \frac{\partial L}{\partial\varphi}: \quad I_1\ddot{\varphi} = -I_1(\Omega\cos\lambda)^2\sin\varphi\,\cos\varphi + p_\psi\Omega\cos\lambda\cos\varphi \tag{12.2.41}$$

or

$$I_1\ddot{\varphi} = -\frac{\partial}{\partial\varphi}V_{\text{eff}}(\varphi) = -V'_{\text{eff}}(\varphi) \tag{12.2.42}$$

with the effective potential energy

$$V_{\text{eff}}(\varphi) = \frac{1}{2I_1}(I_1\Omega\cos\lambda\sin\varphi - p_\psi)^2\,. \tag{12.2.43}$$

The φ values for which the torque $-V'_{\text{eff}}(\varphi)$ vanishes are those with $\cos\varphi = 0$ or $p_\psi = I_1\Omega\cos\lambda\sin\varphi$. Now, the gyrocompass is prepared with a rotation about the figure axis so fast that $|p_\psi| \gg I_1\Omega$ and, therefore, $\cos\varphi = 0$ is the only option.

Of the two equilibrium orientations at $\varphi = \frac{1}{2}\pi$ and $\varphi = -\frac{1}{2}\pi$, one must be the maximum of the effective potential energy, the other the minimum. In view of

$$V''_{\text{eff}}(\varphi = \pm\tfrac{1}{2}\pi) = \pm p_\psi\Omega\cos\lambda - I_1(\Omega\cos\lambda)^2 \cong \pm p_\psi\Omega\cos\lambda\,, \qquad (12.2.44)$$

the minimum is at $\varphi = \frac{1}{2}\pi$ if $p_\psi \gg I_1\Omega$ and at $\varphi = -\frac{1}{2}\pi$ if $-p_\psi \gg I_1\Omega$. Whether we have the stable equilibrium at $\varphi = \frac{1}{2}\pi$ or at $\varphi = -\frac{1}{2}\pi$ makes no difference, however, as in both cases the figure axis is aligned with the y axis of the local coordinate system — aligned with the south-north direction of the horizontal component of $\boldsymbol{\Omega}$. So, let us assume that $p_\psi \gg I_1\Omega$ is the case and write $\varphi = \frac{1}{2}\pi + \alpha$. Then, $\sin\varphi = \cos\alpha \cong 1$ and $\cos\varphi = -\sin\alpha \cong -\alpha$ apply for small-amplitude oscillations, for which

$$I_1\ddot{\alpha} = -\Big(p_\psi\Omega\cos\lambda - I_1(\Omega\cos\lambda)^2\Big)\alpha \qquad (12.2.45)$$

implies harmonic oscillations around $\alpha = 0$ with the period

$$T = 2\pi\sqrt{\frac{I_1}{p_\psi\Omega\cos\lambda - I_1(\Omega\cos\lambda)^2}} \cong 2\pi\sqrt{\frac{I_1}{p_\psi\Omega\cos\lambda}}\,. \qquad (12.2.46)$$

The unavoidable ubiquitous friction damps these oscillations (the critical damping of Section 2.2.5 could be engineered) and ensures that the figure axis points north after the oscillations have ceased. This, of course, is the purpose of a compass.

Exercises with Hints

Chapter 1

1 Three vectors $\boldsymbol{a}$, $\boldsymbol{b}$, $\boldsymbol{c}$ are linearly dependent if there are coefficients α, β, γ, not all equal to zero, such that $\alpha\boldsymbol{a}+\beta\boldsymbol{b}+\gamma\boldsymbol{c}=0$. Show that $\boldsymbol{a}\cdot(\boldsymbol{b}\times\boldsymbol{c})=0$ if $\boldsymbol{a}$, $\boldsymbol{b}$, $\boldsymbol{c}$ are linearly dependent.

2 Vectors $\boldsymbol{a}$, $\boldsymbol{b}$, and $\boldsymbol{c}=\boldsymbol{a}\times\boldsymbol{b}$ have cartesian coordinates a_j, b_k, and c_l, respectively, with each of the subscripts j,k,l taking values x,y,z. Verify that

$$\boldsymbol{a}\cdot\boldsymbol{b}=\sum_{j,k}\delta_{jk}a_jb_k \quad \text{and} \quad c_l=\sum_{j,k}\epsilon_{jkl}a_jb_k\,,$$

and show that

$$\sum_{l}\epsilon_{jkl}\epsilon_{j'k'l}=\delta_{jj'}\delta_{kk'}-\delta_{jk'}\delta_{kj'}\,.$$

Which statement in Section 1.1 is equivalent to this identity? What do you get for $\sum_{k,l}\epsilon_{jkl}\epsilon_{j'kl}$?

3 Four vectors $\boldsymbol{a}$, $\boldsymbol{b}$, $\boldsymbol{c}$, $\boldsymbol{d}$ are such that

$$\boldsymbol{e}=(\boldsymbol{a}\times\boldsymbol{b})\times(\boldsymbol{c}\times\boldsymbol{d})\neq 0\,.$$

How is $\boldsymbol{e}$ related to the two planes spanned by $\boldsymbol{a}$ and $\boldsymbol{b}$ and by $\boldsymbol{c}$ and $\boldsymbol{d}$?

4 The three vectors $\boldsymbol{a}$, $\boldsymbol{b}$, $\boldsymbol{c}$ are linearly independent, but not necessarily pairwise orthogonal or of unit length. Express the coefficients α, β, γ in

$$\boldsymbol{r}=\alpha\boldsymbol{a}+\beta\boldsymbol{b}+\gamma\boldsymbol{c}$$

in terms of $\boldsymbol{a}$, $\boldsymbol{b}$, $\boldsymbol{c}$, and $\boldsymbol{r}$.

5 Show that the Jacobi identity (1.1.50) is obeyed by any three vectors $\boldsymbol{a}$, $\boldsymbol{b}$, and $\boldsymbol{c}$. Verify the identity for the three vectors with the cartesian coordinates

$$\boldsymbol{a} \mathrel{\widehat{=}} \begin{pmatrix} 1 \\ 0 \\ 2 \end{pmatrix}, \quad \boldsymbol{b} \mathrel{\widehat{=}} \begin{pmatrix} 0 \\ 2 \\ -1 \end{pmatrix}, \quad \boldsymbol{c} \mathrel{\widehat{=}} \begin{pmatrix} 3 \\ -1 \\ 0 \end{pmatrix}$$

by explicitly computing all vector products.

6 Four vectors $\boldsymbol{a}_1$, $\boldsymbol{a}_2$, $\boldsymbol{a}_3$, $\boldsymbol{a}_4$ are such that $\boldsymbol{a}_1 + \boldsymbol{a}_2 + \boldsymbol{a}_3 + \boldsymbol{a}_4 = 0$ and

$$\boldsymbol{a}_j \cdot \boldsymbol{a}_k = \begin{cases} 1 & \text{if } j = k\,, \\ -\frac{1}{3} & \text{if } j \neq k\,. \end{cases}$$

What is the geometrical situation? Determine the values of $\boldsymbol{a}_1 \cdot (\boldsymbol{a}_2 \times \boldsymbol{a}_3)$, $\boldsymbol{a}_2 \cdot (\boldsymbol{a}_3 \times \boldsymbol{a}_4)$, $\boldsymbol{a}_3 \cdot (\boldsymbol{a}_4 \times \boldsymbol{a}_1)$, and $\boldsymbol{a}_4 \cdot (\boldsymbol{a}_1 \times \boldsymbol{a}_2)$.

7 Find the solution of the coupled differential equations

$$\frac{\mathrm{d}}{\mathrm{d}t} x(t) = \gamma_1 y(t)\,, \quad \frac{\mathrm{d}}{\mathrm{d}t} y(t) = \gamma_2 x(t)$$

with constant γ_1 and γ_2 and $\gamma_1\gamma_2 > 0$ to the initial values $x(t = 0) = x_0$, $y(t = 0) = y_0$.

8 Consider two different infinitesimal rotations by $\boldsymbol{\delta\phi}_1$ and $\boldsymbol{\delta\phi}_2$. What is the result of first rotating by $\boldsymbol{\delta\phi}_1$, then by $\boldsymbol{\delta\phi}_2$, next by $-\boldsymbol{\delta\phi}_1$, and finally by $-\boldsymbol{\delta\phi}_2$? Note that $\boldsymbol{\delta\phi}_1$ and $\boldsymbol{\delta\phi}_2$ are *independently* infinitesimal, so that their product is not of second order, while their squares are.

9 The net effect of a sequence of rotations is another rotation. Why? — Consider the following sequence of rotations around the x axis and the z axis: First around the z axis by 90°, then around the x axis by 90°, then around the z axis by −90°, finally around the x axis by −90°. The net effect is a rotation around which axis by which angle?

10 Consider the following sequence of three rotations: First around the x axis by 90°, then around the y axis by 180°, finally around the z axis by 90°. The net effect is a rotation around which axis by which angle?

11 The mirror image of $\boldsymbol{r}$ with respect to the plane through $\boldsymbol{r} = 0$ that is perpendicular to unit vector $\boldsymbol{e}$ is $\boldsymbol{r} - 2\boldsymbol{e}\,\boldsymbol{e} \cdot \boldsymbol{r}$. If we take two successive mirror images, first for e_1 and then for e_2, the overall effect is a rotation. Show that the axis of rotation is specified by $e_1 \times e_2$. Express the angle of rotation ϕ in terms of the angle α between e_1 and e_2, that is $\cos\alpha = e_1 \cdot e_2$.

12 For vectors in spherical coordinates,

$$\boldsymbol{A} = (\boldsymbol{e}_r\ \boldsymbol{e}_\vartheta\ \boldsymbol{e}_\varphi)\begin{pmatrix} a_r \\ a_\vartheta \\ a_\varphi \end{pmatrix},$$

find the 3×3-matrix differential operator $\mathcal{D}$ that is needed in

$$\frac{\mathrm{d}}{\mathrm{d}t}\boldsymbol{A} = (\boldsymbol{e}_r\ \boldsymbol{e}_\vartheta\ \boldsymbol{e}_\varphi)\mathcal{D}\begin{pmatrix} a_r \\ a_\vartheta \\ a_\varphi \end{pmatrix},$$

and establish its square $\mathcal{D}^2$. Then use this for $\boldsymbol{A} = \boldsymbol{r}$ to confirm the expressions for the velocity vector in (1.1.137) and for the acceleration vector in (1.1.138).

13 Assume that the surface of the earth is a sphere with radius 6 378 km. When traveling from Singapore (1° 17′ north, 103° 50′ east) first to Munich (48° 8′ north, 11° 34′ east), then to Los Angeles (34° 03′ north, 118° 15′ west), finally back to Singapore on the shortest routes, what distances are covered between the three cities?

14 The so-called *parabolic coordinates* ξ, η, φ with $\xi \geq 0$, $\eta \geq 0$, and $0 \leq \varphi < 2\pi$ (or φ periodic) are defined by

$$\begin{aligned} x &= a\,\xi\,\eta\,\cos\varphi\,, \\ y &= a\,\xi\,\eta\,\sin\varphi\,, \\ z &= \frac{a}{2}(\xi^2 - \eta^2)\,, \end{aligned}$$

where a is a constant length. What are the two-dimensional surfaces of constant ξ, of constant η, of constant φ? What are the singularities of the parabolic coordinates?

15 Determine the local unit vectors $\boldsymbol{e}_\xi$, $\boldsymbol{e}_\eta$, $\boldsymbol{e}_\varphi$ for parabolic coordinates, and verify that they are pairwise orthogonal. In which order are they a right-handed trio of unit vectors?

16 For constant vectors $\boldsymbol{k}$, $\boldsymbol{a}$, and $\boldsymbol{\omega}$, find the gradients of

$$f_1(\boldsymbol{r}) = \boldsymbol{k}\cdot\boldsymbol{r}\,,\quad f_2(\boldsymbol{r}) = \frac{1}{|\boldsymbol{r}-\boldsymbol{a}|}\,,\quad f_3(\boldsymbol{r}) = \frac{1}{2}(\boldsymbol{\omega}\times\boldsymbol{r})^2$$

without using a coordinate representation.

17 What is the expression for the gradient in terms of the parabolic coordinates of Exercise 14? For each of the four coordinate systems — cartesian, cylindrical, spherical, and parabolic — verify that $\boldsymbol{\nabla}\cdot\boldsymbol{r} = 3$ and $\boldsymbol{\nabla}\times\boldsymbol{r} = 0$.

18 Consider a vector field $\boldsymbol{A}$ that is given in terms of cylindrical coordinates, that is: $\boldsymbol{A}(\boldsymbol{r}) = \boldsymbol{e}_s A_s(s,\varphi,z) + \boldsymbol{e}_\varphi A_\varphi(s,\varphi,z) + \boldsymbol{e}_z A_z(s,\varphi,z)$. Express the divergence $\boldsymbol{\nabla}\cdot\boldsymbol{A}$ and the curl $\boldsymbol{\nabla}\times\boldsymbol{A}$ of $\boldsymbol{A}(\boldsymbol{r})$ in terms of the component functions A_s, A_φ, and A_z. Then verify that your expressions give the right answers for $\boldsymbol{A}(\boldsymbol{r}) = \boldsymbol{e}_x x$, $\boldsymbol{A}(\boldsymbol{r}) = \boldsymbol{e}_x y - \boldsymbol{e}_y x$, and $\boldsymbol{A}(\boldsymbol{r}) = \boldsymbol{r}$.

19 Repeat Exercise 18 for spherical coordinates.

20 For the parabolic coordinates of Exercise 14, find the surface elements for the surfaces on which one of the coordinates is constant. Express the volume element $(\mathrm{d}\boldsymbol{r})$ in parabolic coordinates.

21 Consider a system of coordinates a, b, c with local unit vectors $\boldsymbol{e}_a$, $\boldsymbol{e}_b$, $\boldsymbol{e}_c$, which are right-handed in this order — that is: $(\boldsymbol{e}_a\times\boldsymbol{e}_b)\cdot\boldsymbol{e}_c > 0$ — so that

$$\mathrm{d}\boldsymbol{r} = \boldsymbol{e}_a h_a\,\mathrm{d}a + \boldsymbol{e}_b h_b\,\mathrm{d}b + \boldsymbol{e}_c h_c\,\mathrm{d}c$$

states the differential line element with h_a, h_b, $h_c > 0$. State the gradient, the surface elements, and the volume element for this coordinate system.

Chapter 2

22 With reference to (2.2.41) and (2.2.42), compare the early $(0 < t \ll \tau)$ and the late $(t \gg \tau)$ forms of

$$z(t) = g\tau^2\log\Bigl(\cosh(t/\tau)\Bigr) \quad\text{and}\quad z(t) = g\tau^2\left(\frac{t}{\tau} - 1 + \mathrm{e}^{-t/\tau}\right),$$

that is: determine first the leading correction to $z(t) \cong \frac{1}{2}gt^2$ for $t \ll \tau$ as well as the constant c in, and the leading correction to, $z(t) \cong g\tau t + c$ for $t \gg \tau$, and then comment on what you found.

23 A stone ($\equiv$ point mass m) is thrown upward from $\boldsymbol{r}(t=0) = 0$ with initial velocity $\boldsymbol{v}(t=0) = v_0\,\boldsymbol{e}_z$, $v_0 > 0$, and then moves under the influence of its weight $m\boldsymbol{g} = -mg\boldsymbol{e}_z$ and the Newtonian frictional force $-m\gamma\boldsymbol{v}$. The point mass is highest above ground at time t_1, and it is back at the initial height $z = 0$ at time t_2. Explain by qualitative arguments, why $t_2 > 2t_1$, that is: it takes longer to fall down than to fly up. Find expressions that relate t_1 and t_2 to $\gamma v_0/g$ and then use them to demonstrate that $t_2 > 2t_1$ is indeed the case.

24 A stone is thrown upward from ground level with initial velocity $\boldsymbol{v}(t=0) = v_0\,\boldsymbol{e}_z$, $v_0 > 0$, and then moves under the influence of its weight $m\boldsymbol{g} = -mg\boldsymbol{e}_z$ and the air-drag frictional force $-m\kappa v\boldsymbol{v} = -mgv\boldsymbol{v}/v_\infty^2$ with $v_\infty = \sqrt{g/\kappa} > 0$. What height h above ground does the stone reach?

25 When the stone of Exercise 24 is back at ground level, what is its speed v_1? Express v_1 in terms of v_0 and v_∞.

26 An electron with mass m and charge $-q$ is moving in the constant magnetic field $\boldsymbol{B} = B\,\boldsymbol{e}_z$, so that the Lorentz force

$$\boldsymbol{F} = -q\,\boldsymbol{v} \times \boldsymbol{B} = m\boldsymbol{\omega}_{\mathrm{c}} \times \boldsymbol{v}$$

is acting, where $\omega_{\mathrm{c}} = |\boldsymbol{\omega}_{\mathrm{c}}|$ is the so-called cyclotron frequency. Use cartesian coordinates to find $\boldsymbol{v}(t)$ for $\boldsymbol{v}(t=0) = \boldsymbol{v}_0$ from Newton's equation of motion. Then determine $\boldsymbol{r}(t)$ with the initial position $\boldsymbol{r}(t=0) = \boldsymbol{r}_0$. Give a verbal description of the electron's trajectory, and introduce a fitting choice of cylindrical coordinates for another parameterization of $\boldsymbol{r}(t)$.

27 A homogeneous, linear, nth-order differential equation with constant coefficients has the general form (why?)

$$\left(\frac{\mathrm{d}}{\mathrm{d}t} - \lambda_1\right)^{n_1} \left(\frac{\mathrm{d}}{\mathrm{d}t} - \lambda_2\right)^{n_2} \cdots \left(\frac{\mathrm{d}}{\mathrm{d}t} - \lambda_k\right)^{n_k} x(t) = 0$$

where the n_js are positive integers and their sum is $n_1 + n_2 + \cdots + n_k = n$, and $\lambda_j \neq \lambda_{j'}$ if $j \neq j'$. Verify that

$$x(t) = (c_0 + c_1 t + c_2 t^2 + \cdots + c_{n_j-1} t^{n_j-1})\,\mathrm{e}^{\lambda_j t}$$

with $1 \leq j \leq k$ and constant coefficients $c_0, \ldots, c_{n_j-1}$ is a solution.

28 A general construction for a model of the delta function is this: Take a function $D(\phi)$ with the properties (i) $D(\phi) \to 0$ as $|\phi| \to \infty$, (ii) $D(\phi) \to 1$ as $\phi \to 0$. Then

$$\delta_\tau(t) = \int_{-\infty}^{\infty} \frac{\mathrm{d}\Omega}{2\pi}\, D(\Omega\tau)\, \mathrm{e}^{\mathrm{i}\Omega t}$$

with $\tau > 0$ is a model for the delta function (it is understood that $\tau \to 0$ eventually). Which model do you get for $D(\phi) = \mathrm{e}^{-|\phi|}$, and what is the corresponding model for the step function?

29 Integration by parts tells us (how?) that the derivative of the delta function has the defining property

$$\int_{-\infty}^{\infty} \mathrm{d}t\, f(t) \frac{\mathrm{d}}{\mathrm{d}t}\delta(t - t_0) = -\frac{\mathrm{d}f}{\mathrm{d}t}(t_0)\,.$$

Show that $-t\frac{\mathrm{d}}{\mathrm{d}t}\delta(t) = -t\delta'(t) = \delta(t)$. Use this and a model for the delta function to construct another model, and comment on what you get.

30 Explain why

$$\begin{aligned} f(t)\delta(t - t_0) &= f(t_0)\delta(t - t_0)\,, \\ f(t)\delta'(t - t_0) &= f(t_0)\delta'(t - t_0) - f'(t_0)\delta(t - t_0) \end{aligned}$$

hold whenever $f(t)$ is continuous and has a continuous derivative at $t = t_0$. What is the analogous statement about $f(t)\delta''(t - t_0)$?

31 Explain why the ansatz

$$G(t,t') = \left(a_1\, \mathrm{e}^{\lambda_1 t} + a_2\, \mathrm{e}^{\lambda_2 t}\right)\eta(t' - t) + \left(b_1\, \mathrm{e}^{\lambda_1 t} + b_2\, \mathrm{e}^{\lambda_2 t}\right)\eta(t - t')$$

is appropriate for all Green's functions that solve

$$\left[\left(\frac{\mathrm{d}}{\mathrm{d}t}\right)^2 + \gamma\frac{\mathrm{d}}{\mathrm{d}t} + \omega_0^2\right] G(t,t') = \delta(t - t')\,.$$

Then show that the four coefficients a_1, a_2, b_1, and b_2 must be such that the two equations

$$\begin{aligned} (b_1 - a_1)\, \mathrm{e}^{\lambda_1 t'} + (b_2 - a_2)\, \mathrm{e}^{\lambda_2 t'} &= 0\,, \\ (b_1 - a_1)\lambda_1\, \mathrm{e}^{\lambda_1 t'} + (b_2 - a_2)\lambda_2\, \mathrm{e}^{\lambda_2 t'} &= 1\,, \end{aligned}$$

hold. Which additional requirements identify the retarded and the advanced Green's functions? State them explicitly.

32 The constant force $m\boldsymbol{g}$ and the frictional force $-m\gamma\boldsymbol{v}$ are acting on point mass m. The point mass has velocity $\boldsymbol{v}_0$ at time $t = 0$ and is at position $\boldsymbol{r} = 0$ at time T. What is $\boldsymbol{r}(t)$ for $0 < t < T$?

33 A point mass m is exposed to a time-dependent force $\boldsymbol{F}(t)$ and a frictional force $-m\gamma\boldsymbol{v}(t)$. Determine the position $\boldsymbol{r}(t)$ of the point mass for the initial conditions $\boldsymbol{r}(t = 0) = \boldsymbol{r}_0$ and $\dot{\boldsymbol{r}}(t = 0) = \boldsymbol{v}_0$. Check that you get the familiar results when $\boldsymbol{F}(t) = m\boldsymbol{g}$ is constant in time.

34 For the situation of Exercise 33, what is $\boldsymbol{r}(t)$ for the boundary conditions $\boldsymbol{r}(t = 0) = \boldsymbol{r}_0$ and $\boldsymbol{r}(t = T) = \boldsymbol{r}_1$? What do you get for $\boldsymbol{F}(t) = m\boldsymbol{g}$?

35 In the situation of Exercise 33, consider the periodic force $\boldsymbol{F}(t) = m\boldsymbol{a}\sin(\Omega t)$, with constant acceleration $\boldsymbol{a}$ and positive frequency Ω. What are $\boldsymbol{v}(t)$ and $\boldsymbol{r}(t)$ at late times when $\gamma t \gg 1$?

36 A constant force $F = mg$ is applied to an undamped harmonic oscillator (mass m, circular frequency ω_0, damping constant $\gamma = 0$) for a finite duration T. It so happens that the oscillator is at $x = 0$ at time $t = 0$ and has vanishing velocity at time $t = T$. What are the position $x(t)$ and the velocity $\dot{x}(t)$ for $0 < t < T$?

37 The undamped harmonic oscillator of Exercise 36 is initially at rest, so that $x(t) = 0$ for $t < 0$. Then a time-dependent force $F(t)$ is applied for a finite duration T, that is: $F(t) = 0$ for $t < 0$ and $t > T$. It so happens that the oscillator is again at rest after the force has ceased to act. What does this tell you about $F(t)$?

38 Use (2.2.142) or (2.2.143) for the force of (2.2.91) to derive anew the solution given in Section 2.2.6(a) for late times, $\gamma t \gg 1$.

39 Consider the damped harmonic oscillator, driven by the force $F(t)$. Find the Green's function $G_T(t, t')$ such that

$$x(t) = \int_0^T \mathrm{d}t'\, G_T(t, t') \frac{1}{m} F(t')$$

solves the equation of motion, for $0 < t < T$, with the boundary conditions $x(t=0) = 0$ and $x(t=T) = 0$.

40 Use your favorite plotting program to generate plots of $x^{(\mathrm{css})}(t)$ in (2.2.155) for $\omega_0 = 10\gamma$ (subcritical damping), a fixed value of v_0, and different values of the period T. Can you identify T ranges for which the amplitude is particularly large? Repeat for critical damping.

Chapter 3

41 A point mass m is moving along the x axis under the influence of the force associated with the potential energy

$$V(x) = \frac{1}{2}m\omega_0^2(2x^2 - a^2)\,\mathrm{e}^{-(x/a)^2}$$

with $a > 0$. For which energy ranges do you have motion with one, two, or no turning points? For the oscillatory motion between two turning points, determine the period of small-amplitude oscillations.

42 A point mass m is moving along the x axis under the influence of the force associated with the potential energy $V(x) = \kappa|x|^\nu$ with $\kappa > 0$ and $\nu > 0$. Find the period of oscillatory motion as a function of energy. For this problem, Euler's beta function integral $(\alpha, \beta > -1)$

$$\int_0^1 \mathrm{d}\zeta\, \zeta^\alpha (1-\zeta)^\beta = \frac{\alpha!\,\beta!}{(\alpha+\beta+1)!}$$

is useful. Evaluate also $\displaystyle\int_b^a \mathrm{d}x\,(a-x)^\alpha(x-b)^\beta$ for $a > b$ and recognize the $\alpha = \beta = -\frac{1}{2}$ case of (3.1.40), and establish that $(-\frac{1}{2})! = \sqrt{\pi}$.

43 A point mass m is moving along the x axis under the influence of the force associated with the potential energy $V(x) = \frac{1}{2}m\omega_0^2|x|(a - |x|)$ with $a > 0$. For which energy ranges do you have motion with one, two, or no turning points? For the oscillatory motion between two turning points, determine the period as a function of energy.

44 A point mass m is moving along the x axis under the influence of the force associated with the potential energy

$$V(x) = -\frac{E_0}{\left[\cosh(kx)\right]^2},$$

where E_0 and k are positive constants. For which ranges of energy is the motion of the point mass bounded by two turning points, by one turning point, or not bounded at all? For energy E such that there is periodic motion between two turning points, find the period $T(E)$.

45 For $-\frac{1}{2}\pi a < x < \frac{1}{2}\pi a$, the potential energy for a point mass m is

$$V(x) = V_0\left[\tan(x/a)\right]^2,$$

where a and V_0 are positive constants. What is the period of small-amplitude oscillations? Find the energy-dependent period $T(E)$ for all permissible values of the energy E.

46 A point mass m is moving along the x axis under the influence of the force

$$F = -ma\,\mathrm{sgn}(x) = \left\{\begin{array}{rl} -ma & \text{for} \quad x > 0 \\ ma & \text{for} \quad x < 0 \end{array}\right\}$$

with constant $a > 0$. Show that the energy $E = \frac{1}{2}mv^2 + ma|x|$ is constant in time. What is the energy-dependent period $T(E)$ of the periodic motion?

47 In the situation of Exercise 46, the period-averages of the kinetic energy and the potential energy are

$$\overline{E_{\mathrm{kin}}} = \frac{1}{T}\int_0^T \mathrm{d}t\, \frac{m}{2}v^2 \quad \text{and} \quad \overline{E_{\mathrm{pot}}} = \frac{1}{T}\int_0^T \mathrm{d}t\, ma|x|\,.$$

Express each of them in terms of energy E.

48 A point mass m is moving along the x axis under the influence of the force associated with the potential energy

$$V(x) = E_0 a^2 \frac{x^2 - a^2}{(x^2 + 2a^2)^2} \quad \text{with} \quad E_0 > 0 \quad \text{and} \quad a > 0.$$

For which energy ranges do you have motion with one, two, or no turning points? For the oscillatory motion between two turning points, what is the period of small-amplitude oscillations?

49 Repeat Exercise 48 for the case $E_0 < 0$.

50 A point mass m is moving along the x axis under the influence of the force associated with the potential energy

$$V(x) = E_0 \frac{a^2 x^2}{(x^2 + a^2)^2} \quad \text{with} \quad E_0 > 0 \quad \text{and} \quad a > 0.$$

For which energy ranges do you have motion with one, two, or no turning points? For the oscillatory motion between two turning points, what is the period of small-amplitude oscillations?

51 Repeat Exercise 50 for the case $E_0 < 0$.

52 A point mass m is moving along the x axis under the influence of the force associated with the potential energy

$$V(x) = F\left(\sqrt{|x| + a} - \sqrt{a}\right)^2$$

with constants $F > 0$ and $a > 0$. What are the metrical dimensions of F and a? Which combination of m, F, and a has the metrical dimension of energy? Which combination has the metrical dimension of time? Which simple expressions approximate $V(x)$ for $|x| \ll a$ and $|x| \gg a$? What is the period of small-amplitude oscillations?

53 In the situation of Exercise 52, what is the energy-dependent period $T(E)$ for motion between two turning points?

54 A point mass m moves along the x axis under the influence of the force

$$F(x) = -a(x^2 - x_0^2),$$

where a and x_0 are positive constants. The force vanishes for $x = \pm x_0$. Which of them is the location of a stable equilibrium? What is the period of small-amplitude oscillations about the stable-equilibrium position? Without any calculation: Is the period longer or shorter for oscillations with amplitudes that are not small? Explain.

55 Can you find a symmetric potential energy $V(x) = V(-x)$ for which the motion between two turning points has the energy-dependent period $T(E) = T_0\,\mathrm{e}^{E/E_0}$, where T_0 and E_0 are positive constants? Comment on what you found.

56 Christiaan Huygens's tautochrone: A point mass m moves along the curve $y(x)$ without friction, and the weight $-mg\boldsymbol{e}_y$ (with $g > 0$) is the only external force acting, whereby $y(0) = 0$, $y'(0) = 0$, $y(-x) = y(x)$, and $xy'(x) > 0$ for $x \neq 0$. We want to choose $y(x)$ such that the period of the oscillation is independent of the value of the energy

$$E = \frac{m}{2}\dot{s}^2 + mgy \quad \text{with} \quad s(x) = \int_0^x \mathrm{d}x'\,\sqrt{1 + y'(x')^2}\,.$$

Explain why this requires $y(x) = s(x)^2/R$ with some constant length R. Then find $y(x)$.

57 A point mass m is moving in the xy plane under the influence of the force associated with the potential energy

$$V(x, y) = V_0 \cos(2kx) - 2V_0 \cos(kx)\cos(\sqrt{3}ky)\,,$$

where V_0 and k are positive constants. Determine the positions at which the force vanishes and examine whether the potential energy has a maximum, a minimum, or a saddle point there. Then find the periods of the natural small-amplitude oscillations at the minima.

58 A point mass m is moving in the xy plane under the influence of the force associated with the potential energy

$$V(x, y) = V_0 \cos(k_1 x)\cos(k_2 y)\,,$$

where V_0, k_1, and k_2 are positive constants. Determine the positions at which the force vanishes and examine whether the potential energy has a maximum, a minimum, or a saddle point there. Then find the periods of the natural small-amplitude oscillations at the minima.

59 Consider the force field $\boldsymbol{F}(\boldsymbol{r}) = k_1 yz\,\boldsymbol{e}_x + k_2 zx\,\boldsymbol{e}_y + k_3 xy\,\boldsymbol{e}_z$ with constant coefficients k_1, k_2, and k_3. Are there values of the coefficients for which the force is conservative? For arbitrary values of k_1, k_2, k_3, calculate the line integral of $\boldsymbol{F}(\boldsymbol{r})$ for the closed loop that connects the points with cartesian components $(x, y, z) = (0,0,0)$, $(0,1,2)$, $(1,0,2)$, $(1,2,0)$, and back to $(0,0,0)$ with straight lines.

60 A central-force field has the form $\boldsymbol{F}(\boldsymbol{r}) = f(r)\boldsymbol{e}_r$. Is this a conservative force? If yes, find a potential energy for it.

61 A force field $\boldsymbol{F}(\boldsymbol{r})$ has the form

$$\boldsymbol{F}(\boldsymbol{r}) = f_1(r)\boldsymbol{a} + f_2(r)\boldsymbol{a}\cdot\boldsymbol{r}\,\boldsymbol{r}$$

with a constant vector $\boldsymbol{a}$ and non-singular functions $f_1(r)$ and $f_2(r)$ that depend only on the distance $r = |\boldsymbol{r}|$ from the origin of the coordinate system. How are $f_1(r)$ and $f_2(r)$ related if $\boldsymbol{F}(\boldsymbol{r})$ is a conservative force? What is the potential energy associated with such a conservative force?

62 Which of the following five force fields are conservative?

$$\text{(i)}\quad \boldsymbol{F} \mathrel{\widehat{=}} \begin{pmatrix} 2kx + ky \\ kx + kz \\ ky + 2kz \end{pmatrix} \quad \text{with} \quad k = \text{constant;}$$

$$\text{(ii)}\quad \boldsymbol{F} \mathrel{\widehat{=}} \lambda \begin{pmatrix} y^2 + yz \\ 2xy - z^2 + xz \\ 2yz + xy \end{pmatrix} \quad \text{with} \quad \lambda = \text{constant;}$$

$$\text{(iii)}\quad \boldsymbol{F} = r\boldsymbol{b} \quad \text{with} \quad \boldsymbol{b} = \text{constant;}$$

$$\text{(iv)}\quad \boldsymbol{F} = \frac{r^2\boldsymbol{a} - \boldsymbol{r}\,\boldsymbol{r}\cdot\boldsymbol{a}}{r^3} \quad \text{with} \quad \boldsymbol{a} = \text{constant;}$$

$$\text{(v)}\quad \boldsymbol{F} = \boldsymbol{a}\times(\boldsymbol{r}\times\boldsymbol{a}) \quad \text{with} \quad \boldsymbol{a} = \text{constant.}$$

Find a potential energy for each conservative force.

63 The gradient of the position vector, $\boldsymbol{\nabla}\boldsymbol{r}$, is a dyadic; which one? What is the dyadic double gradient $\boldsymbol{\nabla}\boldsymbol{\nabla}V(r)$ of a potential energy $V(r)$ that depends only on the length r of position vector $\boldsymbol{r}$?

64 In $\boldsymbol{a}\times\boldsymbol{1}$ and $\boldsymbol{1}\times\boldsymbol{b}$ we have the column-type vector $\boldsymbol{a}$ and the row-type vector $\boldsymbol{b}$. Show that the following identities hold:

$$\boldsymbol{a}\times\boldsymbol{1} = -(\boldsymbol{a}\times\boldsymbol{1})^{\mathrm{T}},$$
$$\boldsymbol{a}\times\boldsymbol{1} = \boldsymbol{1}\times\boldsymbol{a},$$
$$\boldsymbol{a}\times\boldsymbol{1}\times\boldsymbol{b} = \boldsymbol{b}\,\boldsymbol{a} - \boldsymbol{a}\cdot\boldsymbol{b}\,\boldsymbol{1}.$$

65 For any three vectors $\boldsymbol{a}$, $\boldsymbol{b}$, and $\boldsymbol{c}$, we have

$$\boldsymbol{a}\,\boldsymbol{b}\times\boldsymbol{c}+\boldsymbol{b}\,\boldsymbol{c}\times\boldsymbol{a}+\boldsymbol{c}\,\boldsymbol{a}\times\boldsymbol{b}=\boldsymbol{a}\cdot(\boldsymbol{b}\times\boldsymbol{c})\,\mathbf{1}\,.$$

Verify this.

66 Verify also that $(\boldsymbol{a}\,\boldsymbol{b}-\boldsymbol{b}\,\boldsymbol{a})^2=\boldsymbol{a}\times\boldsymbol{b}\,\boldsymbol{a}\times\boldsymbol{b}-(\boldsymbol{a}\times\boldsymbol{b})^2\mathbf{1}$ for any two vectors $\boldsymbol{a}$ and $\boldsymbol{b}$, where $\mathsf{A}^2=\mathsf{A}\cdot\mathsf{A}$ for dyadic A. What is $(\boldsymbol{e}\,\boldsymbol{e}-\boldsymbol{e}'\,\boldsymbol{e}')^2$ for two unit vectors $\boldsymbol{e}$ and $\boldsymbol{e}'$?

67 Write the finite rotation in (1.1.77) as $\boldsymbol{r}_0$ multiplied by the rotation dyadic $\mathsf{R}(\boldsymbol{e},\varphi)$, that is: $\boldsymbol{r}(\varphi)=\mathsf{R}(\boldsymbol{e},\varphi)\cdot\boldsymbol{r}_0$, and verify that R is an *orthogonal dyadic*: $\mathsf{R}^{\mathrm{T}}\cdot\mathsf{R}=\mathbf{1}\,.$

68 Explain why $\mathsf{R}(\boldsymbol{e},\varphi_2)\cdot\mathsf{R}(\boldsymbol{e},\varphi_1)=\mathsf{R}(\boldsymbol{e},\varphi_1+\varphi_2)$ must hold, and verify this relation.

69 The eigenvalues, eigencolumns, and eigenrows of the 3×3 matrix that represents a dyadic correspond to eigenvalues, right-sided eigenvectors, and left-sided eigenvectors of the dyadic itself. Find the eigenvalues and eigenvectors of $\boldsymbol{a}\times\mathbf{1}$, $\boldsymbol{a}\times\mathbf{1}\times\boldsymbol{b}$, and $\mathsf{R}(\boldsymbol{e},\varphi)$.

70 The trace and the determinant of a dyadic are equal to the trace and the determinant, respectively, of the 3×3 matrix that represents the dyadic. All the familiar identities for traces and determinants of matrices hold also for traces and determinants of dyadics, such as $\mathrm{tr}\{\mathsf{A}+\mathsf{B}\}=\mathrm{tr}\{\mathsf{A}\}+\mathrm{tr}\{\mathsf{B}\}$ or $\det\{\mathsf{A}\cdot\mathsf{B}\}=\det\{\mathsf{A}\}\det\{\mathsf{B}\}$. Show that $\mathrm{tr}\{\boldsymbol{b}\,\boldsymbol{a}\}=\boldsymbol{a}\cdot\boldsymbol{b}$ for any two vectors $\boldsymbol{a}$ and $\boldsymbol{b}$. Further, demonstrate that

$$\mathsf{A}^3-\mathsf{A}^2\mathrm{tr}\{\mathsf{A}\}+\frac{1}{2}\mathsf{A}\Big(\mathrm{tr}\{\mathsf{A}\}^2-\mathrm{tr}\{\mathsf{A}^2\}\Big)-\mathbf{1}\det\{\mathsf{A}\}=0$$

for all dyadics A, where $\mathsf{A}^3=\mathsf{A}^2\cdot\mathsf{A}=\mathsf{A}\cdot\mathsf{A}\cdot\mathsf{A}$, and conclude that

$$\det\{\mathsf{A}\}=\frac{1}{3}\mathrm{tr}\{\mathsf{A}^3\}-\frac{1}{2}\mathrm{tr}\{\mathsf{A}\}\,\mathrm{tr}\{\mathsf{A}^2\}+\frac{1}{6}\mathrm{tr}\{\mathsf{A}\}^3\,.$$

71 For given dyadic A and its transpose A^{T}, the products $\mathsf{A}^{\mathrm{T}}\cdot\mathsf{A}$ and $\mathsf{A}\cdot\mathsf{A}^{\mathrm{T}}$ are *positive dyadics*, that is: $\boldsymbol{a}\cdot\mathsf{A}^{\mathrm{T}}\cdot\mathsf{A}\cdot\boldsymbol{a}\geq0$ and $\boldsymbol{a}\cdot\mathsf{A}\cdot\mathsf{A}^{\mathrm{T}}\cdot\boldsymbol{a}\geq0$ for all nonzero vectors $\boldsymbol{a}$. Why? Show that

$$\mathsf{A}=\sum_{j=1}^{3}\boldsymbol{e}_j\lambda_j\boldsymbol{e}'_j\,,\qquad(*)$$

where the $\boldsymbol{e}_j$s are the right-handed trio of eigenvectors of $\mathsf{A}\cdot\mathsf{A}^{\mathrm{T}}$ and the $\boldsymbol{e}_j'$s are the right-handed trio of eigenvectors of $\mathsf{A}^{\mathrm{T}}\cdot\mathsf{A}$, and the common eigenvalues of $\mathsf{A}^{\mathrm{T}}\cdot\mathsf{A}$ and $\mathsf{A}\cdot\mathsf{A}^{\mathrm{T}}$ are the λ_j^2s. Conclude that $\det\{\mathsf{A}\} = \lambda_1\lambda_2\lambda_3$. — Note: If $\det\{\mathsf{A}\} \geq 0$, we can choose the eigenvectors such that all $\lambda_j \geq 0$, and then $(*)$ is the *Schmidt* decomposition* of A; if $\det\{\mathsf{A}\} < 0$, the Schmidt decomposition with $\lambda_j > 0$ requires that one of the eigenvector trios is left-handed.

72 A particular dyadic-valued product of two dyadics, the two-fold vector product $\mathsf{A} \times\!\!\times \mathsf{B}$, is defined by

$$(\boldsymbol{a}\,\boldsymbol{b}) \times\!\!\times (\boldsymbol{a}'\,\boldsymbol{b}') = \boldsymbol{a}\times\boldsymbol{a}'\;\boldsymbol{b}'\times\boldsymbol{b}$$

and the distributive law for both factors. Show that $\mathsf{A} \times\!\!\times \mathsf{B} = \mathsf{B} \times\!\!\times \mathsf{A}$ and $(\mathsf{A} \times\!\!\times \mathsf{B})^{\mathrm{T}} = \mathsf{A}^{\mathrm{T}} \times\!\!\times \mathsf{B}^{\mathrm{T}}$ for all dyadics A and B.

73 Dyadics A and B are arbitrary, dyadic R is a rotation dyadic as in Exercise 67. Show that

$$(\mathsf{R}\cdot\mathsf{A}) \times\!\!\times (\mathsf{R}\cdot\mathsf{B}) = \mathsf{R}\cdot(\mathsf{A} \times\!\!\times \mathsf{B}) \quad\text{and}\quad (\mathsf{A}\cdot\mathsf{R}) \times\!\!\times (\mathsf{B}\cdot\mathsf{R}) = (\mathsf{A} \times\!\!\times \mathsf{B})\cdot\mathsf{R}\,.$$

74 Verify that

$$\mathsf{A}^{\mathrm{T}}\cdot(\mathsf{A} \times\!\!\times \mathsf{A}) = -2\det\{\mathsf{A}\}\,\mathbf{1}$$

for all dyadics A. Then conclude that $\mathsf{A}^{-1} = -\dfrac{(\mathsf{A} \times\!\!\times \mathsf{A})^{\mathrm{T}}}{2\det\{\mathsf{A}\}}$ is the reciprocal of A if $\det\{\mathsf{A}\} \neq 0$, that is: $\mathsf{A}\cdot\mathsf{A}^{-1} = \mathsf{A}^{-1}\cdot\mathsf{A} = \mathbf{1}$.

75 Harking back to Section 3.2.5, consider the following realization of a Paul trap. A charged particle is in a time-varying electric field, so that — in the relevant region of space — the force $\boldsymbol{F} = -\boldsymbol{\nabla}V$ derives from the time-dependent potential energy

$$V(\boldsymbol{r},t) = \frac{m}{2}f(t)\omega_0^2(x^2+y^2-2z^2)$$

where $f(t) = f(t+T)$ is periodic with

$$f(t) = \begin{cases} 1 \text{ for } 0 < t < \frac{1}{2}T\,, \\ -1 \text{ for } \frac{1}{2}T < t < T\,. \end{cases}$$

Express $\boldsymbol{r}(T)$ and $\boldsymbol{v}(T)$ in terms of $\boldsymbol{r}(0)$ and $\boldsymbol{v}(0)$. For which values of $\omega_0 T$ is the particle trapped?

---

*Erhard SCHMIDT (1876–1959)

Chapter 4

76 As in Section 4.5.3, point masses $m_1, m_2, \ldots, m_J$ have conservative line-of-sight pair forces among them and are exposed to the external forces $\boldsymbol{F}_j^{(\text{ext})} = m_j\boldsymbol{g}$ where $\boldsymbol{g}$ is the same gravitational acceleration for all masses. At time t_0, the initial conditions give values $\boldsymbol{R}_0$, $\boldsymbol{P}_0$, E_0, and $\boldsymbol{L}_0$ to the center-of-mass position $\boldsymbol{R}$, the total momentum $\boldsymbol{P}_{\text{tot}}$, the total energy E_{tot}, and the total angular momentum $\boldsymbol{L}_{\text{tot}}$. Find $\boldsymbol{R}(t)$, $\boldsymbol{P}_{\text{tot}}(t)$, $E_{\text{tot}}(t)$, and $\boldsymbol{L}_{\text{tot}}(t)$.

77 Consider a closed system of point masses m_j with pair forces among them. Show that

$$\frac{\mathrm{d}}{\mathrm{d}t}\sum_j \boldsymbol{r}_j \cdot m_j\boldsymbol{v}_j = 2E_{\text{kin}} + \sum_j \boldsymbol{r}_j \cdot \boldsymbol{F}_j$$

with the total kinetic energy $E_{\text{kin}} = \frac{1}{2}\sum_j m_j\boldsymbol{v}_j^2$. Now, assume that the forces and initial conditions are such that all the masses remain within some finite region of space and also their velocities remain in a finite region of velocity space (example: our solar system), and introduce long-time averages in accordance with

$$\overline{f(t)} = \frac{1}{T}\int_0^T \mathrm{d}t\, f(t)\bigg|_{T\to\infty} .$$

Show that

$$\overline{E_{\text{kin}}} = -\frac{1}{2}\overline{\sum_j \boldsymbol{r}_j \cdot \boldsymbol{F}_j} ,$$

which is the so-called *virial theorem*, a term introduced by Clausius.*

78 Continuing with the situation of Exercise 77, assume in addition that the pair forces are conservative. Conclude that

$$\overline{E_{\text{kin}}} = \frac{1}{2}\overline{\sum_j \boldsymbol{r}_j \cdot \boldsymbol{\nabla}_j E_{\text{pot}}} ,$$

where $E_{\text{pot}} = \sum_{j<k} V_{(jk)}(\boldsymbol{r}_j - \boldsymbol{r}_k)$ is the total potential energy.

*Rudolf Julius Emanuel (born Rudolf Gottlieb) CLAUSIUS (1822–1888)

79 Continuing with the situation of Exercises 77 and 78, consider now the case of conservative line-of-sight pair forces with a power law,

$$V_{(jk)}(r) = A_{(jk)} r^n,$$

with a *common* value of n for all pairs (example: $n = -1$ in the solar system). Show that

$$\overline{E_{\text{kin}}} = \frac{n}{n+2} E_{\text{tot}} \quad \text{and} \quad \overline{E_{\text{pot}}} = \frac{2}{n+2} E_{\text{tot}}$$

under these circumstances. Comment on the special values of $n = -1$ and $n = 2$.

Chapter 5

80 During a solar eclipse, is the net force on the moon toward the earth or toward the sun?

81 A planet moves on his Kepler's ellipse with numerical eccentricity ϵ. The minor axis of the ellipse cuts the orbit into two:

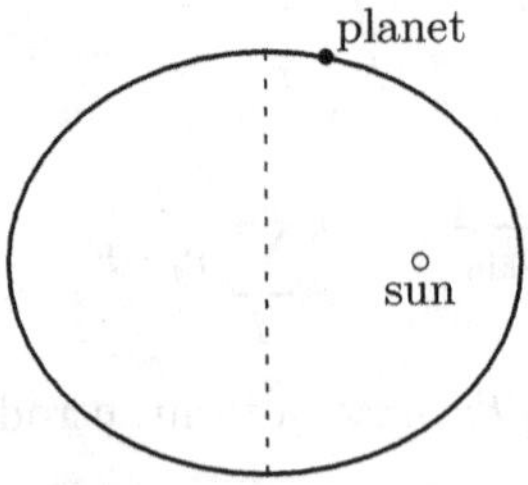

What fraction of the round-trip time does the planet spend in each half?

82 A planet of mass m moves on his Kepler's ellipse with major half-axis a, numerical eccentricity ϵ, angular momentum $|\boldsymbol{l}| = m\kappa > 0$, and period $T = 2\pi\sqrt{1-\epsilon^2}a^2/\kappa$. Show that the so-called *axis vector*

$$\boldsymbol{A} = \frac{\boldsymbol{r}}{r} - \frac{\boldsymbol{v} \times (\boldsymbol{r} \times \boldsymbol{v})}{Gm_\odot}$$

is a constant of motion; the axis vector is also known as the *Laplace–Runge*-Lenz*† vector*. Then consider $\frac{\boldsymbol{r}}{r}\cdot\boldsymbol{A}$ and $\frac{\boldsymbol{r}}{r}\times\boldsymbol{A}$, infer the length $|\boldsymbol{A}|$ of the axis vector, and establish the geometrical meaning of $\boldsymbol{A}$.

83 For the planet of Exercise 82, what does the virial theorem say about the time averages of the kinetic and the potential energy, averaged over one period? By an explicit evaluation of the time integral, show that the time average of the potential energy is $-Gmm_{\odot}/a$. Derive Kepler's Third Law by combining the two observations.

84 Apply (5.3.17) to the situation of a Kepler's ellipse and so re-derive Kepler's Third Law.

85 A point mass is moving, with total energy $E<0$ and positive κ parameter, in the force field associated with the potential energy $V(s)=-C/s+B/s^2$, where B and C are positive constants. Verify that the orbit is a precessing ellipse, that is:

$$s(\varphi)=\frac{(1-\epsilon^2)a}{1+\epsilon\cos(\lambda\varphi)}$$

with $a>0$, $0\le\epsilon<1$, and $\lambda>0$. Determine the values of a,ϵ,λ and check that you get the right answer for $B=0$. Is the precession forward or backward?

86 Confirm your answer to this question by a calculation of the angular period with the aid of (5.3.21).

87 Point mass m moves in the central-force field associated with the potential energy $V(r)$. The force is attractive, $V'(r)>0$, and the motion is confined to the radial range $s_1\le r\le s_2$. As usual the bounds are determined by the energy E and the angular momentum $\boldsymbol{l}$ of the orbit. Explain why the circular orbits ($s_1=s_2$) have smallest energy for given angular momentum and largest angular momentum for given energy.

88 For the potential energy $V(r)=-\dfrac{A}{r(a+r)^2}$ with constants $A>0$ and $a>0$ — the so-called *Tietz* potential* — one can have bound orbits for $E=0$. For which values of $\kappa=|\boldsymbol{l}|/m$ is this possible? For such an orbit with $E=0$, find the angular period and determine how much time elapses

*Carl David Tolmé RUNGE (1856-1927) †Wilhelm LENZ (1888–1957)
*Tadeusz TIETZ (1920–1996)

between two successive instants at which the point mass is closest to the force center.

89 Isotropic harmonic oscillator: Point mass m moves in the force field $\boldsymbol{F} = -m\omega_0^2\boldsymbol{r} = -\boldsymbol{\nabla}\frac{1}{2}m\omega_0^2r^2$ with energy E and angular momentum $\boldsymbol{l} \neq 0$. What is the angular period and how much time elapses between two successive instants at which the point mass is closest to the force center? Do you get what you expect?

90 Consider the isotropic harmonic oscillator of Exercise 89. Show that the dyadic $\mathsf{D} = \boldsymbol{v}\,\boldsymbol{v} + \omega_0^2\boldsymbol{r}\,\boldsymbol{r}$ does not depend on time.

91 The isotropic harmonic oscillator of Exercises 89 and 90: By generalizing the familiar $x(t)$ of the one-dimensional harmonic oscillator, state $\boldsymbol{r}(t)$ for initial position $\boldsymbol{r}(t=0) = \boldsymbol{r}_0$ and initial velocity $\boldsymbol{v}(t=0) = \boldsymbol{v}_0$. Express E and $\boldsymbol{l}$ in terms of $\boldsymbol{r}_0$ and $\boldsymbol{v}_0$. Consider $(\boldsymbol{l} \times \boldsymbol{r}_0) \cdot \boldsymbol{r}(t)$ and $(\boldsymbol{v}_0 \times \boldsymbol{l}) \cdot \boldsymbol{r}(t)$ and use them to show that the planar orbit is an ellipse centered at $\boldsymbol{r} = 0$.

92 In analogy to the expression (5.3.21) for the angular period Φ, express the scattering angle θ as an integral over s from the single turning radius s_1 to $s = \infty$. Apply your expression to $V(s) = C/s$ to derive anew the Rutherford cross section and verify that it does not matter whether C is positive or negative.

93 A projectile of mass m_1 is scattered by a target of mass m_2, whereby a conservative line-of-sight force is acting. Before the scattering event, the target is at rest, and the projectile is approaching with the asymptotic speed v_∞ and the impact parameter b. For the *relative motion*, the scattering angle is θ. What is the scattering angle Θ that is actually observed?

94 In the situation of Exercise 93, what is the range of possible Θ values when $m_1 < m_2$, $m_1 = m_2$, or $m_1 > m_2$? If the differential cross section in the center-of-mass frame is $\dfrac{\mathrm{d}\sigma}{\mathrm{d}\Omega} = f(\theta)$, what is the differential cross section observed in the laboratory frame when $m_1 = m_2$?

95 Point masses that have energy E are scattered by the force field associated with the potential energy $V(r) = B/r^2$ with $B > 0$. Find the differential scattering cross section.

96 A point mass m is scattered by the velocity-dependent force

$$\boldsymbol{F} = m\kappa_0 \frac{\boldsymbol{r} \times \boldsymbol{v}}{r^3},$$

where the parameter $\kappa_0 > 0$ measures the strength of the force. What is the differential cross section $\dfrac{d\sigma}{d\Omega}$?

97 In ray optics the path of a light ray through a ball of water with radius R, without any internal reflection, is as depicted here:

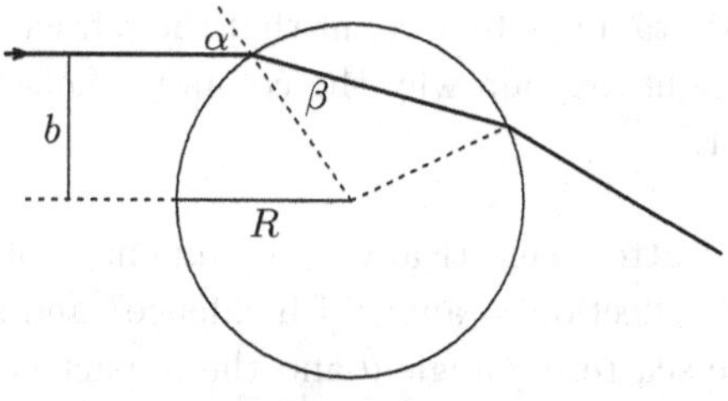

where *Snell's** *Law* of refraction applies in the form $3\sin\alpha = 4\sin\beta$. The ray has impact parameter b and is deflected by angle θ. Express $y = \cos(\frac{1}{2}\theta)$ as a function of $x = (b/R)^2$ and sketch the graph of $y(x)$. Show that the differential cross section is $\dfrac{d\sigma}{d\Omega} = -\dfrac{R^2}{8y}\dfrac{dx}{dy}$ and find $\dfrac{d\sigma}{d\Omega}$.

98 For the situation of Exercise 97, calculate the total cross section $\sigma = \displaystyle\int d\Omega\, \frac{d\sigma}{d\Omega}$. Do you get the expected result?

99 The optical phenomenon of a rainbow requires one internal reflection:

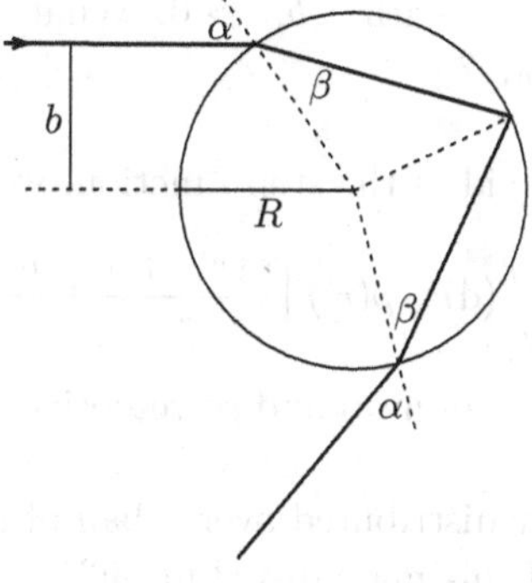

*Willebrord SNELLIUS (1580–1626)

Express the cosine of the scattering angle θ as a function of the ratio b/R of the impact parameter b and the radius R of the ball of water, and plot this function. You should observe that $\frac{\mathrm{d}\cos\theta}{\mathrm{d}b^2}$ is positive for some values of b, negative for other values, and zero for a critical scattering angle θ_{RB}. What is the consequence thereof for the differential scattering cross section $\frac{\mathrm{d}\sigma}{\mathrm{d}\Omega}$, and what is the value of θ_{RB}?

100 Repeat for two internal reflections; they give rise to the secondary rainbow which is usually not as bright as the primary rainbow from one internal reflection. By taking into account that the refractive index depends on the color of the light, explain why the colors in the secondary rainbow are in reversed order.

101 A point mass is scattered elastically by an impenetrable ball with radius R. Invoke "angle of reflection = angle of incidence" and so determine the relation between the scattering angle θ and the impact parameter b. Then find the differential scattering cross section $\frac{\mathrm{d}\sigma}{\mathrm{d}\Omega}$ and the total cross section $\sigma = \int \mathrm{d}\Omega \, \frac{\mathrm{d}\sigma}{\mathrm{d}\Omega}$.

102 Apply the integral expression of Exercise 92 to the situation of Exercise 101.

Chapter 6

103 The surface of a homogeneous body of mass M is the ellipsoid $(x/a)^2 + (y/b)^2 + (z/c)^2 = 1$ with $a, b, c > 0$. What is the quadrupole moment dyadic $\mathbf{Q}$ of this body?

104 Show that, with the aid of the step function, we can write (6.3.6) as

$$\Phi(r) = -G \int (\mathrm{d}\boldsymbol{r}')\, \rho(r') \left[\frac{\eta(r - r')}{r} + \frac{\eta(r' - r)}{r'}\right].$$

Then evaluate the gradient of $\Phi(r)$ and so re-derive (6.3.13).

105 Mass M is uniformly distributed over a ball of radius R. What is the gravitational potential of this mass distribution?

106 A point mass m is at distance r from the center of a planet. The planet has mass M and the shape of a ball of radius R. The ball has uniform mass density ρ_0 for $R/2 < r < R$, and the mass density is twice as big in the planet's inner core of radius $R/2$. What is the gravitational force that the planet exerts on the point mass?

107 Two homogeneous balls, each of mass $\frac{1}{2}M$ and radius R, are centered at $\boldsymbol{r}_1$ and $\boldsymbol{r}_2 = \boldsymbol{r}_1 + \boldsymbol{a}$ with $|\boldsymbol{a}| > 2R$. What is the quadrupole moment dyadic $\mathbf{Q}$ of this mass distribution? The balls are released from rest and then accelerated toward each other by their gravitational attraction. How much time elapses before the balls touch if $|\boldsymbol{a}| \gg R$?

Chapter 7

108 In ray optics, the ray from (x_0, y_0) to (x_1, y_1) follows the path of shortest optical length L that is given by

$$L = \int_0^1 \mathrm{d}\ell\, n \qquad (L \text{ minimal for the actual path})$$

where $\mathrm{d}\ell = \sqrt{(\mathrm{d}x)^2 + (\mathrm{d}y)^2}$ is a path element and n is the position-dependent refractive index. For a refractive index that depends only on x, that is $n = n(x)$, consider rays with trajectories parameterized by $\bigl(x, y(x)\bigr)$ and show that

$$n(x)\sin\bigl(\alpha(x)\bigr) = \text{const} \quad \text{with} \quad \frac{\mathrm{d}}{\mathrm{d}x}y(x) = \tan\bigl(\alpha(x)\bigr).$$

This is Snell's Law, of course.

109 Now consider the more general situation in which the refractive index $n(x, y)$ can also depend on y. What about the three-dimensional case with $\mathrm{d}\ell = |\mathrm{d}\boldsymbol{r}|$ and a position-dependent $n(\boldsymbol{r})$?

110 What does (7.3.11) tell you about the brachistochrone parameters ϕ_0 and R when $a \ll b$ or $a \gg b$?

111 Under the gravitational pull $\boldsymbol{g} = g\boldsymbol{e}_y$, a point mass is moving along the brachistochrone to get from $(x_1, y_1) = (0, 0)$ to $(x_2, y_2) = (a, 0)$. What is the average speed?

112 A point mass gets from its initial position$(x, y) = (0, 0)$ to the final position at $(a, b) = R(\phi_0 - \sin\phi_0, 1 - \cos\phi_0)$ along the brachistochrone:

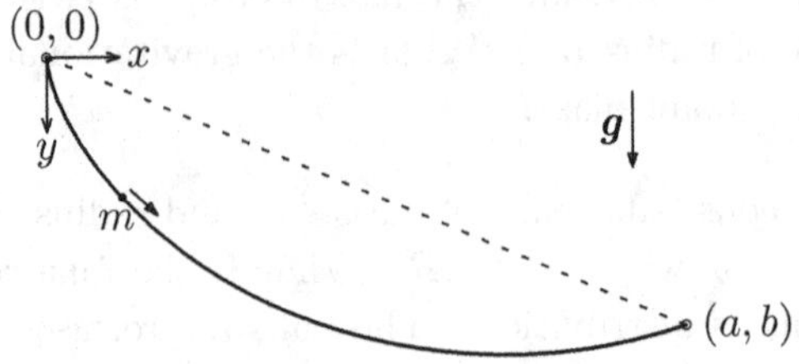

It reaches (a, b) in shorter time than it would along the straight line although the path is longer for the brachistochrone. It follows that the average speed for the brachistochrone is larger than that for the straight line. To confirm this, find the average speed for each path and compare the two speeds.

113 In the situation of Exercise 112, is there a path for which the average speed is largest?

114 Verify that (7.4.16) and (7.4.18) imply (7.4.19).

115 The functional

$$F[y] = \frac{2}{5}\int_0^\infty \mathrm{d}x\, \frac{y(x)^{5/2}}{x^{1/2}} + \frac{1}{2}\int_0^\infty \mathrm{d}x\, y'(x)^2$$

with $y(x = 0) = 1$ and $y(x) \to 0$ for $x \to \infty$ plays a certain role in the *Thomas*-Fermi† model* of atomic physics. The extremum of $F[y]$ is a minimum. Which differential equation is obeyed by the $y(x)$ that minimizes $F[y]$? Confirm that $y(x) \cong 144/x^3$ for sufficiently large x values.

116 The functional of Exercise 115: Denote the values of the two integrals for the minimizing $y(x)$ by F_1 and F_2, so that

$$F_{\min} = \min_{y(x)}\{F[y]\} = \frac{2}{5}F_1 + \frac{1}{2}F_2\,,$$

and show that $F_1 + F_2 = B$ where $B = -y'(x = 0)$ is known as *Baker's‡ constant.*

*Llewellyn Hilleth THOMAS (1903–1992) †Enrico FERMI (1901–1954)
‡Edward B. BAKER (work of 1930)

117 The functional of Exercises 115 and 116: Consider the derivative of $4x^{1/2}y(x)^{5/2} - 5xy'(x)^2$ and conclude that $2F_1 = 5F_2$. Express F_1, F_2, and F_{min} in terms of B.

Chapter 8

118 Mass m is moving under the influence of its weight $-mg\boldsymbol{e}_z$ on the cone specified by $z = \beta\sqrt{x^2 + y^2}$ with $\beta > 0$. State the Lagrange function in terms of cylindrical coordinates and then derive the equations of motion for $z(t)$ and $\varphi(t)$. Find an effective potential $V_{\text{eff}}(z)$ for the z motion, and give a description of the typical trajectories.

119 Mass m_1 is hanging from the ceiling to which it is attached by a spring with spring constant k and natural length l; mass m_2 is attached to mass m_1 by a string of length a. Neglect the masses of the spring and the string, restrict the motion of mass m_1 to vertical up-down, and the motion of mass m_2 to a vertical plane. State the Lagrange function for this coupled system and the implied equations of motion.

120 Mass m is hanging vertically at the bottom end of a spring with spring constant k and natural length l; the mass of the spring can be disregarded. A device moves the top end of the spring up and down so that its height above a certain reference level is given by $z_{\text{top}}(t)$, which is an externally-determined given function of time. Denote by $z(t)$ the height of mass m above the reference level, state the Lagrange function for $z(t)$ and $\dot{z}(t)$, derive the equation of motion for $z(t)$, and solve it with the initial conditions $z(0) = z_0$, $\dot{z}(0) = 0$.

121 A point mass m is moving without friction in the horizontal xy plane. Three equal springs (spring constant $k = m\omega_0^2$, natural length a) are used to attach the point mass to $(x_1, y_1) = (-a, 0)$, $(x_2, y_2) = (0, -a)$, and $(x_3, y_3) = (a\cos\theta_0, a\sin\theta_0)$ with $0 < \theta_0 < \frac{1}{2}\pi$. The masses of the springs are negligibly small. What is the Lagrange function for this situation? Verify that the potential energy has its minimum at $(x, y) = (0, 0)$.

122 Point mass m is moving along the horizontal x axis. A spring of natural length a and spring constant k connects the mass to point $(0, a)$ on the y axis. State the Lagrange function $L(t, x, \dot{x})$ and derive the equation of motion for $x(t)$. Is there a conserved quantity? For the parameterization $x = a\sinh\vartheta$, state the Lagrange function $L(t, \vartheta, \dot{\vartheta})$, and derive the equation of motion for $\vartheta(t)$. Which approximate equations of motion apply for $|x| \ll a$ and $|\vartheta| \ll 1$?

123 The motion of a point mass m is described by the Lagrange function

$$L = \frac{m}{2}\boldsymbol{v}^2 + \boldsymbol{v} \cdot \boldsymbol{f}(\boldsymbol{r}),$$

where $\boldsymbol{f}(\boldsymbol{r})$ is a given vector field. Derive the Newton's equation of motion obeyed by $\boldsymbol{r}(t)$, and then identify the physical situation.

124 The dynamics of a point particle (mass m, position $\boldsymbol{r}(t)$, velocity $\boldsymbol{v}(t)$) is described by the Lagrange function

$$L = \frac{m}{2}\boldsymbol{v}^2 - V(\boldsymbol{r}) + \boldsymbol{v} \cdot \boldsymbol{\nabla} u(\boldsymbol{r}),$$

where $V(\boldsymbol{r})$ is the potential energy and $u(\boldsymbol{r})$ is some given function of position $\boldsymbol{r}$. Derive the Newton's equation of motion for the particle and so show that the same physical system is described, irrespective of which $u(\boldsymbol{r})$ is chosen.

125 What physical system would be described by Bateman's* Lagrange function $L = m\dot{x}\dot{y} + m\gamma x\dot{y} - kxy$?

126 The Lagrange function for a particle of mass m in relativistic motion is

$$L = mc^2 - mc\sqrt{c^2 - \boldsymbol{v}^2} - V(\boldsymbol{r}),$$

where c is the speed of light and $V(\boldsymbol{r})$ is the potential energy. State the implied equation of motion and explain in which sense there is a "velocity-dependent mass." Do you get the familiar nonrelativistic expressions when $|\boldsymbol{v}| \ll c$?

Chapter 9

127 Show that ${X^{(\ell)}}^{\mathrm{T}} M X^{(\ell')} = 0$ if $\omega_\ell^2 \neq \omega_{\ell'}^2$ for the normal modes of Section 9.2, and then express the coefficients a_l and b_l in (9.2.4) in terms of $X_0 = X(t=0)$ and $\dot{X}_0 = \dot{X}(t=0)$.

*Harry BATEMAN (1882–1946)

128 Express the Lagrange function in (9.3.24) in terms of the cartesian coordinates y_1, y_2 and their velocities, that is $L(y_1, \dot{y}_1, y_2, \dot{y}_2)$. Then find the normal modes for $a_1 = a_2 = g/\omega_0^2$ and $m_1 = 3m_2 = 3m$.

129 Consider the situation of (8.3.19) for

$$m_1 = \frac{16}{25}M\,, \quad m_2 = \frac{9}{25}M\,, \quad k_1 = \frac{4}{5}M\omega_0^2\,, \quad \text{and} \quad k_2 = \frac{36}{125}M\omega_0^2\,,$$

and find the normal modes.

130 Two point masses m are coupled to three springs as shown in the figure, with the far ends of the outer springs fixed to nearby walls:

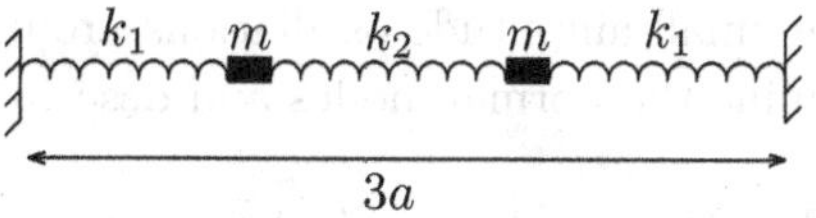

Here, k_1 and k_2 are the spring constants for the outer springs and the inner spring, respectively. Each spring has length a when it is relaxed. The masses of the springs are negligibly small. The two point masses can only move along the horizontal line specified by the figure, and no forces other than those of the springs are acting. Find the normal modes and their characteristic frequencies. Use words and suitable sketches to describe the normal modes.

131 For a single plane pendulum, we have the Lagrange function (9.3.23) with $m_1 = m$, $a_1 = a$, and $m_2 = 0$. Show that an oscillation with amplitude α_0 has the period

$$T(\alpha_0) = \sqrt{\frac{2a}{g}} \int_{-\alpha_0}^{\alpha_0} \frac{\mathrm{d}\alpha}{\sqrt{\cos\alpha - \cos\alpha_0}}\,.$$

Then find the leading correction to $T(\alpha_0) \cong T(0) = 2\pi\sqrt{a/g}$, which is the period of (9.3.17) for small-amplitude oscillations.

132 Two equal point masses m are moving without friction in the vertical xy plane, with the gravitational acceleration $\boldsymbol{g} = -g\boldsymbol{e}_y$; the top mass moves along the cycloid parameterized by $(x, y) = R(\phi + \sin\phi, 1 - \cos\phi)$ with $-\pi < \phi < \pi$ and $R > 0$; the bottom mass is connected to the top mass by a massless string of length $3R$; the string is always fully stretched and has angle θ with the vertical direction:

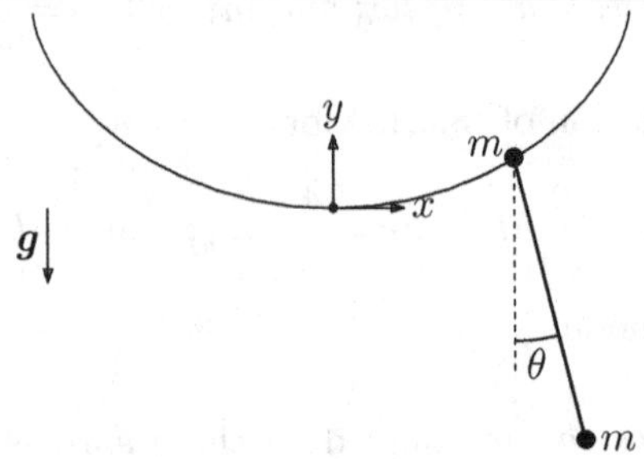

State the Lagrange function $L(\phi, \dot{\phi}, \theta, \dot{\theta})$ and also find the approximate Lagrange function for small-amplitude oscillations around the equilibrium configuration. Determine the normal modes and describe them.

133 Find the normal modes for the physical systems of Exercises 119 and 121. Use words and suitable sketches to describe them.

134 Two equal point masses m can move without friction in the horizontal xy plane. A spring (natural length a, spring constant $k = m\omega_0^2$) connects the two masses. Another spring of the same kind connects one mass to point $(x, y) = (-3a, 0)$, and yet another spring of that kind connects the other mass to point $(x, y) = (3a, 0)$. Taking into account that the masses can move both in the x direction and in the y direction, find the characteristic frequencies of small-amplitude oscillations and describe the normal modes.

135 Find the period of small-amplitude oscillations in the situation of Exercise 122.

136 Consider a triatomic molecule as in Section 9.3.3, but now the two masses m_1 are at x_2 and x_3, and the mass m_2 is at x_1 with $x_1 < x_2 < x_3$. The right spring between the two masses m_1 has spring constant k_1 and natural length a_1; the left spring between the mass m_2 and the central mass m_1 has spring constant k_2 and natural length a_2. State the Lagrange function, assume that the center-of-mass is at rest, and find the normal modes for $m_1 = 4m_2$ and $k_1 = 2k_2$.

137 A simple diatomic molecule, such as CO or HCl, can be modeled by two point masses m_1 and m_2 at a fixed distance a and an electric dipole moment $\boldsymbol{d}$ that is along the line of sight between the atoms. The potential energy of the dipole moment in an electric field $\boldsymbol{E}$ is $-\boldsymbol{d}\cdot\boldsymbol{E}$. Ignore gravity, assume that $\boldsymbol{E}$ is homogeneous (that is: does not depend on position $\boldsymbol{r}$), and consider a molecule that is at rest at $t=0$ with $\boldsymbol{d}_0=\boldsymbol{d}(t=0)$ neither parallel nor antiparallel to $\boldsymbol{E}$. Derive the equation of motion for $\boldsymbol{d}(t)$ and solve it for small-amplitude oscillations.

Chapter 10

138 Point mass m moves along the x axis with the constant acceleration a, so that

$$L(x,\dot{x})=\frac{1}{2}m\dot{x}^2+max$$

is the Lagrange function. Find $x(t)$ for given initial $x(t_0)=x_0$ and final $x(t_1)=x_1$. Then determine the action

$$W(x_0,t_0;x_1,t_1)=\int\limits_{t_0}^{t_1}\mathrm{d}t\,L(x,\dot{x})\,.$$

Now differentiate $W(x_0,t_0;x_1,t_1)$ with respect to each of its arguments. Do you get what you expect?

139 As usual, $X^{\mathrm{T}}=(x_1,x_2,\dots,x_n)$ is the row of coordinates, and $\dot{X}^{\mathrm{T}}=(\dot{x}_1,\dot{x}_2,\dots,\dot{x}_n)$ is the row of velocities. The Lagrange function has the form

$$L(t,X,\dot{X})=\frac{1}{2}\dot{X}^{\mathrm{T}}M(t,X)\dot{X}-V(t,X)$$

where $M(t,X)$ is a symmetric, invertible $n\times n$ mass matrix that could depend on the coordinates X and on time t. Show that the corresponding Hamilton function is

$$H(X,P,t)=\frac{1}{2}P^{\mathrm{T}}M(t,X)^{-1}P+V(t,X)\,,$$

where $P^{\mathrm{T}}=(p_1,p_2,\dots,p_n)$ is the row of momenta, and then verify that $\dfrac{\partial H}{\partial x_k}=-\dfrac{\partial L}{\partial x_k}$.

140 Starting with the Lagrange function (9.3.23), find the Hamilton function for the double pendulum. Which approximate Hamilton function applies when small-amplitude oscillations are considered? Compare it with the Hamilton function that is associated with the Lagrange function (9.3.24).

141 What are the Hamilton functions for the Lagrange functions in Exercises 118–120 and 123–126?

142 With the axis vector $\boldsymbol{A}$ of Exercise 82 and an infinitesimal vector $\boldsymbol{\epsilon}$, take $\boldsymbol{\epsilon}\cdot\boldsymbol{A}$ for "ϵF" in (10.7.3) and find the resulting infinitesimal variations $\delta\boldsymbol{r}$ and $\delta\boldsymbol{p}$ of the position and the momentum.

143 Two unit vectors $\boldsymbol{e}_1$ and $\boldsymbol{e}_2$, which need not be orthogonal, pick out one component of the dyadic D of Exercise 90. Use this $F = \boldsymbol{e}_1\cdot\mathsf{D}\cdot\boldsymbol{e}_2$ in (10.7.3) and find the resulting infinitesimal variations $\delta\boldsymbol{r}$ and $\delta\boldsymbol{p}$ of the position and the momentum.

144 For the Hamilton function

$$H(x,p,t) = \frac{p^2}{2m} - \frac{F_0 x}{\cosh(t/T)^2}$$

with constant positive F_0 and T, find the phase-space density $\rho(x,p,t)$ in terms of $\rho_0(x,p) = \rho(x,p,t=0)$.

145 For phase-space density $\rho(X,P,t)$, the *mean value* $\langle F\rangle$ of a phase-space function $F(X,P,t)$ is its ρ-weighted average over phase space,

$$\langle F\rangle = \frac{\displaystyle\int(\mathrm{d}X)(\mathrm{d}P)\,\rho(X,P,t)F(X,P,t)}{\displaystyle\int(\mathrm{d}X)(\mathrm{d}P)\,\rho(X,P,t)},$$

where the integration covers all of phase space. Show that

$$\frac{\mathrm{d}}{\mathrm{d}t}\langle F\rangle = \left\langle\frac{\mathrm{d}F}{\mathrm{d}t}\right\rangle = \left\langle\frac{\partial F}{\partial t} + \{F,H\}\right\rangle.$$

146 The *variance* $\sigma(F)$ of a phase-space function $F(X,P,t)$ is the mean value of $(F-\langle F\rangle)^2$, that is: $\sigma(F) = \langle(F-\langle F\rangle)^2\rangle$. Verify that

$$\sigma(F) = \langle F^2\rangle - \langle F\rangle^2,$$

which is the more frequently used expression.

147 For point mass m in force-free motion along the x axis, find the time-dependent mean values of position x and momentum p as well as their variances in terms of expected values at the initial time $t = 0$.

148 Repeat Exercise 147 for a one-dimensional harmonic oscillator.

Chapter 11

149 A very thin and flat body has a mass density that is well approximated by

$$\rho(\boldsymbol{r}) = \sigma(x, y)\delta(z),$$

where $\sigma(x, y)$ is the mass per unit area in the xy plane. Show that, in this situation, we have $I_{zz} = I_{xx} + I_{yy}$ for the diagonal matrix elements of the inertia dyadic.

150 A body has mass density $\rho(\boldsymbol{r})$ with the center-of-mass at $\boldsymbol{r} = 0$. How are the inertia dyadic $\mathbf{I}$ and the quadrupole moment dyadic $\mathbf{Q}$ related to each other?

151 Body 1 has center-of-mass position $\boldsymbol{R}_1$, mass M_1, and inertia dyadic $\mathbf{I}_1$; likewise, there are $\boldsymbol{R}_2$, M_2, and $\mathbf{I}_2$ for body 2. What is the inertia dyadic $\mathbf{I}$ of this two-body system? What is the corresponding statement about quadrupole moment dyadics?

152 Four equal point masses m are placed at the non-adjacent corners of a cube with volume a^3. What is the inertia dyadic of this four-body system? One of the four masses is removed. What is the inertia dyadic of the remaining three-body system? State your answer in terms of $\boldsymbol{r}_4$, the vector from the center of the cube to the corner of the removed fourth mass.

153 What is the inertia dyadic for the body of Exercise 103; for the ball of Exercise 105; for the planet of Exercise 106; for the pair of balls of Exercise 107?

154 The surface of a rigid body of mass M is the oblate ellipsoid given by $x^2 + y^2 + 4z^2 = 4R^2$ with $R > 0$. The body has a homogeneous mass density ρ_0, except for a ball-shaped core of radius a that has density $2\rho_0$.

Find the inertia dyadic in terms of M and R. The body rotates with angular velocity $\boldsymbol{\omega} = \omega(\boldsymbol{e}_x \sin\theta + \boldsymbol{e}_z \cos\theta)$ about an axis through its center. What is the angular momentum?

155 A homogeneous body has the shape of a conical frustum, that is: its surface is made up by the disks with $z = 0, s \le a$ and $z = h, s \le b < a$ as well as the truncated cone with $0 \le z \le h, s = a - (a-b)z/h$. Find the inertia dyadic for rotations around the center-of-mass, and for rotations around the tip of the cone at $z = ah/(a-b)$. The body rotates with angular velocity $\boldsymbol{\omega} = \omega(\boldsymbol{e}_x \sin\theta + \boldsymbol{e}_z \cos\theta)$ about an axis through its center-of-mass. What is the angular momentum?

156 A physical pendulum is a body that can rotate freely about a horizontal axis that does not pass through the center-of-mass. Suppose you have found two parallel axes, one at distance a from the center-of-mass, the other at distance b, such that the period of small-amplitude oscillations is the same for both axes. Show that $a+b = g\left(\frac{T}{2\pi}\right)^2$, where T is that common period. This observation is exploited when the so-called *Kater's* pendulum* is used to measure g rather precisely. How?

157 According to Euler, any rotation can be realized by the following sequence of three rotations by the Euler angles φ, ϑ, and ψ: first rotate around the z axis by angle φ, this turns the unit vectors $(\boldsymbol{e}_x, \boldsymbol{e}_y, \boldsymbol{e}_z)$ into $(\boldsymbol{e}_{x'}, \boldsymbol{e}_{y'}, \boldsymbol{e}_{z'})$ with $\boldsymbol{e}_{z'} = \boldsymbol{e}_z$; then rotate around the y' axis by angle ϑ, this turns the unit vectors $(\boldsymbol{e}_{x'}, \boldsymbol{e}_{y'}, \boldsymbol{e}_{z'})$ into $(\boldsymbol{e}_{x''}, \boldsymbol{e}_{y''}, \boldsymbol{e}_{z''})$ with $\boldsymbol{e}_{y''} = \boldsymbol{e}_{y'}$; finally rotate around the z'' axis by angle ψ. Show that this is equivalent to this sequence: first rotate around the z axis by angle ψ, then around the y axis by angle ϑ, finally around the z axis by angle φ.

158 Upside-down pendulum: A thin rigid homogeneous rod of length ℓ and mass m is standing upright, but is slightly off the exact vertical position and tends to fall over because the gravitational acceleration g is pulling it down. The bottom end of the rod is periodically moved vertically up and down along the z axis, such that its acceleration switches between λg and $-\lambda g$ at regular instants separated by the half-period $T/2$, whereby λ is a positive constant parameter. With the top of the rod at distance $s(t) = \sqrt{x(t)^2 + y(t)^2}$ from the z axis, state the equations of motion for $x(t)$ and $y(t)$ when $s \ll \ell$. Which condition must be met by λ and T, so that the rod stays upright and does not fall over?

*Henry KATER (1777–1835)

Chapter 12

159 Is the centrifugal force $\boldsymbol{F}(\boldsymbol{r}) = -m\boldsymbol{\omega} \times (\boldsymbol{\omega} \times \boldsymbol{r})$ conservative? If yes, state its potential energy.

160 You are at the equator and throw a stone upward from ground level. Neglect the frictional forces and also the centrifugal force, and account for the Coriolis force to first order. Is there a deflection (east or west?) when the stone is back at ground level, or do the Coriolis forces of the upward and downward motion compensate for each other?

161 An anisotropic oscillator with potential energy $V(\boldsymbol{r}) = \frac{1}{2}k_1x^2 + \frac{1}{2}k_2y^2 + \frac{1}{2}k_3z^2$ is in a frame that rotates with angular velocity $\boldsymbol{\Omega} = \Omega\boldsymbol{e}_z$. Including the Coriolis force and the centrifugal force, the equation of motion is

$$m\ddot{\boldsymbol{r}} = -\boldsymbol{\nabla}V(\boldsymbol{r}) - 2m\boldsymbol{\Omega} \times \dot{\boldsymbol{r}} - m\boldsymbol{\Omega} \times (\boldsymbol{\Omega} \times \boldsymbol{r}) .$$

The exponential ansatz $\boldsymbol{r}(t) = \boldsymbol{c}\,\mathrm{e}^{\mathrm{i}\omega t}$ with constant vector $\boldsymbol{c}$ yields a third-degree characteristic polynomial in ω^2. How do you have to choose k_1, k_2, k_3, and Ω to ensure that all roots of the polynomial are positive?

162 A point mass m is near $\boldsymbol{r} = 0$, where we have the center of a ring with a large diameter; the ring and the point mass carry electric charge of the same sign, so that the force on the point mass derives from the potential energy

$$V(\boldsymbol{r}) = \frac{1}{2}m\omega_0^2\left[r^2 - 3(\boldsymbol{n}\cdot\boldsymbol{r})^2\right],$$

where $\boldsymbol{n}$ is the unit vector normal to the plane of the ring. The ring rotates about a diameter with angular velocity Ω, so that

$$\boldsymbol{n} = \boldsymbol{e}_x\cos(\Omega t) + \boldsymbol{e}_y\sin(\Omega t),$$

if we choose the z axis as the axis of rotation and have the ring in the yz plane at $t = 0$. What is the equation of motion for the point mass in the laboratory frame? What is the equation of motion in the rotating frame in which the ring does not move? How do we need to choose Ω to ensure that the point mass remains near $\boldsymbol{r} = 0$?

Hints

1 Express either one of the three vectors in terms of the other two.

2 Recall the defining properties of the delta symbol and the epsilon symbol; glance at page 10.

3 The geometrical meaning of $\boldsymbol{a} \times \boldsymbol{b}$ and $\boldsymbol{c} \times \boldsymbol{d}$ matters.

4 Consider $(\boldsymbol{a} \times \boldsymbol{b}) \cdot \boldsymbol{r}$ and the like. The trio of vectors $\boldsymbol{a}'$, $\boldsymbol{b}'$, $\boldsymbol{c}'$ that appear in $\alpha = \boldsymbol{a}' \cdot \boldsymbol{r}$, $\beta = \boldsymbol{b}' \cdot \boldsymbol{r}$, $\gamma = \boldsymbol{c}' \cdot \boldsymbol{r}$ are called the *reciprocal vectors* to the trio $\boldsymbol{a}$, $\boldsymbol{b}$, $\boldsymbol{c}$.

5 Use (1.1.37) three times.

6 The $\boldsymbol{a}_j$s could be the normal vectors of the four triangular faces of a tetrahedron, or the vectors pointing from the center of a cube to non-adjacent corners.

7 Either one of the three methods of Section 1.1.8 works when suitably adapted to the new circumstances.

8 The Jacobi identity will be helpful. You should find that the net result is an infinitesimal rotation by $\boldsymbol{\delta\phi} = \boldsymbol{\delta\phi}_2 \times \boldsymbol{\delta\phi}_1$.

9–11 Remember that the vector specifying the axis of the over-all rotation is not affected by that rotation.

12 Modify the procedure that produced (1.1.108) for cylindrical coordinates.

13 Distance on the surface of the earth matters, and the shortest routes are along great circles.

14, 15 This is closely analogous to the corresponding questions about cylindrical and spherical coordinates that are answered in Sections 1.1.10 and 1.1.11, respectively.

16 Recall the definition of the gradient in (1.2.2).

17 The method of Section 1.2, where expressions for the gradient in cartesian, cylindrical, and spherical coordinates are found, applies also to parabolic coordinates.

18, 19 Remember that the local unit vectors depend on the coordinates.

20 Follow the procedures of Section 1.3.

21 The procedures of Sections 1.2 and 1.3 apply here, too, and the lesson of Exercise 4 is useful.

22 In each of the four cases, identify what is the small quantity for the expansion. No, it is not always t/τ or τ/t.

23 Note that on the way up, both the weight and the frictional force decelerate the stone, on the way down the frictional force decelerates and the weight accelerates; remember that $\vartheta < \sinh\vartheta$ for $\vartheta > 0$.

24, 25 The replacing of $\dot{v} = \pm\ddot{z}$ by $\pm v\dfrac{dv}{dz} = \pm\dfrac{1}{2}\dfrac{dv^2}{dz}$ simplifies the calculation, but it is also fine to find first $v(t)$ and then $z(t)$. For the latter, the identities $\tan\alpha = -\dfrac{1}{2}\dfrac{d}{d\alpha}\log\Bigl((\cos\alpha)^2\Bigr)$ and $\tanh\alpha = \dfrac{1}{2}\dfrac{d}{d\alpha}\log\Bigl((\cosh\alpha)^2\Bigr)$ are useful.

26 You will encounter another version of the pair of differential equations dealt with in Section 1.1.8.

27 Show first that the differential operators $\dfrac{d}{dt} - \lambda$ and $e^{\lambda t}\dfrac{d}{dt}e^{-\lambda t}$ are the same, that is: they have the same effect on a function of t.

28 The required integration over Ω is easy when the integration is split into two parts, $-\infty < \Omega < 0$ and $0 < \Omega < \infty$.

29 Remember that a delta function is meant to be multiplied by another function to form an integrand. If you start with a model of the kind depicted in (2.2.122), the new model will have $\delta_\tau(t = 0) = 0$ and two narrow peaks on both sides of $t = 0$.

30 Remember that expressions involving $\delta(t-t_0)$ are meant to be multiplied by some function of t and integrated.

31 Note that $G(t,t')$ is a solution of the homogeneous differential equation for $t<t'$ and $t>t'$, and is continuous at $t=t'$. The retarded Green's function vanishes for $t<t'$, the advanced for $t>t'$.

32 Which initial position $\boldsymbol{r}_0$ do you need in (2.2.19)?

33 The version of (2.2.142) with $\omega_0=0$, $\omega=\frac{1}{2}\mathrm{i}\gamma$ does it. Why? Alternatively, derive the Green's function needed here by first considering the time derivative of $\mathrm{e}^{\gamma t}\boldsymbol{v}(t)$, or by any other method.

34 You could determine the initial velocity $\boldsymbol{v}_0$ that yields the final position $\boldsymbol{r}(T)=\boldsymbol{r}_1$. Alternatively, find the needed Green's function directly.

35 At very late times, the solution of the homogeneous differential equation does not matter.

36 A change of variables $x(t)=x'(t)+\bar{x}$ with a suitably chosen value for the constant displacement $\bar{x}$ yields a familiar differential equation for $x'(t)$.

37 State your answer in terms of two integrals that must vanish for such a force $F(t)$.

38 No hint needed.

39 What are the additional equations for a_1, a_2, b_1, b_2 in Exercise 31? For critical damping, you should get

$$G_T(t,t')=-\mathrm{e}^{-\frac{1}{2}\gamma(t-t')}\,\frac{t_<(T-t_>)}{T}$$

with $t_<=\min\{t,t'\}$ and $t_>=\max\{t,t'\}$.

40 Make sure that your code can reproduce (2.2.148).

41 First sketch a graph of $V(x)$: It is symmetric in x, negative at $x=0$, positive for large $|x|$, changes sign at $x=\pm a/\sqrt{2}$, ...

42 Just apply (3.1.19). Check that you get the familiar harmonic-oscillator result: $T(E)$ is constant for $\nu=2$.

43 As in Exercise 41, first sketch a graph of $V(x)$; remember (3.1.19).

44 The identity $(\cosh\vartheta_1)^2 - (\cosh\vartheta_2)^2 = (\sinh\vartheta_1)^2 - (\sinh\vartheta_2)^2$ could be useful.

45 The identity $(\tan\alpha)^2 - (\tan\beta)^2 = \dfrac{(\sin\alpha)^2 - (\sin\beta)^2}{(\cos\alpha\,\cos\beta)^2}$ could be useful.

46 This is partly a repeat of Exercise 42 for $\nu = 1$.

47 It is sufficient to average over the half-period during which $x(t) > 0$. This is a preview of the virial theorem of Exercises 77–79.

48, 49 Remember the hints for Exercise 41.

50, 51 Ditto.

52, 53 Ditto. Perhaps try the substitution $x = (y^2 + 2y)a$.

54 Regarding the last question: Consider the limiting situation of an energy so large that you still have periodic motion, but not if the energy is a bit larger.

55 Since no lower bound is stated for E, apply (3.1.35) for $x_0 = 0$ and $V(x_0) = -\infty$.

56 Remember that harmonic oscillators have periods that do not depend on the energy. The parameterization of the cycloid in (7.3.9) is useful.

57, 58 Keep in mind that both $\dfrac{\partial}{\partial x}V(x,y)$ and $\dfrac{\partial}{\partial y}V(x,y)$ must vanish at maxima, minima, and saddle points of the potential energy.

59 You can simplify the integration over the closed loop by first writing $\boldsymbol{F}(\boldsymbol{r})$ as the sum of a gradient and a remainder with a nonzero curl.

60 Glance at page 36.

61 The curl of a scalar field $b(\boldsymbol{r})$ times a vector field $\boldsymbol{B}(\boldsymbol{r})$ is given by the product rule, $\boldsymbol{\nabla} \times (b\boldsymbol{B}) = \boldsymbol{\nabla} b \times \boldsymbol{B} + b\boldsymbol{\nabla} \times \boldsymbol{B}$.

62 Note that (iii)–(v) are particular cases of the force field in Exercise 61.

63 Recall what you found in Exercise 16: $\boldsymbol{k} = \boldsymbol{\nabla} \boldsymbol{k} \cdot \boldsymbol{r} = \boldsymbol{\nabla} \boldsymbol{r} \cdot \boldsymbol{k} = (\boldsymbol{\nabla} \boldsymbol{r}) \cdot \boldsymbol{k}$; apply the chain rule and the product rule of differentiation.

64 Two dyadics are equal if they are represented by the same 3×3 matrices.

65 Remember Exercise 4.

66 What is the scalar product with $\boldsymbol{a}$, $\boldsymbol{b}$, and $\boldsymbol{a} \times \boldsymbol{b}$?

67 Both $\boldsymbol{a} \times \boldsymbol{b} = (\boldsymbol{a} \times \mathbf{1}) \cdot \boldsymbol{b}$ and $\boldsymbol{a} \times (\boldsymbol{b} \times \boldsymbol{c}) = (\boldsymbol{a} \cdot \boldsymbol{c}\,\mathbf{1} - \boldsymbol{c}\,\boldsymbol{a}) \cdot \boldsymbol{b}$ could be useful.

68 Recall Exercise 9.

69 For each of the three dyadics, there are some vectors that suggest themselves; try them out. Alternatively, you can exploit the matrix representations.

70 Keep in mind that the trace and the determinant are the sum and the product, respectively, of the eigenvalues, and remember that the characteristic polynomial equals zero, $(\mathsf{A} - \alpha_1 \mathbf{1})(\mathsf{A} - \alpha_2 \mathbf{1})(\mathsf{A} - \alpha_3 \mathbf{1}) = 0$, where α_1, α_2, α_3 are the eigenvalues of A.

71 The positive dyadics are represented by positive 3×3 matrices, which are symmetric.

72 Both identities are immediate consequences of the definition.

73 Remember that the vector product of two vectors is another vector and responds to rotations accordingly.

74 Write $\mathsf{A} = \sum_j \boldsymbol{e}_j \lambda_j \boldsymbol{e}_j'$ as in Exercise 71 and note that $\epsilon_{jkl}\epsilon_{klm} = \delta_{jm}\epsilon_{jkl}{}^2$.

75 Being "trapped" means that $\boldsymbol{r}(t)$ and $\boldsymbol{v}(t)$ are restricted to some finite range around $\boldsymbol{r} = 0$ and $\boldsymbol{v} = 0$, and it is enough to look at, say, $(\omega_0 \boldsymbol{r})^2 + \boldsymbol{v}^2$ for $t = T, 2T, 3T, \ldots$.

76 No hint needed.

77 There is a finite range of values for $\boldsymbol{r}_j(t) \cdot \boldsymbol{v}_j(t)$.

78 Exploit the familiar relation between force and potential energy.

79 The differential operator $\boldsymbol{r}\cdot\boldsymbol{\nabla}$ counts the powers of r, $\boldsymbol{r}\cdot\boldsymbol{\nabla}r^n = nr^n$.

80 The moon is between the sun and the earth during a solar eclipse.

81 Remember Kepler's Second Law.

82 When calculating $\frac{\mathrm{d}}{\mathrm{d}t}\boldsymbol{A}$, keep in mind that $\boldsymbol{r}\times\boldsymbol{v}$ is constant; remember Kepler's Third Law; the relation $\frac{\boldsymbol{r}}{r}\cdot\boldsymbol{v} = -\frac{\kappa\epsilon\sin\varphi}{(1-\epsilon^2)a}$ is useful (establish it before you use it).

83 Recall Exercise 79; establish $\overline{V(r)} = -\frac{Gmm_\odot}{\kappa T}\int_0^{2\pi}\mathrm{d}\varphi\, r$ and use (5.3.28).

84 First show $\int_{s_1}^{s_2}\mathrm{d}s\,\frac{s}{\sqrt{(s_2-s)(s-s_1)}} = \frac{\pi}{2}(s_1+s_2)$, then use it.

85, 86 As compared with the Kepler's ellipses, $\kappa^2 + 2B/m$ replaces κ^2 in the effective potential energy.

87 Consider the graph of $V_{\mathsf{eff}}(s)$.

88 Observe that $V_{\mathsf{eff}}(s)$ is positive for very small and for very large distances s; conclude that the minimum of $V_{\mathsf{eff}}(s)$ is negative if there is a bound orbit for $E = 0$; use (5.3.17) and (5.3.21).

89 Use (5.3.17) and (5.3.21) once more.

90 Find $\frac{\mathrm{d}}{\mathrm{d}t}\mathsf{D}$.

91 $x(t), y(t)$ trace out a centered ellipse if $(x\;y)A\begin{pmatrix}x\\y\end{pmatrix} = 1$ with a positive, symmetric 2×2 matrix A.

92 You should find the right-hand side of (5.3.21) with s_2 replaced by ∞; by setting $\kappa^2 = \frac{2}{m}Eb^2$ you express θ in terms of E and b. The integral

$$\int_{s_1}^{\infty}\frac{\mathrm{d}s}{s\sqrt{(s+s_0)(s-s_1)}} = \frac{2}{\sqrt{s_0 s_1}}\tan^{-1}\left(\sqrt{s_0/s_1}\right) \quad \text{for} \quad s_1\,,\ s_1+s_0>0$$

could be useful; derive this as an exercise; the substitution $s = \frac{1}{2}(s_1 - s_0) + \frac{1}{2}(s_1 + s_0)\cosh\vartheta$ is worth trying.

93 The velocities in the lab frame (target at rest) differ from the velocities in the center-of-mass frame (center-of-mass at rest) by the center-of-mass velocity in the lab frame. The relation between Θ and θ looks simplest as a statement about $\tan\Theta$.

94 Note that $\mathrm{d}\sigma = \frac{\mathrm{d}\sigma}{\mathrm{d}\Omega}\mathrm{d}\Omega = \left(\frac{\mathrm{d}\sigma}{\mathrm{d}\Omega}\right)_{\mathrm{lab}}\mathrm{d}\Omega_{\mathrm{lab}}$ with $\mathrm{d}\Omega = \mathrm{d}\phi\,\mathrm{d}\theta\,\sin\theta$ and $\mathrm{d}\Omega_{\mathrm{lab}} = \mathrm{d}\phi\,\mathrm{d}\Theta\,\sin\Theta$.

95 Remember Exercise 92.

96 Note that $\boldsymbol{v}^2$, $\boldsymbol{r}\times\boldsymbol{v} + \kappa_0\frac{\boldsymbol{r}}{r}$, and $(\boldsymbol{r}\times\boldsymbol{v})^2$ are constant in time, and conclude that the point mass moves with constant speed v on a cone. Find $r(t)$, $\boldsymbol{r}(t)$, and $\boldsymbol{v}(t)$, then the scattering angle as a function of $\kappa_0/(vb)$. Pay attention to the remarks on pages 142 and 143. Use numerical tools to produce a plot of the differential cross section.

97 You need to use some simple trigonometric identities and should find that $y(x)$ is a monotonically decreasing function of x.

98 Convert the integral over $\cos\theta$ into an integral over x.

99, 100 Keep paying attention to the remarks on pages 142 and 143.

101 No hint needed.

102 The potential energy exceeds energy E inside the ball, and is zero outside.

103 The integration is simplified by noting that $\mathbf{Q} = 3\,\mathbf{Q}' - \mathrm{tr}\{\mathbf{Q}'\}\,\mathbf{1}$ with $\mathbf{Q}' = \int(\mathrm{d}\boldsymbol{r})\,\rho(\boldsymbol{r})\,\boldsymbol{r}\,\boldsymbol{r}$ and that substitutions such as $x = \left(\frac{a^2}{bc}\right)^{1/3} x'$ turn the integral over the ellipsoid into an integral over a ball.

104 Use (1.2.19) and (2.2.114).

105, 106 Newton's shell theorem is useful here.

107 Keep in kind that, as a consequence of Newton's shell theorem, the forces that the balls exert on each other are equal to those between point masses; Euler's beta function integral (see Exercise 42) for $\alpha = \frac{1}{2}$, $\beta = -\frac{1}{2}$ could be useful.

108 Bring the functional into the standard form of (7.2.1) and note that there is no dependence on $y(x)$.

109 Adopt a local coordinate system such that $\boldsymbol{\nabla} n \propto \boldsymbol{e}_x$; note that $\mathrm{d}\ell = \mathrm{d}t\, v = \mathrm{d}t\, |\boldsymbol{v}|$ and observe that $\dot{\boldsymbol{v}}$ is in the plane spanned by $\boldsymbol{v}$ and $\boldsymbol{\nabla} n$.

110 Consider $0 \leq \phi_0 \ll 1$ and $0 \leq 2\pi - \phi_0 \ll 1$.

111, 112 Just like when driving a car, the average speed is the distance traveled divided by the duration of the trip.

113 Consider a path composed of straight-line segments.

114 Consider $L^2 - (y_1 - y_0)^2$.

115 Think of the right-hand side as a single integral of the kind met in (7.2.1).

116 Exploit the differential equation found in Exercise 115 for an integration by parts.

117 The differential equation found in Exercise 115 is useful.

118–122 In each case, use the suggested coordinates or choose them appropriately, and state the kinetic energy and the potential energy. Check for cyclic coordinates.

123, 124 These are special cases of (10.10.21).

125 The equation of motion for $x(t)$ is familiar; that for $y(t)$ is not.

126 No hint needed.

127 When $\omega_\ell^2 = \omega_{\ell'}^2$, choose $X^{(\ell)}$ such that ${X^{(\ell)}}^{\mathrm{T}} M X^{(\ell')} \propto \delta_{\ell,\ell'}$. Why is this always possible?

128–130 Just apply the procedure of Section 9.2.

131 Apply (3.1.19); establish

$$\left(\cos\alpha - \cos\alpha_0\right)\Big|_{\alpha\,=\,\alpha_0\sin\varphi} = \frac{1}{2}(\alpha_0\cos\varphi)^2\left(1 - \frac{\alpha_0^2}{12}\left[1 + (\sin\varphi)^2\right] + \cdots\right)$$

and use this.

132–133 Just apply the procedure of Section 9.2.

134 The situation is similar to, but not identical with, that of Exercise 130.

135 Here, the period depends on the energy; recall Exercise 42.

136 Follow the example of Section 9.3.3.

137 Torque is important here, not force.

138 The PSA tells you what to expect.

139 No hint needed.

140, 141 Keep in mind that you need to express the velocities in terms of the momenta.

142, 143 Remember to express $\boldsymbol{A}$ and D in terms of $\boldsymbol{r}$ and $\boldsymbol{p}$ for proper evaluation of the Poisson brackets.

144 Follow the procedure that established (10.9.16).

145 Integrate by parts.

146 Establish first that $\langle\ \rangle$ is linear: $\langle\lambda_1F_1 + \lambda_2F_2\rangle = \lambda_1\langle F_1\rangle + \lambda_2\langle F_2\rangle$ for numbers λ_1 and λ_2 and phase-space functions F_1 and F_2.

147, 148 Exploit the lesson of Exercise 145 to set up differential equations; then solve them. On the way, you will also need $\langle xp\rangle$ as a function of time.

149 This is an immediate consequence of the definition of I.

150 Dyadic Q' of Exercise 103 is a useful auxiliary quantity.

151 Remember that the inertia dyadic of a body refers to its center-of-mass; use Steiner's theorem.

152 The geometry is that of Exercise 6; apply the lesson of Exercise 151.

153 No hints needed.

154 Think of the body as composed of two homogeneous bodies, both centered at $\boldsymbol{r} = 0$.

155 Note that the frustum is a cone with its tip removed.

156 Apply Steiner's theorem; at an intermediate step you should find that the moment of inertia for the axis through the center-of-mass is Mab; for the measurement one uses equipment with sufficient symmetry to ensure that the parallel axis through center-of-mass is between the two axes with the same period T.

157 Recall how rotations affect scalar products and vector products and show that

$$\mathsf{R}' \cdot \mathsf{R}(\boldsymbol{e}, \alpha) = \mathsf{R}(\mathsf{R}' \cdot \boldsymbol{e}, \alpha) \cdot \mathsf{R}',$$

where $\mathsf{R}(\boldsymbol{e}, \alpha)$ is the rotation dyadic of Exercise 67 for axis $\boldsymbol{e}$ and angle α, and R' is any other rotation dyadic; then look at the over-all dyadic of the three subsequent rotations.

158 The equations of motion are most easily found from the Lagrange function for this problem; regarding the stability criterion, remember Exercise 75.

159 This repeats Exercise 62(v).

160 You could use (12.2.21) with the appropriate $\boldsymbol{r}_0(t)$ and extract the correction to order $\boldsymbol{\Omega}$; or use any other method.

161 The differential equations for $x(t)$ and $y(t)$ are coupled, but that for $z(t)$ is all by itself; therefore, one root of the third-degree polynomial is easy to find.

162 In the rotating frame, there is no parametric time dependence; try an exponential ansatz; the stability criteria are analogous to those of Exercises 75 and 158; make use of your findings in Exercise 161.

Appendix

A On conic sections

A.1 *Foci; vertices; cartesian coordinates; ray optics*

The conic sections are the planar curves that we call ellipse, hyperbola, and parabola — terminology introduced by Appolonius* more than two millenia ago with reference to the Greek words for "falling short," "excessive," and "comparable." Rather than characterizing them as the intersections of planes and circular cones (see Section A.3), we use the standard cartesian-coordinate equations,

$$\begin{aligned} \text{ellipse:} \quad & \left(\frac{x}{a}\right)^2 + \left(\frac{y}{b}\right)^2 = 1 \quad \text{with} \quad a \geq b > 0\,, \\ \text{hyperbola:} \quad & \left(\frac{x}{a}\right)^2 - \left(\frac{y}{b}\right)^2 = 1 \quad \text{with} \quad a, b > 0\,, \\ \text{parabola:} \quad & 4ax = y^2 \quad \text{with} \quad a > 0\,. \end{aligned} \tag{A.1}$$

With $(x, y) = (0, 0)$ at the center of the respective drawing, the x axis horizontal, and the y axis vertical, the thick lines trace out an ellipse on the left and a hyperbola on the right:

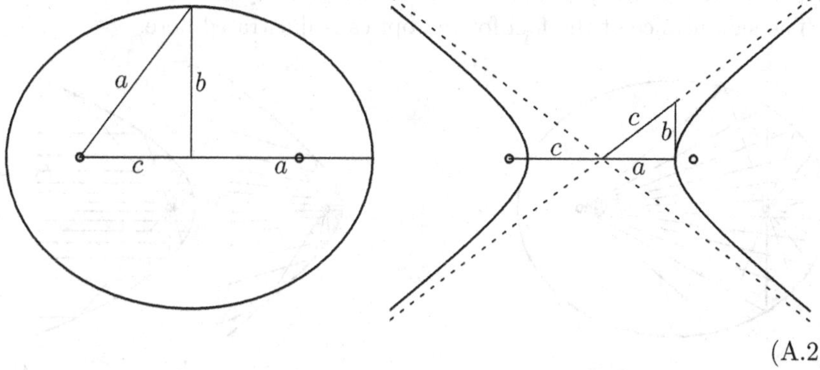

(A.2)

*Appolonius of Perga (262–190 BC)

For both ellipse and hyperbola, the points of largest curvatures, the two *vertices*, are at $(x,y) = (\pm a, 0)$ and the two *foci* (marked by $\circ$) are at $(x,y) = (\pm c, 0)$ with $c = \sqrt{a^2 - b^2} < a$ for the ellipse and $c = \sqrt{a^2 + b^2} > a$ for the hyperbola, respectively. The ellipse is a circle when $a = b$ and the two foci coincide. The hyperbola has two branches, one for each vertex, the right branch with $x \geq a$ and the left branch with $x \leq -a$; the two *asymptotes* of the hyperbola are the dashed straight lines where $\dfrac{x}{a} = \pm\dfrac{y}{b}$.

The Kepler's ellipse of Section 5.2 has a privileged focus, where the sun is located; the nearer vertex is the perihelion, the farther vertex is the aphelion. For the hyperbolic orbit of (5.3.41), there is only the perihelion.

The parabola has a single vertex at $(x,y) = (0,0)$ and a single focus at $(x,y) = (a,0)$:

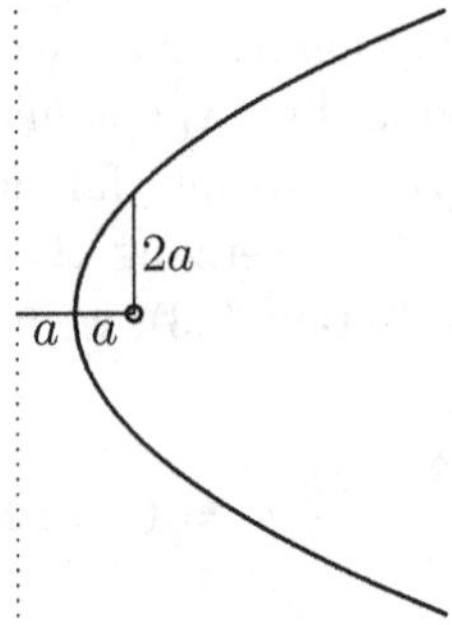

(A.3)

Here, too, the x axis is horizontal and the y axis vertical but, as in (A.2), they are not indicated in the figure. The dotted line at $x = -a$ is the *directrix*; more about it in Section A.2.

In the case of the ellipse, one calls a the *major half-axis* and b the *minor half-axis*; some prefer to speak of the semimajor axis and the semiminor axis, respectively. For a parabola, a is known as the *focal distance.*

The significance of the foci for ray optics is illustrated here:

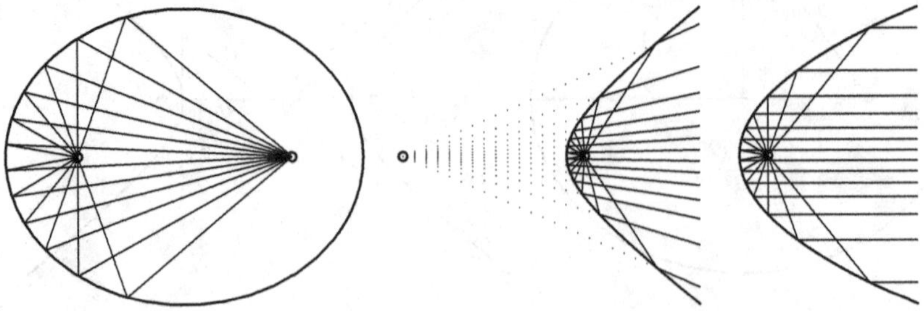

(A.4)

If we put a point source of light into a focus of an ellipse, then all rays reflected at the ellipse go through the other focus; rays originating in a focus of a hyperbola and reflected off one branch appear as if they were emitted by a point source in the other focus; and rays from the focus of a parabola become parallel after reflection off the parabola.

For both the ellipse and the hyperbola, the reflected rays are approximately parallel when the other focus is far away on the scale set by the distance between the focus with the light source and the adjacent vertex. This suggests that the parabola is the limiting shape of an ellipse and also of a hyperbola branch. Indeed, when $a \to \infty$ and $b \to \infty$ such that the locations of a vertex and the nearby focus are fixed, the ellipse or one branch of the hyperbola become a parabola. We leave it to the reader to verify these statements about ray optics and the ellipse/hyperbola → parabola limit.

A.2 *Eccentricity; directrices; polar coordinates*

Convenient parameterizations of the cartesian-coordinate equations in (A.1) are

$$\begin{aligned} \text{ellipse:} \quad & \begin{pmatrix} x \\ y \end{pmatrix} = \begin{pmatrix} a\cos\alpha \\ b\sin\alpha \end{pmatrix}, \\ \text{hyperbola:} \quad & \begin{pmatrix} x \\ y \end{pmatrix} = \begin{pmatrix} a\sec\alpha \\ b\tan\alpha \end{pmatrix} = \begin{pmatrix} \pm a\cosh u \\ b\sinh u \end{pmatrix}, \\ \text{parabola:} \quad & \begin{pmatrix} x \\ y \end{pmatrix} = \begin{pmatrix} au^2 \\ 2au \end{pmatrix}, \end{aligned} \tag{A.5}$$

where α is a 2π-periodic angle parameter whereas u takes on all real values. The α parameterization of the hyperbola covers both branches with $-\frac{1}{2}\pi < \alpha < \frac{1}{2}\pi$ for the branch with $x \geq a$ and $\frac{1}{2}\pi < \alpha < \frac{3}{2}\pi$ for the other; each of the two u parameterizations is for one branch.

In particular, these parameterizations give simple expressions for the distances $f_\pm$ of a point on an ellipse or hyperbola to the foci,

$$f_\pm = \sqrt{(x \mp c)^2 + y^2} \qquad \text{for the focus at } x = \pm c. \tag{A.6}$$

For an ellipse, we have

$$\begin{aligned} f_\pm^2 &= (a\cos\alpha \mp c)^2 + (b\sin\alpha)^2 \\ &= (a\cos\alpha)^2 \mp 2ac\cos\alpha + \underbrace{c^2+b^2}_{=a^2} - \underbrace{(b\cos\alpha)^2}_{=(a^2-c^2)(\cos\alpha)^2} \\ &= a^2 \mp 2ac\cos\alpha + (c\cos\alpha)^2 = (a \mp c\cos\alpha)^2 \end{aligned} \tag{A.7}$$

or

$$f_\pm = a \mp c\cos\alpha\,, \tag{A.8}$$

since $a > c$ for an ellipse. Likewise, we find

$$f_\pm^2 = (a \mp c\sec\alpha)^2 \tag{A.9}$$

for a hyperbola or

$$f_\pm = \begin{cases} \mp a + c\sec\alpha & \text{for the branch with} \quad \sec\alpha > 1\,, \\ \pm a - c\sec\alpha & \text{for the branch with} \quad \sec\alpha < -1\,, \end{cases} \tag{A.10}$$

since $a < c$ for a hyperbola. It follows that $f_+ + f_- = 2a$ for all points of an ellipse, $f_- - f_+ = 2a$ for the hyperbola branch with $x \geq a$, and $f_+ - f_- = 2a$ for the other branch.

Accordingly, an ellipse is the locus of a point that moves such that the distances from the two foci have a constant sum; we exploited this property of an ellipse in the context of (5.2.33)–(5.2.36), and it is also behind the usual way of drawing an ellipse by a pencil that stretches a string with fixed ends. For each branch of a hyperbola, the distances from the foci have a constant difference.

For points of a parabola, the distance f to the single focus,

$$f = \sqrt{(x-a)^2 + y^2} = \sqrt{(x-a)^2 + 4ax} = x + a = d\,, \tag{A.11}$$

is equal to the distance d from the directrix, the dotted line with $x = -a$ in (A.3). Therefore, a parabola is the locus of all points that are equally distant from a line and a point — the directrix and the focus, respectively.

There is also a directrix for each focus of an ellipse or hyperbola because (A.8) and both cases of (A.10) are summarized in

$$f_\pm = \left|\frac{c}{a}x \mp a\right| = \frac{c}{a}\left|x - (\pm a^2/c)\right| \tag{A.12}$$

or

$$f_\pm = \epsilon d_\pm \quad \text{with} \quad \epsilon = \frac{c}{a} \quad \text{and} \quad d_\pm = |x - (\pm a/\epsilon)|\,. \tag{A.13}$$

Here, ϵ is the *numerical eccentricity*, or simply the eccentricity, with $0 < \epsilon < 1$ for an ellipse and $\epsilon > 1$ for a hyperbola; and $d_\pm$ is the distance from the directrix at $x = \pm a/\epsilon$. The distance c from the center of the ellipse or hyperbola to a focus is called the *linear eccentricity*. A circle has $\epsilon = 0$ and no useful directrix since it would have to be infinitely far away.

We assign unit eccentricity to a parabola, $\epsilon = 1$, and then

$$f = \epsilon d \tag{A.14}$$

applies to ellipse, hyperbola, and parabola alike: A conic section with given eccentricity is the locus of all points for which the ratio of the distance from a point (= focus) and the distance from a line (= directrix) equals the eccentricity. The focus must not be on the directrix, unless we want to regard the point $(x, y) = (0, 0)$ as a special case of the ellipse and the pair of asymptotes as a special case of the hyperbola in (A.2), and the ray with $x > 0$ and $y = 0$ as a special case of the parabola in (A.3).

Distance from the focus = eccentricity times distance from the directrix is illustrated here:

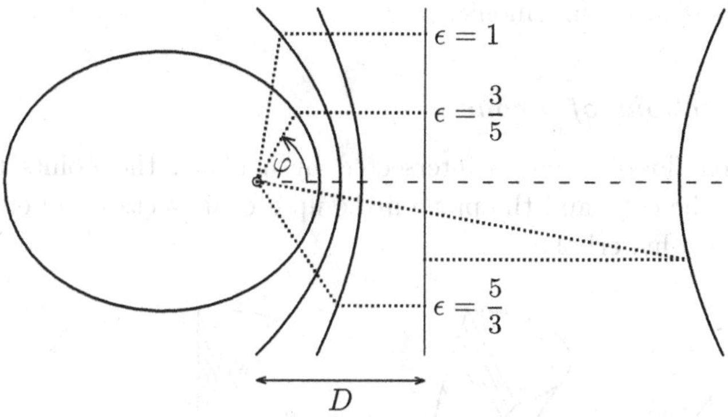

(A.15)

for the ellipse with $\epsilon = \frac{3}{5}$, the parabola ($\epsilon = 1$), and the hyperbola with $\epsilon = \frac{5}{3}$. The vertex is on the dashed horizontal line that goes through the focus and is normal to the directrix. If we denote the distance from focus to vertex by $s_{\min}$, and by D the distance from focus to directrix, then

$$s_{\min} = \frac{\epsilon D}{\epsilon + 1} \tag{A.16}$$

for the ellipse, the parabola, and the near branch of the hyperbola, and

$$s_{\min} = \frac{\epsilon D}{\epsilon - 1} \tag{A.17}$$

for the far branch.

Regarding the conic sections as these loci that refer to focus and directrix is particularly useful for finding the distance $s(\varphi)$ from the focus as a function of the azimuth φ that is indicated in (A.15) for the ellipse. With polar coordinates centered at the focus, we have $f = s$ and $d = D - x = D - s\cos\varphi$ in (A.14) and find

$$s(\varphi) = \frac{1+\epsilon}{1+\epsilon\cos\varphi} s_{\min} \tag{A.18}$$

after eliminating D in favor of $s_{\min}$ with the aid of (A.16). This relation holds for the ellipse ($0 \le \epsilon < 1$), the parabola ($\epsilon = 1$), and the near branch of the hyperbola ($\epsilon > 1$); see (5.2.36) and (5.3.30). For the far branch, we have $d = x - D = s\cos\varphi - D$ and get

$$s(\varphi) = \frac{\epsilon - 1}{\epsilon\cos\varphi - 1} s_{\min} \tag{A.19}$$

after using (A.17). Formally, the replacement $\epsilon \to -\epsilon$ turns (A.18) into (A.19), but this is just a mnemonic.

A.3 *Plane sections of a cone*

When a circular double cone is intersected by a plane, the points that belong both to the cone and the plane make up a conic section, an ellipse or a parabola or a hyperbola:

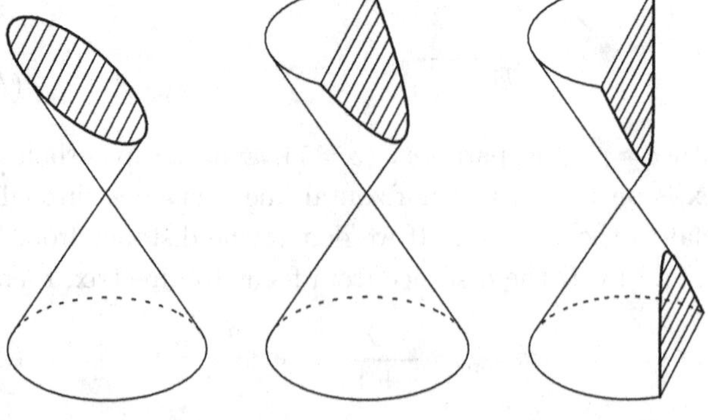

(A.20)

To describe this situation, we specify the double cone by the unit vector $\boldsymbol{e}$ of its symmetry axis and the opening angle θ, so that a point with position vector $\boldsymbol{r}$ is on the double cone if

$$(\boldsymbol{e}\cdot\boldsymbol{r})^2 = r^2(\cos\theta)^2\,, \tag{A.21}$$

with the tip of the cone at $\boldsymbol{r}=0$. Recall, in this context, that cones are the surfaces on which the polar angle of spherical coordinates is constant; see Section 1.1.10. We are here dealing with a double cone and combine the cones for θ and $\pi-\theta$ by squaring $\boldsymbol{e}\cdot\boldsymbol{r}$, rather than keeping track of the sign of $\cos\theta$. By convention, then, we have

$$0<\theta<\frac{1}{2}\pi\,,\qquad 0<\sin\theta,\,\cos\theta<1\,. \tag{A.22}$$

The plane that cuts through the double cone is specified by the vector $\boldsymbol{w}$ that points from the tip of the double cone to the nearest point in the plane. Since $\boldsymbol{r}-\boldsymbol{w}$ is perpendicular to $\boldsymbol{w}$ if $\boldsymbol{r}$ is in the plane, we have

$$\boldsymbol{w}\cdot(\boldsymbol{r}-\boldsymbol{w})=0 \tag{A.23}$$

for all points in the plane. Whether we obtain ellipse, parabola, or hyperbola at the intersection of cone and plane depends on the angle χ between $\boldsymbol{e}$ and $\boldsymbol{w}$,

$$\boldsymbol{e}\cdot\boldsymbol{w}=w\cos\chi\,. \tag{A.24}$$

We are free to choose $\boldsymbol{e}$ such that

$$0<\sin\chi,\,\cos\chi<1\,, \tag{A.25}$$

since $-\boldsymbol{e}$ and $\boldsymbol{e}$ specify the symmetry axis of the double cone equally well.

We supplement $\boldsymbol{w}$ with two unit vectors $\boldsymbol{e}_1$ and $\boldsymbol{e}_2$, which we choose such that the three vectors are pairwise orthogonal and right-handed in this order,

$$(\boldsymbol{e}_1\times\boldsymbol{e}_2)\cdot\boldsymbol{w}=w\,. \tag{A.26}$$

Then,

$$\boldsymbol{r}=\boldsymbol{e}_1u+\boldsymbol{e}_2v+\boldsymbol{w} \tag{A.27}$$

parameterizes all vectors $\boldsymbol{r}$ in the plane with the in-plane cartesian coordinates u and v since (A.23) holds for all pairs (u,v). It is expedient to choose

$$e_1 = \frac{w \times (e \times w)}{w^2 \sin\chi} \quad \text{and} \quad e_2 = \frac{w \times e}{w \sin\chi} \tag{A.28}$$

because then $e \cdot e_2 = 0$, and we have

$$e \cdot r = e \cdot e_1 u + e \cdot w = u \sin\chi + w \cos\chi \tag{A.29}$$

on the left-hand side of (A.21).

With $r^2 = u^2 + v^2 + w^2$ on the right-hand side, the (u, v) pairs of the points of intersection solve a quadratic equation,

$$(u \sin\chi + w \cos\chi)^2 = (u^2 + v^2 + w^2)(\cos\theta)^2 \tag{A.30}$$

or

$$\left[(\cos\theta)^2 - (\sin\chi)^2\right]u^2 - \sin(2\chi)wu + (\cos\theta)^2 v^2 = \left[(\cos\chi)^2 - (\cos\theta)^2\right]w^2. \tag{A.31}$$

The sign of $(\cos\theta)^2 - (\sin\chi)^2 = (\cos\chi)^2 - (\sin\theta)^2$, the coefficient multiplying u^2, decides which conic section results:

$$\begin{aligned} &\text{ellipse for } \cos\theta > \sin\chi\,, \\ &\text{parabola for } \cos\theta = \sin\chi\,, \\ &\text{hyperbola for } \cos\theta < \sin\chi\,. \end{aligned} \tag{A.32}$$

In the case of the parabola, we replace $\cos\chi$ by $\sin\theta$ and $\sin\chi$ by $\cos\theta$ in (A.31) and solve for v^2,

$$v^2 = 2w \tan\theta \left[u - w\cot(2\theta)\right], \tag{A.33}$$

which is $4ax = y^2$ of (A.1) for $a = \frac{1}{2}w\tan\theta$, $x = u - w\cot(2\theta)$, and $y = v$. The parabola has its vertex at the point with $(u, v) = \bigl(w\cot(2\theta), 0\bigr)$ and its focus at $(u, v) = \bigl(\frac{1}{2}w\cot\theta, 0\bigr)$.

After completing a square, the $\cos\theta \neq \sin\chi$ case of (A.31) reads

$$\left(\frac{(\cos\theta)^2 - (\sin\chi)^2}{w\cos\theta\,\sin\theta}\right)^2 \left(u - \frac{w\cos\chi\,\sin\chi}{(\cos\theta)^2 - (\sin\chi)^2}\right)^2 + \frac{(\cos\theta)^2 - (\sin\chi)^2}{(w\sin\theta)^2} v^2 = 1\,. \tag{A.34}$$

In the ellipse case of $\cos\theta > \sin\chi$, this is $\frac{x^2}{a^2}+\frac{y^2}{b^2}=1$ of (A.1) with

$$x = u - \frac{w\cos\chi\,\sin\chi}{(\cos\theta)^2-(\sin\chi)^2} \quad\text{and}\quad y = v\,,$$

$$a = \frac{w\cos\theta\,\sin\theta}{(\cos\theta)^2-(\sin\chi)^2} \quad\text{and}\quad b = \frac{w\sin\theta}{\sqrt{(\cos\theta)^2-(\sin\chi)^2}}\,. \tag{A.35}$$

The ellipse has linear eccentricity $c = \frac{w\sin\theta\,\sin\chi}{(\cos\theta)^2-(\sin\chi)^2}$, numerical eccentricity $\epsilon = \frac{c}{a} = \frac{\sin\chi}{\cos\theta}$, and the foci at

$$(u,v) = \left(\frac{w\sin\chi}{\cos\chi\mp\sin\theta}, 0\right). \tag{A.36}$$

In the hyperbola case of $\cos\theta < \sin\chi$, (A.34) is $\frac{x^2}{a^2}-\frac{y^2}{b^2}=1$ of (A.1) with

$$x = u + \frac{w\cos\chi\,\sin\chi}{(\sin\chi)^2-(\cos\theta)^2} \quad\text{and}\quad y = v\,,$$

$$a = \frac{w\cos\theta\,\sin\theta}{(\sin\chi)^2-(\cos\theta)^2} \quad\text{and}\quad b = \frac{w\sin\theta}{\sqrt{(\sin\chi)^2-(\cos\theta)^2}}\,. \tag{A.37}$$

The hyperbola has linear eccentricity $c = \frac{w\sin\theta\,\sin\chi}{(\sin\chi)^2-(\cos\theta)^2}$, numerical eccentricity $\epsilon = \frac{c}{a} = \frac{\sin\chi}{\cos\theta}$, and the foci at

$$(u,v) = \left(\frac{\pm w\sin\chi}{\sin\theta\pm\cos\chi}, 0\right). \tag{A.38}$$

We note that $\epsilon = \frac{\sin\chi}{\cos\theta}$ is the numerical eccentricity in all three cases. The distinction of (A.32) is just that of $\epsilon < 1$, $= 1$, or > 1.

A closing remark: Owing to the choice of unit vectors in (A.28), we do not have a uv product term in the quadratic equation (A.31). More generally, the solutions of a quadratic equation

$$\begin{pmatrix} u & v \end{pmatrix} M \begin{pmatrix} u \\ v \end{pmatrix} + \begin{pmatrix} \lambda & \kappa \end{pmatrix}\begin{pmatrix} u \\ v \end{pmatrix} + \mu = 0 \tag{A.39}$$

with a symmetric 2×2 matrix M and constants κ, λ, μ are (u,v) pairs that trace out an ellipse if $\det\{M\} > 0$, a parabola if $\det\{M\} = 0$, and a hyperbola if $\det\{M\} < 0$. While it is possible that, for the given M, κ, λ, and μ, there are no (u,v) pairs that obey (A.39), it is hardly necessary to state explicitly the conditions that prevent this from happening.

B On the exercise for the reader in Section 5.2

When expressing time increments as fractions of the period T of the Kepler's ellipse, see (5.2.48), the separation of variables in (5.2.51) yields

$$\frac{\mathrm{d}t}{T} = \frac{\mathrm{d}\varphi}{2\pi} \frac{\sqrt{1-\epsilon^2}^3}{(1+\epsilon\cos\varphi)^2} = \frac{1}{2\pi}\sqrt{1-\epsilon^2}^3 \frac{\partial}{\partial\epsilon}\frac{\epsilon\,\mathrm{d}\varphi}{1+\epsilon\cos\varphi}, \tag{B.1}$$

where $\dfrac{1}{(1+\epsilon y)^2} = \dfrac{\partial}{\partial\epsilon}\dfrac{\epsilon}{1+\epsilon y}$ is used for $y = \cos\varphi$ to simplify the φ dependence somewhat. With the substitution $u = \tan\dfrac{\varphi}{2} = \dfrac{\sin\varphi}{1+\cos\varphi}$ of (5.2.52) we then have

$$\begin{aligned}\frac{\mathrm{d}\varphi}{1+\epsilon\cos\varphi} &= \mathrm{d}u\,\frac{2}{1+u^2}\,\frac{1}{1+\epsilon\dfrac{1-u^2}{1+u^2}} = \mathrm{d}u\,\frac{2}{(1+\epsilon)+(1-\epsilon)u^2}\\ &= \frac{2}{\sqrt{1-\epsilon^2}}\mathrm{d}\tan^{-1}\left(\sqrt{\frac{1-\epsilon}{1+\epsilon}}\,u\right)\end{aligned} \tag{B.2}$$

or

$$\frac{\mathrm{d}\varphi}{1+\epsilon\cos\varphi} = \frac{2}{\sqrt{1-\epsilon^2}}\mathrm{d}\tan^{-1}\left(\sqrt{\frac{1-\epsilon}{1+\epsilon}}\tan\frac{\varphi}{2}\right) \tag{B.3}$$

so that

$$\frac{\mathrm{d}t}{T} = \frac{1}{\pi}\sqrt{1-\epsilon^2}^3\frac{\partial}{\partial\epsilon}\frac{\epsilon}{\sqrt{1-\epsilon^2}}\mathrm{d}\tan^{-1}\left(\sqrt{\frac{1-\epsilon}{1+\epsilon}}\tan\frac{\varphi}{2}\right). \tag{B.4}$$

The identities

$$\sqrt{1-\epsilon^2}^3\frac{\partial}{\partial\epsilon}\frac{\epsilon}{\sqrt{1-\epsilon^2}} = 1 + (1-\epsilon^2)\epsilon\frac{\partial}{\partial\epsilon} \tag{B.5}$$

and

$$\frac{\partial}{\partial\epsilon}\tan^{-1}\left(\sqrt{\frac{1-\epsilon}{1+\epsilon}}\tan\frac{\varphi}{2}\right) = -\frac{1}{2\sqrt{1-\epsilon^2}}\frac{\sin\varphi}{1+\epsilon\cos\varphi} \tag{B.6}$$

take us to

$$\frac{\mathrm{d}t}{T} = \frac{1}{\pi}\mathrm{d}\tan^{-1}\left(\sqrt{\frac{1-\epsilon}{1+\epsilon}}\tan\frac{\varphi}{2}\right) - \frac{1}{2\pi}\epsilon\sqrt{1-\epsilon^2}\,\mathrm{d}\frac{\sin\varphi}{1+\epsilon\cos\varphi}. \tag{B.7}$$

Accordingly,

$$\frac{t_2 - t_1}{T} = \left[\frac{1}{\pi}\tan^{-1}\left(\sqrt{\frac{1-\epsilon}{1+\epsilon}}\tan\frac{\varphi}{2}\right) - \frac{1}{2\pi}\epsilon\sqrt{1-\epsilon^2}\,\frac{\sin\varphi}{1+\epsilon\cos\varphi}\right]\Bigg|_{\varphi=\varphi(t_1)}^{\varphi(t_2)} \tag{B.8}$$

relates the lapse of time between two positions on the Kepler's ellipse to their azimuth values. Since $-\frac{1}{2}\pi < \tan^{-1}(\;) < \frac{1}{2}\pi$ by convention, this relation applies when

$$-\pi < \varphi(t_1) < \varphi(t_2) < \pi\,, \tag{B.9}$$

that is: between two successive visits to the aphelion. The corresponding expression for

$$0 < \varphi(t_1) < \varphi(t_2) < 2\pi\,, \tag{B.10}$$

that is: between two successive visits to the perihelion, is

$$\frac{t_2 - t_1}{T} = \left[\frac{1}{\pi}\cot^{-1}\left(\sqrt{\frac{1+\epsilon}{1-\epsilon}}\cot\frac{\varphi}{2}\right) - \frac{1}{2\pi}\epsilon\sqrt{1-\epsilon^2}\,\frac{\sin\varphi}{1+\epsilon\cos\varphi}\right]\Bigg|_{\varphi=\varphi(t_1)}^{\varphi(t_2)} \tag{B.11}$$

since $0 < \cot^{-1}(\;) < \pi$. As a check, we note that (B.8) gives the answer to Exercise 81 for $\cos\varphi(t_1) = \cos\varphi(t_2) = -\epsilon$ and $\sin\varphi(t_1) < 0 < \sin\varphi(t_2)$.

Although we cannot solve (B.8) or (B.11) for $\varphi(t_2)$ as an explicit function of t_1, t_2, and $\varphi(t_1)$, it is easy to exploit these relations numerically. For example, we can find successive values of $\varphi(t)$ such that the respective time difference is $\frac{1}{12}T$, thereby breaking the period T of the Kepler's ellipse — the "year" for the planet considered — into twelve "months." For the ellipse of (5.2.20), which has $\epsilon = \frac{3}{5}$, we so get this picture:

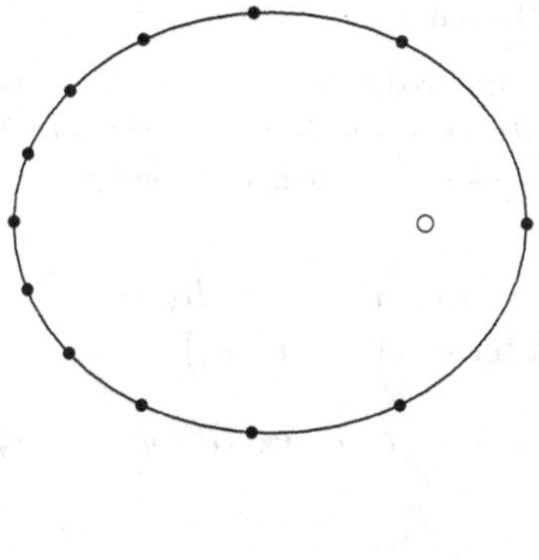

(B.12)

C On the exercise for the reader in Section 11.2.2

We relate the row $\bigl(\boldsymbol{e}_1\ \boldsymbol{e}_2\ \boldsymbol{e}_3\bigr)$ of principal-axis unit vectors to the row $\bigl(\boldsymbol{e}_x\ \boldsymbol{e}_y\ \boldsymbol{e}_z\bigr)$ of fixed unit vectors by three rotations that are analogous to those mentioned after (11.3.37). First, we rotate about $\boldsymbol{e}_x$ by angle φ,

$$\begin{aligned}\bigl(\boldsymbol{e}_x\ \boldsymbol{e}_y\ \boldsymbol{e}_z\bigr) \to \bigl(\boldsymbol{e}_x\ \boldsymbol{e}_y\ \boldsymbol{e}_z\bigr)&\begin{pmatrix}1 & 0 & 0\\ 0 & \cos\varphi & -\sin\varphi\\ 0 & \sin\varphi & \cos\varphi\end{pmatrix}\\ &= \bigl(\boldsymbol{e}_x\ \boldsymbol{e}_y\ \boldsymbol{e}_z\bigr)R_1(\varphi) = \bigl(\boldsymbol{e}_x\ \boldsymbol{e}_y\ \boldsymbol{e}_z\bigr)',\end{aligned} \tag{C.1}$$

then about $\boldsymbol{e}'_z$ by angle ϑ,

$$\begin{aligned}\bigl(\boldsymbol{e}_x\ \boldsymbol{e}_y\ \boldsymbol{e}_z\bigr)' \to \bigl(\boldsymbol{e}_x\ \boldsymbol{e}_y\ \boldsymbol{e}_z\bigr)'&\begin{pmatrix}\cos\vartheta & -\sin\vartheta & 0\\ \sin\vartheta & \cos\vartheta & 0\\ 0 & 0 & 1\end{pmatrix}\\ &= \bigl(\boldsymbol{e}_x\ \boldsymbol{e}_y\ \boldsymbol{e}_z\bigr)'R_3(\vartheta) = \bigl(\boldsymbol{e}_x\ \boldsymbol{e}_y\ \boldsymbol{e}_z\bigr)'',\end{aligned} \tag{C.2}$$

finally about $\boldsymbol{e}''_x$ by angle ψ,

$$\begin{aligned}\bigl(\boldsymbol{e}_x\ \boldsymbol{e}_y\ \boldsymbol{e}_z\bigr)'' \to \bigl(\boldsymbol{e}_x\ \boldsymbol{e}_y\ \boldsymbol{e}_z\bigr)''&\begin{pmatrix}1 & 0 & 0\\ 0 & \cos\psi & -\sin\psi\\ 0 & \sin\psi & \cos\psi\end{pmatrix}\\ &= \bigl(\boldsymbol{e}_x\ \boldsymbol{e}_y\ \boldsymbol{e}_z\bigr)''R_1(\psi) = \bigl(\boldsymbol{e}_1\ \boldsymbol{e}_2\ \boldsymbol{e}_3\bigr),\end{aligned} \tag{C.3}$$

where $R_1(\ )$ and $R_3(\ )$ are the matrices for the rotation about the current first and third axes. Taken together, we have

$$\bigl(\boldsymbol{e}_1\ \boldsymbol{e}_2\ \boldsymbol{e}_3\bigr) = \bigl(\boldsymbol{e}_x\ \boldsymbol{e}_y\ \boldsymbol{e}_z\bigr)R(\varphi,\vartheta,\psi) \tag{C.4}$$

for the rotation by the three Euler angles, with the orthogonal 3×3 matrix

$$\begin{aligned}&R(\varphi,\vartheta,\psi) = R_1(\varphi)R_3(\vartheta)R_1(\psi)\\ &= \begin{pmatrix}\cos\vartheta & -\sin\vartheta\cos\psi & \sin\vartheta\sin\psi\\ \cos\varphi\sin\vartheta & \cos\varphi\cos\vartheta\cos\psi-\sin\varphi\sin\psi & -\cos\varphi\cos\vartheta\sin\psi-\sin\varphi\cos\psi\\ \sin\varphi\sin\vartheta & \sin\varphi\cos\vartheta\cos\psi+\cos\varphi\sin\psi & -\sin\varphi\cos\vartheta\sin\psi+\cos\varphi\cos\psi\end{pmatrix}.\end{aligned} \tag{C.5}$$

Its inverse $R(\varphi,\vartheta,\psi)^{-1} = R(\varphi,\vartheta,\psi)^{\mathrm{T}} = R(-\psi,-\vartheta,-\varphi)$ is needed when expressing $\bigl(\boldsymbol{e}_x\ \boldsymbol{e}_y\ \boldsymbol{e}_z\bigr)$ in terms of $\bigl(\boldsymbol{e}_1\ \boldsymbol{e}_2\ \boldsymbol{e}_3\bigr)$,

$$\bigl(\boldsymbol{e}_x\ \boldsymbol{e}_y\ \boldsymbol{e}_z\bigr) = \bigl(\boldsymbol{e}_1\ \boldsymbol{e}_2\ \boldsymbol{e}_3\bigr)R(-\psi,-\vartheta,-\varphi)\,. \tag{C.6}$$

The time dependence of $\boldsymbol{e}_1$, $\boldsymbol{e}_2$, and $\boldsymbol{e}_3$ results from the time dependence of φ, ϑ, and ψ, whereby

$$\frac{\mathrm{d}}{\mathrm{d}t}\bigl(\boldsymbol{e}_1\ \boldsymbol{e}_2\ \boldsymbol{e}_3\bigr) = \boldsymbol{\omega}\times\bigl(\boldsymbol{e}_1\ \boldsymbol{e}_2\ \boldsymbol{e}_3\bigr) \tag{C.7}$$

with

$$\begin{aligned}\boldsymbol{\omega} &= \dot{\varphi}\boldsymbol{e}_x + \dot{\vartheta}\boldsymbol{e}_z' + \dot{\psi}\boldsymbol{e}_x'' \\ &= \bigl(\boldsymbol{e}_1\ \boldsymbol{e}_2\ \boldsymbol{e}_3\bigr)\left[\dot{\varphi}\begin{pmatrix}\cos\vartheta\\ -\cos\psi\sin\vartheta\\ \sin\psi\sin\vartheta\end{pmatrix} + \dot{\vartheta}\begin{pmatrix}0\\ \sin\psi\\ \cos\psi\end{pmatrix} + \dot{\psi}\begin{pmatrix}1\\0\\0\end{pmatrix}\right],\end{aligned} \tag{C.8}$$

which is the analog of (11.3.37). The comparison with (11.2.6),

$$\boldsymbol{\omega} = \bigl(\boldsymbol{e}_1\ \boldsymbol{e}_2\ \boldsymbol{e}_3\bigr)\begin{pmatrix}\omega_1\\ \omega_2\\ \omega_3\end{pmatrix}, \tag{C.9}$$

tells us that

$$\begin{aligned}\omega_1 &= \dot{\varphi}\cos\vartheta + \dot{\psi}\,,\\ \omega_2 &= -\dot{\varphi}\cos\psi\sin\vartheta + \dot{\vartheta}\sin\psi\,,\\ \omega_3 &= \dot{\varphi}\sin\psi\sin\vartheta + \dot{\vartheta}\cos\psi\,,\end{aligned} \tag{C.10}$$

for the chosen parameterization of $\boldsymbol{e}_1$, $\boldsymbol{e}_2$, and $\boldsymbol{e}_3$.

Before proceeding, we must now decide how we orient the trio of fixed unit vectors. Since $\boldsymbol{L} = \mathbf{I}\cdot\boldsymbol{\omega}$ is constant in time, this conserved vector singles out one direction in space and invites us to choose $\boldsymbol{e}_x$ in this direction: $\boldsymbol{L} = |\boldsymbol{L}|\,\boldsymbol{e}_x$. Further, the rotating body has $\boldsymbol{e}_1$ as its figure axis and, therefore, any choice of orientation for $\boldsymbol{e}_2$ and $\boldsymbol{e}_3$ at $t=0$ is fine. In particular, we can opt for $\boldsymbol{e}_z = \boldsymbol{e}_3(t=0) \propto \boldsymbol{e}_x\times\boldsymbol{e}_1(t=0)$. Then

$$\begin{aligned}&\varphi(t=0) = 0\,,\\ &\vartheta(t=0) = \vartheta_0 \quad\text{with}\quad \boldsymbol{e}_x\cdot\boldsymbol{e}_1(t=0) = \cos\vartheta_0\,,\\ &\psi(t=0) = 0\,,\end{aligned} \tag{C.11}$$

and

$$\omega_3(t=0) = 0 \tag{C.12}$$

is a consequence of these conventions about $\bigl(\boldsymbol{e}_x\ \boldsymbol{e}_y\ \boldsymbol{e}_z\bigr)$ and $\bigl(\boldsymbol{e}_1\ \boldsymbol{e}_2\ \boldsymbol{e}_3\bigr)\big|_{t=0}$.

A look at

$$\boldsymbol{L} = \begin{pmatrix}\boldsymbol{e}_1 & \boldsymbol{e}_2 & \boldsymbol{e}_3\end{pmatrix} \begin{pmatrix} I_1\omega_1 \\ I_2\omega_2 \\ I_2\omega_3 \end{pmatrix} = |\boldsymbol{L}|\,\boldsymbol{e}_x = |\boldsymbol{L}|\begin{pmatrix}\boldsymbol{e}_1 & \boldsymbol{e}_2 & \boldsymbol{e}_3\end{pmatrix} \begin{pmatrix} \cos\vartheta \\ -\cos\psi\sin\vartheta \\ \sin\psi\sin\vartheta \end{pmatrix} \tag{C.13}$$

reveals

$$\omega_2 \sin\psi + \omega_3 \cos\psi = 0\,, \tag{C.14}$$

which we combine with (C.10) to arrive at

$$\dot{\vartheta} = 0 \quad \text{or} \quad \vartheta(t) = \vartheta_0\,. \tag{C.15}$$

We can also combine (C.14) with (11.2.17), which gives

$$\omega_2(0)\sin(\Omega t + \psi) + \omega_3(0)\cos(\Omega t + \psi) = 0\,. \tag{C.16}$$

This implies $\Omega t + \psi = \text{constant}$ or, with the initial value in (C.11),

$$\psi(t) = -\Omega t\,. \tag{C.17}$$

Finally, we recall that $\omega_1(t) = \omega_1(0)$ and use (C.10), (C.15), and (C.17) to conclude that

$$\dot{\varphi} = \frac{\omega_1(0) + \Omega}{\cos\vartheta_0}\,. \tag{C.18}$$

With Ω of (11.2.16), this establishes

$$\varphi(t) = \frac{I_1\omega_1(0)t}{I_2\cos\vartheta_0} = \Omega' t\,. \tag{C.19}$$

In summary, we have

$$\begin{pmatrix}\boldsymbol{e}_1 & \boldsymbol{e}_2 & \boldsymbol{e}_3\end{pmatrix} = \begin{pmatrix}\boldsymbol{e}_x & \boldsymbol{e}_y & \boldsymbol{e}_z\end{pmatrix} R(\Omega' t, \vartheta, -\Omega t) \tag{C.20}$$

for the time-dependent principal axes. We note that the angles between the angular momentum $\boldsymbol{L} = |\boldsymbol{L}|\,\boldsymbol{e}_x$, the angular velocity $\boldsymbol{\omega}$, and the unit vector $\boldsymbol{e}_1$ of the figure axis are constant in time,

$$\begin{aligned} \boldsymbol{e}_x \cdot \boldsymbol{\omega} &= \omega_1(0)\cos\vartheta_0 - \omega_2(0)\sin\vartheta_0\,, \\ \boldsymbol{\omega} \cdot \boldsymbol{e}_1 &= \omega_1(0)\,, \\ \boldsymbol{e}_1 \cdot \boldsymbol{e}_x &= \cos\vartheta_0\,. \end{aligned} \tag{C.21}$$

Accordingly, the figure axis and the angular-velocity vector precess around the angular-momentum vector, and these two precessions are coordinated such that, as seen from the body-fixed system, the angular-momentum vector precesses around the figure axis.

Hereby, the constancy of $\boldsymbol{\omega} \cdot \boldsymbol{L}$ is just the conservation of the kinetic energy of the rotational motion,

$$\boldsymbol{\omega} \cdot \boldsymbol{L} = \boldsymbol{\omega} \cdot \mathsf{I} \cdot \boldsymbol{\omega} = 2E_{\mathrm{rot}}\,, \tag{C.22}$$

which is the only contribution to the energy in this situation.

Index

Printed in the United States
By Bookmasters

Printed in the United States
By Bookmasters